Van Til
April 20

DICTIONARY OF
ACOUSTICS

DICTIONARY OF ACOUSTICS

CHRISTOPHER L MORFEY

Institute of Sound and Vibration Research
University of Southampton
Southampton, UK

ACADEMIC PRESS

A Harcourt Science and Technology Company

San Diego   San Francisco   New York
Boston   London   Sydney   Tokyo

This book is printed on acid-free paper.

Copyright © 2001 by ACADEMIC PRESS

All Rights Reserved.
No part of this publication may be reproduced or transmitted in any form or by any means, electronic or mechanical, including photocopying, recording, or any information storage and retrieval system, without permission in writing from the publisher.

Academic Press
A Harcourt Science and Technology Company
Harcourt Place, 32 Jamestown Road, London NW1 7BY, UK
http://www.academicpress.com

Academic Press
A Harcourt Science and Technology Company
525 B Street, Suite 1900, San Diego, California 92101-4495, USA
http://www.academicpress.com

ISBN 0-12-506940-5

Library of Congress Catalog Number: 00-102542

A catalogue record for this book is available from the British Library

Typeset by Paston Prepress Ltd, Beccles, Suffolk, UK
Printed and bound in Great Britain by Redwood Books Ltd, Trowbridge, UK

01 02 03 04 05 06 RB 9 8 7 6 5 4 3 2 1

Contents

Subject Consultants and Reviewers	vii
Preface	ix
Acknowledgments	xi
Dictionary Guide and Abbreviations Used	xiii
DICTIONARY A–Z	1
Select Bibliography	427

Subject Consultants and Reviewers

Consultants

Michael Fisher, ISVR, University of Southampton, UK, *Aeroacoustics* [1]

Douglas Robinson†, formerly of ISVR, University of Southampton, UK, *Audiology and Human Effects* [2]

Mike Barron, Dept of Architecture & Civil Engineering, University of Bath, UK, *Auditorium Acoustics* [3]

Ian Flindell, ISVR, University of Southampton, UK
Mike Lower, ISVR Consultancy Services, University of Southampton, UK, *Environmental Noise* [4]

Marc Bedford, Aerospace Engineering and Engineering Mechanics, University of Texas at Austin, USA
Maurice Petyt, ISVR, University of Southampton, UK, *Mathematics and Applied Mechanics* [5]

Alan Cummings, School of Engineering, University of Hull, UK, *Porous Materials for Sound Absorption* [6]

Paul White, ISVR, University of Southampton, UK
Garvin Wills, PA Consulting Group, Cambridge Technology Centre, UK, *Signal Processing and Linear Systems* [7]

Stanley Ehrlich, Stan Ehrlich Associates, USA
Keith Holland, ISVR, University of Southampton, UK, *Transducers and Measurements* [8]

Tim Leighton, ISVR, University of Southampton, UK, *Ultrasonics, Physical Acoustics* [9]

Michael Ainslie, Principal Consultant (Ocean Acoustics), CORDA Ltd, UK, *Underwater Acoustics* [10]

Frank Fahy, ISVR, University of Southampton, UK, *Vibroacoustics, Acoustics* [11]

Reviewers

Michael Ainslie, *10**, Principal Consultant (Ocean Acoustics), CORDA Ltd, UK
Mike Barron, *3*, Dept of Architecture & Civil Engineering, University of Bath, UK

† Deceased.
* Numbers refer to subject areas reviewed, cross-referenced to the preceding list of consultants.

John Chapman, *5*, Dept of Mathematics, University of Keele, UK
Robert Chivers, *9*, *11*, ISVR, University of Southampton and DAMTP, University of Cambridge, UK
Alan Cummings, *6*, *11*, School of Engineering, University of Hull, UK
Stephen Elliott, *7*, ISVR, University of Southampton, UK
Frank Fahy, *11*, ISVR, University of Southampton, UK
Stuart Gatehouse, *2*, MRC Institute of Hearing Research, Glasgow, UK
Joe Hammond, *7*, ISVR, University of Southampton, UK
Tony Herbert, *4*, *11*, Anthony Herbert Acoustics, Lyndhurst, UK
Finn Jacobsen, *7*, Dept of Acoustic Technology, Technical University of Denmark, Lyngby, DK
Tim Leighton, *9*, *10*, ISVR, University of Southampton, UK
Geoffrey Lilley, *1*, School of Engineering Sciences, University of Southampton, UK
Mark Lutman, *2*, ISVR, University of Southampton, UK
John Post, *7*, Applied Research Laboratories, University of Texas at Austin, USA
Victor Sparrow, *3*, *5*, *11*, Graduate Program in Acoustics, Pennsylvania State University, USA
Wayne Wright, *11*, Physics Dept, Kalamazoo College, Michigan, USA

Additional technical contributions

Anthony Atchley, Yves Aurégan, Joseph Beaman, Philippe Béquin, Bernard Berry, Ioannis Besieris, David Blackstock, Niven Brown, Murray Campbell, David Chapman, Argy Chatzipetris, Stuart Colam, Gary Coleman, Leszek Demkowicz, John Dixon, Jim Greenleaf, Michael Griffin, Mark Hamilton, Dean Hewlett, Michael Howe, Phillip Joseph, Jian Kang, Sarosh Kapadia, Ben Lawton, Denys Mead, Pennia Menounou, Kevin Munro, Philip Nelson, Brian Olsen, Allan Pierce, Roger Pinnington, Stewart Radcliffe, Sjoerd Rienstra, Nicholas Rott, Neil Sandham, Christine Shadle, Viggo Tarnow, Ron Verrillo, Erik Vigran, Evan Westwood, Bob White, Matthew Wright, Philippe Young.

Preface

'The language of science is incredibly interesting; it's a natural language under stress. The language is put under stress to explain things that are difficult to explain, that perhaps are explainable in terms of mathematical equations and structures. In this language simple words like power, energy, force, stable, unstable, acquire a host of alternative meanings. . . .'

> Roald Hoffmann, joint winner of the 1981 Nobel Prize in chemistry, in conversation with Lewis Wolpert.*

Although Hoffmann was thinking of theoretical chemistry, his remarks apply with equal force to the cross-disciplinary field of acoustics. It took me several years of teaching the subject to senior engineering undergraduates to recognize that the language of acoustics, however straightforward it might seem to experienced practitioners, represents a significant obstacle to students. Fortunately, when in 1987 I began compiling definitions of acoustical terms for my students, Douglas Robinson had recently retired from the UK National Physical Laboratory and was occupying a nearby office as research professor in the Institute of Sound and Vibration Research (ISVR). When I asked his advice, he lent me his copy of the *Glossary of Acoustical Terms*, produced in 1969 by the British Standards Institution[†]. The BSI Glossary became the model for the present work.

Although well written, the Glossary was already out of date when I first saw it; the technology had moved on, and the theoretical underpinning was that of the acoustics textbooks of the 1950s. The publication in 1981 of Allan Pierce's *Acoustics*[‡] had set a new standard for graduate texts in engineering acoustics, and any dictionary of the subject would need to match that standard to do justice to the rapidly growing field. I also decided it was desirable to try to define specialist acoustical terms in as informative a manner as possible, so that the student of acoustics would recognize interconnections among the various branches of the subject.

The project had not advanced beyond a hundred or so terms when, in 1994, two events provided a fresh impetus: the IEC published Chapter 801 of its International Electrotechnical Vocabulary, and the Acoustical Society of America published its revised American National Standard on Acoustical Terminology. The scope of these documents was impressive, with 431 and 601 entries respectively; they made the task of assembling the dictionary a great deal easier, and I acknowledge the dedicated efforts of the standards committees

---

* Reported in *Passionate Minds: The Inner World of Scientists*, by Lewis Wolpert and Alison Richards (1997). By permission of Oxford University Press.
† BSI 661:1969, long since withdrawn.
‡ See Bibliography for references.

responsible. Wherever possible I have sought to be consistent in my expanded definitions with what they wrote. Over the following two years, prompted by the enthusiastic interest of my younger ISVR colleagues Phillip Joseph and Matthew Wright, a draft of the present dictionary was assembled (covering A–L) that addressed the aims set out in the previous paragraph.

In 1997 Gioia Ghezzi of Academic Press decided that the dictionary project was worth supporting. The University of Southampton provided me with a timely four-month sabbatical, and a generous invitation from the University of Texas at Austin (UT) provided the ideal environment for converting a rough version of A–L into a more respectable A–Z. Nevertheless a further 21 months of sustained effort were needed to turn the resulting draft into a finished product. I am profoundly grateful to David Blackstock, Mark Hamilton and ARL Director Mike Pestorius for their support from the UT side, and to my ISVR colleagues for their encouragement and forbearance as the project approached completion.

Robert T Beyer, writing a few years ago as Series Editor of the *AIP Series on Modern Acoustics and Signal Processing*, penned some wise words[§] on the interdisciplinary nature of modern acoustics. Anyone who has attended meetings of the Acoustical Society of America will be aware of the fascinating variety of topics covered. Yet, despite this variety, communication among acousticians remains possible. Although one can see a tendency for the terminology to fragment between subdisciplines, we still share a largely common vocabulary, one that I have sought to document in the following pages. The attempt has been partial, in that terms relating to speech and music have been omitted entirely. This failing apart, the scope of the dictionary is fairly broad. Altogether nearly 3000 terms are included, ranging from the elementary to the highly specialized. My hope is that the result will be of real benefit to the international acoustical community.

The knowledgeable reader will inevitably notice mistakes, despite my own best efforts and those of the reviewers. Please send comments and corrections to Chris Morfey, ISVR, University of Southampton, Southampton SO17 1BJ, fax + 44 23 8059 3190, or email me at < dict@isvr.soton.ac.uk >.

Southampton, 10 May 2000
Christopher L Morfey

---

§ See *Computational Ocean Acoustics*, Jensen *et al.* (1994), p. xiii.

Acknowledgments

Specific technical contributions are acknowledged elsewhere, but there is more to be said. Thanks first to friends in Austin who took care of me during my four-month visit – your love and care are appreciated; and also to my wife – who I think knows this anyway – for unfailingly providing other topics of conversation over the past 3 years. The task of entering my handwritten notes on the keyboard, complete with equations, was cheerfully undertaken by Sue Brindle at ISVR who has seen the project through from the beginning. Joe Hammond, ISVR Director, and my colleague Phil Nelson, Fluid Dynamics and Acoustics Group Chairman, have been immensely supportive. My former mentors Philip Doak and Peter Davies loaned me office space among the retired professors, where a quiet hour could be spent and papers could be left undisturbed. Emma Roberts and Roopa Baliga at Academic Press were a pleasure to work with at all times. Finally, it is fair to acknowledge that the subject consultants and reviewers listed in the front of the book have contributed enormously to the value and credibility of the present work. Out of the 27 listed, 20 have some connection with ISVR. The first to offer his services was the late Douglas Robinson, who patiently educated me in the terminology of hearing measurement, and then in January 1998 provided a complete set of draft audiological definitions for inclusion in the dictionary. I am indebted to them all for their enthusiasm, patience and attention to detail; any errors that remain are my responsibility alone.

Dictionary Guide and Abbreviations Used

1. Alphabetization

Entries are arranged in alphabetical order. The order is determined by spelling out the terms letter by letter, with spaces and other characters ignored: for example

acoustic reflex tests
acoustics
acoustic scatterer.

2. Spelling

Terms are spelled according to the *Longman Dictionary of the English Language* (1995) (see the Bibliography for details).

3. Typographic conventions

(1) Headwords are capitalized only where standard usage requires it: e.g.

dB
decibel
Fermat's principle
leaky Rayleigh wave.

(2) Italic labels immediately following the headword help to define the area of meaning: examples are

acoustic bullets – *in ultrasonics* ... ,
bias – *in spectral analysis* ... .

(3) Small capitals are used for cross-references to other entries in the dictionary:

acoustic power – the same as SOUND POWER.

(4) Bold italics highlight subsidiary definitions within the main entry: for example

chi-square distribution – a family of continuous PROBABILITY DENSITY FUNCTIONS, with a parameter N called the ***degrees of freedom***.

(5) Often an equation, theorem, or result is named after an individual, as in HELMHOLTZ EQUATION. The year in which the result first appeared is given

wherever possible, preceded by the name in bold small capitals: (**H L F von Helmholtz** *1860*).

(6) Sometimes a term is named after a famous scientist, but for less specific reasons (as in MACH NUMBER). In such cases the name is followed by dates in italics giving the individual's life span: (**E Mach** *1838–1916*).

(7) To save unnecessary repetition, a swung dash (∼) is sometimes used to represent the headword within a definition: for example

scattering coefficient – see BULK ∼, SURFACE ∼, VOLUME ∼.

4. Units

Entries for quantitative terms generally list the appropriate SI unit symbol*, or combination of unit symbols, at the end:

acoustic impedance – ... *Units* Pa s m^{-3}.

sound power – ... *Units* W.

For generic terms like radiation impedance or cross-spectral density, where the units depend on the context, no units are specified. Some dimensionless quantities are assigned special units: for example

phase angle – ... *Units* rad.

Otherwise, where the quantity is measured by a pure number, the units are listed as 'none':

acoustic Mach number – ... *Units* none.

5. Non-SI units

In a few cases, non-SI units are specified at the end of an entry (e.g. Np, B, h, °, %) or appear as entries in their own right (e.g. phon. octave). Most of these non-SI units are accepted for use with the SI system (% is not); the adoption of the neper and bel as SI units is under discussion in ISO as the dictionary goes to press.

6. Notes

Many of the dictionary definitions are supplemented by additional information appended in the form of notes as in the example below.

diffraction – non-specular reflection or SCATTERING of sound waves by an object or boundary, particularly into the SHADOW ZONE. Diffraction generally involves the distortion of incident wavefronts.

Note: It is not easy to draw a sharp distinction between diffraction and scattering. Roughly speaking, diffraction is limited to wave phenomena

---

* These unit symbols have the useful feature of being the same in all languages.

that cannot be accounted for, even approximately, by RAY ACOUSTICS. On the other hand scattering can refer to any reflection process, whether specular or not.

7. Multiple meanings

Where a term has more than one accepted meaning in acoustics, the separate meanings are listed; for example

damping – (1) the absorption of energy in a propagating wave . . . .

damping – (2) *of a room* the addition of sound-absorbing material to the room surfaces . . . .

damping – (3) *of a structure* the addition of damping material to the vibrating parts of the structure . . . .

damping – (4) abbreviation for DAMPING TREATMENT.

8. Synonyms

Where two or more versions of a term exist they are entered separately, but the definition is usually given at only one entry, and a cross-reference is then supplied at the others: for example

acoustic intensity – *at a point in a time-stationary acoustic field* the time-average rate of energy flow per unit area, ... Also known as **sound intensity**. Units $W\,m^{-2}$.

sound intensity – an equivalent term for ACOUSTIC INTENSITY. Units $W\,m^{-2}$.

9. New terms

During the compilation of the dictionary, the author and his consultants have occasionally suggested new terms where the current acoustics vocabulary seemed imprecise, inconsistent, or lacking. The 'new' terms offered to readers are indicated by a warning sign (◆) following the headword; an example is

undisturbed-field equivalent sound level – ◆ *for an individual exposed to sound through headphones* the sound level of a diffuse sound field, incident on the head, that would produce the same level at the listener's eardrums as is produced by the headphones. *Units* dB re $(20\,\mu Pa)^2$.

Altogether 27 such proposals are included in the present dictionary. Several of our ideas continue the honourable tradition of acoustics borrowing its terminology from other fields (e.g. optics, electrical engineering).

10. Abbreviations and unit symbols

The following unit symbols are used:

Symbol	Unit name
%	percent
B	bel
°C	degree Celsius
d	day
dec	decade
h	hour
Hz	hertz
J	joule
K	kelvin
kg	kilogram
m	metre
mol	mole
N	newton
Np	neper
oct	octave
Pa	pascal
phon	phon
rad	radian
s	second
sone	sone
sr	steradian
V	volt
W	watt
°	degree

Other, non-technical abbreviations used in the dictionary are listed below.

e.g.	for example
etc.	and so on
i.e.	that is
NAm	North American
UK	United Kingdom

A

abnormal audibility zone – *in long-range atmospheric propagation* a region, usually observed at ground level, in which the transmission loss from a distant source (e.g. an explosion near the ground) is abnormally low. The mechanism is similar to that responsible for the SOFAR CHANNEL, in that sound is received via downward reflection from the stratosphere where the sound speed is higher than at source height. At the same time, the usual lapse in temperature with height in the lower atmosphere causes upward refraction close to the ground, with the result that a SHADOW ZONE appears at intermediate ranges. See also STRATOSPHERIC DUCT.

absolute phase – *of a system frequency response function* the UNWRAPPED PHASE of the response. See also PHASE-SHIFT FUNCTION. *Units* rad.

Note: Compare RELATIVE PHASE.

absolute temperature – the thermodynamic temperature T, measured in kelvins from absolute zero. A temperature of $T = 273.15$ K on the thermodynamic scale corresponds to zero on the Celsius scale; a temperature of $T = 273.16$ K corresponds to the triple point of water. *Units* K.

Note: In acoustics, the absolute temperature appears explicitly in the expression for the sound speed of an ideal gas,

$$c = (\gamma R T)^{1/2},$$

where γ is the SPECIFIC-HEAT RATIO and R is the SPECIFIC GAS CONSTANT. It also appears in the differential coefficient that connects temperature changes to pressure changes in a fluid, when the fluid is compressed isentropically:

$$dT = \frac{\alpha T}{\rho C_p} dP.$$

Here α is the VOLUME THERMAL EXPANSIVITY of the fluid, ρ is the fluid density, and C_p the specific heat at constant pressure.

absolute threshold – *for a particular listener presented with a specified acoustic signal* the minimum level at which the acoustic signal (e.g. a pure tone) is detectable by the listener, in a specified fraction of trials (conventionally 50%). The term implies quiet listening conditions: that is, it represents the irreducible absolute threshold. In the presence of a MASKING sound or noise, the term ***masked threshold*** is appropriate. *Units* dB re $(20 \, \mu\text{Pa})^2$.

Note (1): The method of measuring the threshold sound pressure level can vary: see MINIMUM AUDIBLE PRESSURE, MINIMUM AUDIBLE FIELD.

Note (2): An equivalent term is *threshold of hearing*. Compare HEARING THRESHOLD LEVEL.

absolute value – *of a complex number* the quantity $\sqrt{a^2 + b^2} = |z|$, where a and b are the real and imaginary parts of the complex number z. In the complex plane, $|z|$ is the distance of the point representing z from the origin. An alternative term is *modulus*. See also POLAR FORM.

absorbance – *for sound waves incident on a boundary* an equivalent term for ABSORPTION COEFFICIENT (1). *Units* none.

absorber – *in acoustics* abbreviation for SOUND ABSORBER. See also VIBRATION ABSORBER.

absorbing area – (1) *of a room* an older term for ROOM ABSORPTION. *Units* m^2.

absorbing area – (2) *of an object in a room* an older term for EQUIVALENT ABSORPTION AREA. *Units* m^2.

absorbing boundary condition – *in computational acoustics* a condition that is applied at the computational domain boundary to simulate extension of the domain to infinity, i.e. FREE-FIELD RADIATION. The domain boundary should ideally be transparent to incident acoustic waves; although perfect transparency is not generally achievable, absorbing boundary conditions can often provide a practical simulation of free-field conditions. Also known as *anechoic boundary condition*.

absorbing power – *of a room* an older term for ROOM ABSORPTION. *Units* m^2.

absorption – (1) *of sound in a medium* the dissipation of acoustic energy that occurs in a lossy medium; it contributes, along with SCATTERING, to the ATTENUATION (1) of freely-propagating sound waves. Compare VOLUME ABSORPTION.

absorption – (2) *of sound at a boundary* the loss or escape of acoustic energy from a sound field that occurs when the boundary is not perfectly reflective. Compare BOUNDARY ABSORPTION.

absorption coefficient – (1) *at a boundary* the fraction of the incident acoustic power arriving at the boundary that is not reflected, and is therefore regarded as being absorbed by the boundary. Equivalent terms are *absorbance* and *absorption factor*. Compare SABINE ABSORPTION COEFFICIENT. *Units* none.

Note (1): The IEC and ANSI 1994 terminology standards do not recognize this term, preferring *sound power absorption coefficient*. The abbreviation given here is widely used by acousticians, however, and is generally unambiguous. (Shortening SOUND POWER REFLECTION COEFFICIENT or SOUND POWER TRANSMISSION COEFFICIENT in a similar way would lead to problems, since REFLECTION COEFFICIENT and TRANSMISSION COEFFICIENT commonly refer to pressure.)

Note (2): The absorption coefficient is a function of frequency and incident wave direction. For practical purposes it is often quoted in one-third octave bands. Unless otherwise stated, a single plot or table of absorption coefficient as a function of frequency is assumed to refer to the STATISTICAL ABSORPTION COEFFICIENT (i.e. for random incidence).

absorption coefficient – (2) *of an acoustic medium* an abbreviation sometimes used in ultrasonics for BULK ABSORPTION COEFFICIENT; otherwise known as the ENERGY ATTENUATION COEFFICIENT. *Units* m^{-1}.

Note: This abbreviation risks confusion with the first definition of absorption coefficient given above, i.e. the fraction of incident power absorbed at a boundary; it is therefore not recommended.

absorption cross-section – *of an object in an acoustic medium* the area σ in the equation $W_{abs} = \sigma I_{inc}$ that gives the net sound power absorbed (within the object or the immediately surrounding medium), when the object is irradiated by plane progressive waves of intensity I_{inc}. Usually σ depends on the frequency and direction of the incident waves. Compare EQUIVALENT ABSORPTION AREA, which is defined similarly except that the incident field is diffuse. *Units* m^2.

absorption length – *for a parametric array* the effective length of the array as determined by attenuation of the primary beam; it is given by $L_a = 1/(\alpha_1 + \alpha_2 - \alpha_-)$. Here the symbol α denotes the linear plane-wave ATTENUATION COEFFICIENT in the medium; α_1 and α_2 refer to the two primary frequencies, and α_- to the difference frequency. See PARAMETRIC ARRAY. *Units* m.

absorption loss – the component of the TRANSMISSION LOSS between two points that comes from acoustic energy absorption, either within the medium or at absorbing boundaries. Separation of transmission loss into absorption loss and other components (e.g. DIVERGENCE LOSS) is feasible only under conditions where INTERFERENCE phenomena average out within the frequency band concerned, so that ENERGY ACOUSTICS becomes a valid approximation. *Units* dB.

a-c – *in audiology* abbreviation for AIR CONDUCTION.

ac, AC – oscillatory (by analogy with alternating current).

acausal response – *of a system* a response that is not CAUSAL; an acausal response begins before the input. An equivalent term is ***non-causal response***.

ac boundary layer – *for oscillatory relative motion of a fluid parallel to a solid boundary* a region near the boundary where the tangential fluid velocity drops toward zero, measured relative to the boundary. In addition, if the unsteady fluid motion is caused by sound, there will be an oscillatory temperature difference between the fluid and the boundary, falling to zero at the boundary itself. An equivalent term is ***viscothermal unsteady boundary layer***. For further detail, see BOUNDARY LAYER (2), THERMAL UNSTEADY BOUNDARY LAYER.

accelerance – *of a point-excited mechanical system* the complex ratio of acceleration to applied force, at a single frequency; for example, a lumped mass m has accelerance $1/m$. Equivalently, it is a frequency response function in which acceleration is the output and force is the input. The alternative term ***inertance*** is not recommended, since it has a conflicting interpretation. *Units* m s^{-2} N^{-1} ≡ kg^{-1}.

acceleration – if a point has position vector $\mathbf{r}(t)$ at time t, its acceleration is $\ddot{\mathbf{r}} = d^2\mathbf{r}/dt^2$. *Units* m s^{-2}.

acceleration of a fluid element – if a fluid flow has a vector velocity field $\mathbf{u}(\mathbf{x}, t)$, where $\mathbf{x}$ is position and t is time, then the acceleration of the fluid element at $(\mathbf{x}, t)$ is given by the MATERIAL DERIVATIVE

$$\frac{D\mathbf{u}}{Dt} = \frac{\partial \mathbf{u}}{\partial t} + (\mathbf{u}\cdot\nabla)\mathbf{u}.$$

See also LAGRANGE ACCELERATION FORMULA. *Units* m s^{-2}.

acceleration waves – *in applied mechanics* a generic term that covers ELASTIC WAVES in solids and pressure waves in fluids. Small-amplitude motion is not implied, so acceleration waves may be nonlinear.

accelerometer – an ELECTROMECHANICAL TRANSDUCER that generates an electrical output in response to an acceleration input, usually along a single axis but not necessarily: triaxial and rotational accelerometers are also used in vibration and shock response measurements.

accession to inertia – the same as VIRTUAL MASS. *Units* kg.

acoustic – (1) associated with sound, or more generally with mechanical wave propagation in any medium. However, for coupled structural–acoustic waves in fluid-loaded structures, the term VIBROACOUSTIC is preferred. The adjective "acoustic" (rather than acoustical) is used to form technical terms, as in ACOUSTIC INTENSITY, and to describe effects in which sound is the agent, as in ACOUSTIC TRAUMA. An **acoustic device** is one that is driven or actuated by sound, e.g. *acoustic refrigerator*.

acoustic – (2) *describing a musical instrument* not using electronic amplification to enhance the sound produced, as in *acoustic guitar* (as opposed to *electric guitar*); hence a humorous adjective for a device in its original mechanical (pre-electric) form, as in *acoustic typewriter*.

acoustic – (3) *used as a singular noun* the acoustical properties of a concert hall or auditorium, as judged subjectively by either performers or the audience. Compare ACOUSTICS (3).

acoustic, ~al – the meanings of these adjectives overlap. Technical terms are usually modified by **acoustic**, and non-technical terms by **acoustical**; thus ~ *impedance*, ~ *signal*, but ~ *al society*, ~ *al engineer*. Some terms attract either usage: thus ~ or ~ *al properties of matter*, ~ or ~ *al consultant*, and ~ or ~ *al qualities of an auditorium*.

Note: Analogies with *electric/electrical*, or *optic/optical*, do not appear to be helpful in this case.

acoustic ~ – some terms include the prefix "acoustic" as standard terminology, e.g. ACOUSTIC EMISSION, and are listed below in that form. For many other acoustical terms, "acoustic" is an optional qualifier, used only where the context requires it. Such terms are listed without the prefix: e.g. *acoustic particle velocity* (found under PARTICLE VELOCITY), *acoustic interference* (found under INTERFERENCE).

acoustic absorber – an equivalent term for SOUND ABSORBER.

acoustic absorption – see ABSORPTION.

acoustic admittance – *of an acoustic system or transmission line* the complex ratio of volume velocity to pressure, at a single frequency; the reciprocal of ACOUSTIC IMPEDANCE. Equivalently, it is a frequency response function in which volume velocity is the output and pressure is the input. Also known as **acoustic mobility**. Units $m^3 s^{-1} Pa^{-1}$.

acoustic agglomeration – the grouping of suspended particles into larger aggregates by the action of sound waves in the suspending fluid, usually at high intensity.

acoustical – see ACOUSTIC, ~AL.

acoustical ~ – a few terms for which the prefix "acoustical" is essential to the meaning, e.g. ACOUSTICAL CEILING, are listed below in that form. Many other terms can be constructed with "acoustical" as a qualifier, but their definition is generally obvious and they are not given entries in the dictionary.

acoustical ceiling – a ceiling designed to have desirable acoustical properties, e.g. a high ABSORPTION COEFFICIENT.

acoustical consultant – a professional expert who advises clients on issues of noise control, building acoustics, sound reinforcement, or acoustics generally.

acoustical glare – *in a concert hall* a brittle or harsh quality imparted to the sound, which modifies the timbre of music. The cause of acoustical glare is thought to be strong early reflections reaching the listener from smooth, flat surfaces as opposed to diffusing surfaces. See also TONE COLOURATION.

acoustically compact – see COMPACT.

acoustical–mechanical conversion efficiency – *of a mechanically driven fluid-loaded structure* the ratio of the sound power radiated into the fluid, W_{rad}, to the mechanical power input to the structure, W_{in}:

$$\eta_{am} = \frac{W_{rad}}{W_{in}}.$$

As an example, the value of η_{am} for a uniform isotropic plate excited below its CRITICAL FREQUENCY, f_c, by a point normal force, when the plate radiates on both sides into a free field, is

$$\eta_{am} \approx 0.37 \frac{\rho_0 \, c_P}{\rho_s \, c_0} \quad \text{(independent of thickness for } f < f_c\text{)}$$

provided the plate dimensions (in its plane) are many times the free bending wavelength. Here ρ_0 and c_0 are the density and sound speed of the ambient fluid, ρ_s is the density of the plate material, and c_P is the PLATE LONGITUDINAL-WAVE SPEED. *Units* none.

acoustical oceanography – the use of sound to probe the structure of the oceans and their floor, and thus to produce maps showing (for example) temperature, salinity, and current distributions; also near-surface bubble distributions, bottom topography, and so on.

acoustical well logging – the measurement of sound transmission between two points along a drill hole in the Earth's crust, in order to determine properties of the surrounding material. For example, the porosity of the material can be estimated in this way; rock fractures can be located; and the presence of gas or oil pockets established.

acoustic analogy – a representation of the density fluctuations in a real fluid flow as if they were due to acoustic waves in a uniform stationary fluid, driven by an externally-applied fluctuating stress field. This is expressed mathematically (see D'ALEMBERTIAN OPERATOR) by the equation

$$\Box^2 [c_0^2(\rho - \rho_0)] = -\mathcal{F}, \quad \text{with} \quad \mathcal{F} = \frac{\partial^2 T_{ij}}{\partial x_i \partial x_j}.$$

Here T_{ij} are the components of the LIGHTHILL STRESS TENSOR (given by the exact equations of fluid motion), and ρ_0, c_0 are the density and sound speed of a hypothetical uniform stationary fluid through which the acoustic waves propagate.

Note (1): By introducing the approximation that T_{ij} is not significantly dependent on the density fluctuations that it excites, Lighthill was able to derive asymptotic scaling laws for sound radiated by turbulent flows. See LIGHTHILL'S U^8 POWER LAW.

Note (2): Variations in pressure, rather than $c_0^2(\rho - \rho_0)$, may be used as the wave equation variable. The resulting source term $\mathcal{F}$ is then different in minor respects, but the same far-field radiation is obtained.

Note (3): LORD RAYLEIGH used a primitive acoustic analogy to describe the scattering of sound by local variations in fluid compressibility and density: see *Theory of Sound* (**1945**), Chapter XV.

acoustic approximation – an approximation to the dynamical equations of the medium in which terms of second order in the perturbation amplitude are ignored, and the density and temperature perturbations are assumed to obey the same linearized wave equation as the pressure. Compare LINEAR ACOUSTICS, LINEARIZATION.

acoustic boundary layer – see BOUNDARY LAYER (2).

acoustic boundary layer thickness – see BOUNDARY LAYER (2), VISCOUS PENETRATION DEPTH, THERMAL PENETRATION DEPTH. *Units* m.

acoustic brightness – ◆ *of a surface viewed from a given point* the field of acoustics lacks a generally agreed term to describe the ANGULAR INTENSITY DISTRIBUTION produced at a point in a room, as a result of partially-diffuse reflection from the room surfaces. In optics the equivalent term is *radiance*, while illumination engineers use *luminance* or *photometric brightness* for the subjectively weighted equivalent in candelas per square metre. These optical terms describe the power density (unweighted or weighted) produced at a given point by a light-reflecting surface, per unit solid angle subtended by the surface. By analogy, the ***acoustic brightness*** of a diffusely reflecting surface may be similarly defined as the acoustic intensity arriving at *P* from a surface element, per unit solid angle subtended at *P* by the element.

A consequence of this definition is that a surface element of acoustic brightness $B(\theta)$ and area dS contributes intensity $dI = B(\theta) \cos \theta \, dS/r^2$, at distance r from the element in a direction at angle θ to the surface normal. Equivalently, the ANGULAR POWER DISTRIBUTION per unit area of surface is $B(\theta) \cos \theta$. The surface is here regarded as a distribution of incoherently reflecting elements, whose contributions to the intensity at *P* are additive. A similar discussion applies to a sound-emitting surface made up of incoherently radiating elements: in the latter case the acoustic brightness is analogous to the *radiant exitance* (in optics) or *luminous exitance* (in photometry) of a self-radiating surface. *Units* $W \, m^{-2} \, sr^{-1}$.

Note: A non-absorbent DIFFUSELY-REFLECTING surface, irradiated with acoustic energy at a rate Φ per unit area and time, has the same brightness $B = \Phi/\pi$ viewed from any direction; see LAMBERT'S COSINE LAW.

acoustic bullets – *in ultrasonics* another name for LOCALIZED WAVES.

acoustic ceiling – alternative (mainly UK) spelling of ACOUSTICAL CEILING.

acoustic centre – *of a source* the point from which outgoing wavefronts appear to diverge in the far field (under free-field conditions). Its position generally depends on the frequency.

Directional sources have a different centre, in general, for each spherical harmonic component of the radiation field (monopole, dipole, etc.). An alternative definition involves assigning each radiation direction two acoustic centres, one for amplitude and one for phase.

acoustic consultant – alternative (mainly UK) spelling of ACOUSTICAL CONSULTANT.

acoustic coupler – (1) *in audiology* a rigid-walled cavity of specified shape and volume that is used for the calibration of an EARPHONE, in conjunction with a calibrated microphone that measures the sound pressure developed within the cavity. Compared to an EAR SIMULATOR, a coupler embodies only a rough approximation to the acoustic properties of the human ear but has the advantage of simple design and construction. It is typically used to calibrate earphones of supra-aural type that fit against the pinna.

acoustic coupler – (2) a rigid-walled cavity in which the active elements of two transducers are coupled by the pressure field in the contained fluid; for example, a microphone in the cavity may be used to pick up the pressure signal driven by a small loudspeaker. Two reciprocal pressure transducers can be calibrated in this way.

acoustic cross-section – *in active sonar* an alternative term for BACKSCATTERING CROSS-SECTION. *Units* m^2.

acoustic daylight – the naturally occurring incoherent ambient noise field in the ocean.

acoustic daylight technique – use of the underwater acoustic "illumination" provided by ACOUSTIC DAYLIGHT to produce acoustic images of underwater objects.

acoustic distance – *between transducers in an acoustic medium* the effective separation in a given direction between two transducers, as measured by the frequency-dependent phase shift between the transducer responses when the excitation consists of plane waves incident from the direction concerned. *Units* m.

acoustic ecologist – a person who studies and records the sounds of wildlife and natural phenomena, and monitors the impact of human civilization on such sounds.

Note: The spelling is taken from *Time* magazine, but it should arguably be ***acoustical ecologist***.

acoustic efficiency – *of a sound source* the acoustic power output of the device, normalized by the power consumed or dissipated. The term ***mechanoacoustic efficiency*** is used for the ratio of acoustic power out to mechanical power in; ***electroacoustic efficiency*** is used for the ratio of acoustic power out to electrical power in. *Units* none.

Note: In the case of a fan, the appropriate normalizing factor is the product of the stagnation pressure rise through the fan and the volume flowrate. For a jet, it is the total mechanical power dissipated in the jet flow; for a reducing valve, it is the power that could have been extracted in lowering the total pressure of the flow from its value upstream to that downstream.

acoustic emission – *in a statically-loaded solid material* the release of stored elastic energy as transient elastic waves within the material, as a result of spontaneously-appearing cracks or microstructural rearrangement. An alternative term is ***stress-wave emission***.

The term also refers to the radiation of sound that accompanies the energy release, and which may provide early warning of impending fracture. The frequency range of acoustic emission is wide, ranging from a few hertz (as in earthquakes) to several megahertz (phase transformations in steels).

acoustic energy – an energy function, based on second-order products of first-order acoustic variables, that obeys a CONSERVATION LAW. Also known as

sound energy. See ACOUSTIC ENERGY DENSITY, ACOUSTIC ENERGY FLUX VECTOR. *Units* J.

acoustic energy density – the time-averaged sum of the kinetic and potential (or compressional) energy densities at a point in an acoustic field, given by $w = w_{\text{kin}} + w_{\text{pot}}$; also known as ***sound energy density***. The ***kinetic energy density*** is $w_{\text{kin}} = \frac{1}{2}\rho\langle u^2\rangle$, where u is the magnitude of the particle velocity vector, $\langle\ldots\rangle$ denotes a time average, and ρ is the density of the fluid. The ***potential energy density*** is $w_{\text{pot}} = \frac{1}{2}\langle p^2\rangle/B$, where p is the acoustic pressure and B is the isentropic BULK MODULUS; in a fluid of sound speed c, B equals ρc^2. The ***instantaneous acoustic energy density*** is defined as above, but with no time averaging. *Units* J m^{-3}.

Note (1): In a REVERBERANT FIELD that is dominated by the resonant response of lightly-damped acoustic modes in an enclosure, the two terms in the expression above have time-average values that are approximately equal; the mean acoustic energy density is then $w \approx (1/\rho c^2)\langle p^2\rangle$.

Note (2): When a mean flow is present, a different expression applies.

acoustic energy flux density, acoustic energy flux vector – alternative terms for the INSTANTANEOUS ACOUSTIC INTENSITY vector. *Units* W m^{-2}.

acoustic environment – (1) the living environment of humans or other species, viewed from an acoustical aspect. Hence *acoustic environmentalist*, a person concerned with the protection or improvement of such environments.

acoustic environment – (2) the passive acoustic environment that a particular space provides, described in terms of its reflective or modal properties.

acoustic environment – (3) the local sound field to which a person or test object is exposed, usually described in terms of physical measurements (e.g. a spectrum of sound pressure in 1/3-octave bands).

acoustic fatigue – *in structures exposed to intense sound or pseudosound* structural FATIGUE caused by long-term dynamic responses to fluid loading, typically in the range 10 Hz to 10 kHz. The loading may be any kind of unsteady pressure field, including turbulent near-field pressures produced by jets or boundary layers. Acoustic fatigue is typically associated with low stress amplitudes and large numbers of stress reversals (up to 10^9 cycles), implying exposure for at least a day and possibly several months.

acoustic filter – a passive linear device (usually in a one-dimensional system) designed to provide a low acoustic INSERTION LOSS over a specified band of frequencies, and a high insertion loss outside the band. See also FILTER.

Note: An enlarged section (*expansion chamber*) in a rigid-walled pipe acts as a low-pass filter. A Helmholtz resonator connected to the pipe as a side branch acts as a *notch filter*, with a high insertion loss over a narrow band of frequencies.

acoustic fountain – when a sound beam in one fluid propagates across an interface and into a more compressible fluid of less than half the density, reflection of the beam reverses the normal component of the incident momentum flow carried by the beam. The momentum flow reversal is equivalent to a steady normal force on the interface, proportional to the power in the incident beam. A sufficiently powerful ultrasonic beam, transmitted vertically upwards across a horizontal water–air interface, produces a vertical jet of water (acoustic fountain) where the beam emerges into the air.

acoustic–gravity waves – see INTERNAL WAVES. The term is commonly used to refer to planetary-scale internal waves in the Earth's atmosphere.

acoustic holography – the use of phase and amplitude information, recorded over a closed surface in space, to reconstruct the sound field in the region exterior to the surface.

acoustic horn – a passive device in the form of a hard-walled tapering acoustic waveguide of finite length. When a volume-velocity driver is placed at the narrow end or *throat* of the horn, the acoustic coupling between the transducer and the surrounding medium is enhanced. The wide end or *mouth* of the horn is usually arranged to be large enough (in comparison with the acoustic wavelength) that most of the incident energy arriving at the mouth is transmitted into the surrounding medium, with the result that the horn also produces a directional far-field response. For a more general description, applicable also to solid horns, see HORN.

acoustician – a specialist in ACOUSTICS (1).

acoustic immittance – a term used in audiology to mean either ACOUSTIC ADMITTANCE or ACOUSTIC IMPEDANCE.

acoustic impedance – *of an acoustic system or transmission line* the complex ratio of pressure to volume velocity, at a single frequency. Equivalently, it is a frequency response function in which pressure is the output and volume velocity is the input. *Units* Pa s m^{-3}.

acoustic inertance – *corresponding to an acoustic lumped element through which the volume velocity is invariant* the complex ratio of the pressure difference across the element to the time derivative of the volume velocity through the element, at a single frequency. See CONDUCTIVITY, LUMPED ELEMENTS. *Units* kg m^{-4}.

acoustic intensity – *at a point in a time-stationary acoustic field* the time-average rate of energy flow per unit area, denoted by the vector **I**. The component of **I** in any direction is the time-average rate of energy flow per unit area normal to that direction. In the absence of mean flow, $\mathbf{I} = \langle p\mathbf{u} \rangle$ where p is the acoustic pressure, **u** is the particle velocity vector, and angle brackets $\langle \ldots \rangle$ denote a time average. Also known as ***sound intensity***. *Units* W m^{-2}.

Note (1): Compare ACTIVE INTENSITY, REACTIVE INTENSITY, COMPLEX ACOUSTIC INTENSITY.

Note (2): When a mean flow is present, a different expression applies. See INSTANTANEOUS ACOUSTIC INTENSITY, BLOKHINTSEV INVARIANT.

acoustic irradiance – ◆ *of a surface exposed to a sound field* the normal component of ACOUSTIC INTENSITY at the surface due to the incident sound field; i.e. the energy incident per unit time on unit area of the surface. For example, at the boundaries of a room containing a DIFFUSE FIELD with mean ENERGY DENSITY w, the acoustic irradiance is $\Phi = cw/4$, where c is the sound speed. Also known as **(*acoustic*) *irradiation strength***. *Units* W m^{-2}.

Note: This is an adaptation of the optical term ***irradiance***, which refers to electromagnetic radiation incident on a surface and is used in the field of radiative heat transfer.

acoustic irradiation strength – *at a point on a boundary* see ACOUSTIC IRRADIANCE. *Units* W m^{-2}.

acoustic levitation – the use of ACOUSTIC RADIATION FORCES to hold objects in position against the effects of other forces (e.g. gravity, buoyancy).

acoustic lumped elements – see LUMPED ELEMENTS.

acoustic Mach number – *at a point in a sound field* the acoustic particle velocity amplitude (or the rms particle velocity, if the sound field is not at a single frequency), divided by the speed of sound. *Units* none.

acoustic mobility – *of an acoustic system or transmission line* an alternative term for either ACOUSTIC ADMITTANCE or acoustic DIFFERENTIAL ADMITTANCE. See MOBILITY (1) and (2). *Units* m^3 s^{-1} Pa^{-1}.

acoustic nerve – the AUDITORY NERVE.

acoustic ohm – see SI ACOUSTIC OHM.

acoustic perfume – an equivalent term for ACOUSTIC WALLPAPER.

acoustic phase coefficient – *for a single-frequency progressive sound wave* the real part of the PROPAGATION WAVENUMBER k. *Units* rad m^{-1}.

Note: An equivalent definition is the imaginary part of the PROPAGATION COEFFICIENT jk. See also PHASE COEFFICIENT (which is not limited to acoustic waves).

acoustic power – the same as SOUND POWER. *Units* W.

acoustic pressure – *in a fluid* a time-dependent pressure disturbance, usually of small amplitude, superimposed on an ambient state. *Units* Pa.

acoustic pulse reflectometry – an experimental technique for determining the acoustic properties of a one-dimensional acoustic system (e.g. a narrow waveguide), by measuring its response to an incident acoustic pulse of known waveform. Response measurements in the time domain are processed to yield either the frequency response function of the system, or its impulse response. The technique has found numerous industrial and medical applications, and has proved useful in the study of musical wind instruments.

acoustic radiance – ◆ *of a surface* the sound power radiated (or reflected) by the surface per unit surface area; it is the integral of the ACOUSTIC BRIGHTNESS with respect to solid angle, over the entire hemisphere of radiation directions. Also known as ***acoustic radiosity***. *Units* W m^{-2}.

Note: This is an adaptation of the optical term ***radiance***, which refers to electromagnetic radiation emitted (or reflected) by a surface and is used in the field of radiative heat transfer.

acoustic radiation force – the net force on an object in a sound field due to the action of ACOUSTIC RADIATION PRESSURE. *Units* N.

acoustic radiation impedance – *of an opening* the complex ratio of the acoustic pressure averaged over the opening to the volume velocity outflow across the opening, at a single frequency:

$$Z_{\text{rad}} = \frac{\bar{p}}{U}, \quad (\bar{p} = \text{average pressure}, \ U = \text{volume velocity}).$$

In this definition, the opening is treated as a LUMPED ELEMENT, and its dimensions are assumed to be small compared with the acoustic wavelength. For an alternative definition of radiation impedance that avoids these limitations, see MODAL RADIATION IMPEDANCE. For a complete list of related terms, see under RADIATION IMPEDANCE. *Units* Pa s m^{-3}.

Note (1): The lumped-element approach ignores the fact that, even at low frequencies, $\bar{p}$ is not determined solely by U; the above definition of Z_{rad} is therefore not precise. One can make it rigorous by prescribing the velocity or pressure distribution: e.g. in the idealized case where *uniform pressure* is imposed over the opening, one obtains the low-frequency asymptotic expression

$$Z_{\text{rad}} \approx \frac{j\omega\rho}{C} + \frac{\omega^2\rho}{\Omega c}, \quad (p \text{ uniform over opening}).$$

Here C is Rayleigh's CONDUCTIVITY, ω is the angular frequency, the opening radiates into a solid angle Ω, and ρ, c are the density and sound speed of the radiating medium.

Note (2): For the idealized case where *uniform normal velocity* is imposed over the plane of the opening, see PISTON RADIATION IMPEDANCE. However neither this model, nor the uniform-pressure model of note (1), is sufficient to describe the low-frequency acoustic properties of an opening to better than 10 percent accuracy when the actual variation of pressure (or normal velocity) over the opening is unknown.

Note (3): An alternative definition of the average pressure $\bar{p}$ is $\bar{p} = \langle pu_n \rangle / \langle u_n \rangle$; here the angle brackets denote an average over the opening area, and u_n is the local normal velocity over the opening. This allows the SOUND POWER crossing the opening to be expressed without approximation as the time-average product of $\bar{p}$ and U.

Note (4): One practical case for which precise description is possible is the reflection of plane waves at the open end of a long tube; see END CORRECTION.

acoustic radiation pressure – the time-average excess pressure on a material surface due to a sound field, expressed by $\langle P^L - P_0 \rangle$. Here P^L is the Lagrangian pressure, i.e. the pressure of the fluid at a point moving with the surface, and P_0 is the steady ambient pressure, i.e. with the sound field absent. The angle brackets $\langle \ldots \rangle$ denote a time average. *Units* Pa.

Note (1): The excess pressure $\langle P^L - P_0 \rangle$ is not straightforward to evaluate from linear acoustic theory, since it is a second-order quantity. It depends in a subtle way on the Eulerian excess density, $\langle \rho^E - \rho_0 \rangle$, at a fixed point next to the surface. This in turn depends on whether the sound field is confined by rigid boundaries, or exists in an unconfined expanse of fluid. For details, see notes (2) and (3).

Note (2): The ***Rayleigh radiation pressure*** applies in circumstances where $\langle \rho^E - \rho_0 \rangle$ averages to zero over the entire field, which is the case when the sound field is confined by rigid boundaries. Its value at a rigid boundary is βw_{pot}, where β is the COEFFICIENT OF NONLINEARITY of the fluid and w_{pot} is the mean POTENTIAL ENERGY DENSITY.

Note (3): The ***Langevin radiation pressure*** applies when the fluid is unconfined. Its value at any boundary is $w_{\text{pot}} - w_{\text{kin}}$, i.e. the difference between the mean potential energy density and the mean KINETIC ENERGY DENSITY.

acoustic radiation resistance – *of an opening* the real part of the ACOUSTIC RADIATION IMPEDANCE. For an opening whose dimensions are small compared with the external acoustic wavelength, the acoustic radiation resistance $R_{a,\text{rad}}$ is defined without ambiguity by

$$R_{a,\text{rad}} = \frac{W_{\text{rad}}}{U_{\text{rms}}^2};$$

here W_{rad} is the radiated sound power, and U_{rms} is the rms volume velocity through the opening. *Units* Pa s m^{-3}.

acoustic radiation stress tensor – *in a fluid* the second-order quantity defined by

$$\Pi_{ij} = \langle P - P_0 \rangle \delta_{ij} + \rho_0 \langle u_i u_j \rangle,$$

which represents the time-average flux of j-component momentum, per unit surface area, across a fixed surface whose normal points in the i direction. Here P is the pressure, ρ is the fluid density, and u_i, u_j are particle velocity components. Subscript 0 denotes steady ambient conditions with the sound field absent, and angle brackets $\langle \ldots \rangle$ denote a time average at a fixed point. *Units* Pa.

Note: The standard definition, as given above, excludes viscous stresses.

acoustic reflex – *in response to loud sounds* a reflex contraction of the muscle attached to the stapes (third in the chain of three OSSICLES). The effect is to limit the vibration amplitude of the stapes and thus to reduce the transmission of loud sounds to the inner ear, particularly at low frequencies. Also known as the *stapedial reflex* or *middle ear muscle reflex*.

acoustic reflex tests – audiological tests in which the acoustic admittance of the ear is monitored during presentation of a moderately intense acoustic stimulus. In normal subjects, the stimulus alters the admittance via reflex contraction of muscles attached to the ossicles. See ACOUSTIC REFLEX, OTOADMITTANCE TESTS.

acoustics – (1) the science and technology of sound in all its aspects. Covers its production, propagation and control; its interaction with materials; its reception by the ear, and its effects on the hearer. Also used to denote a discipline or field of study; hence *acoustics professor/student*.

Note: A useful chart due to R B Lindsay, showing the broad subdivisions of acoustics, is reproduced in Chapter 1 of A D Pierce's *Acoustics* (**1989**). Also informative is the subject classification scheme (PACS 43) published in each volume of the *Journal of the Acoustical Society of America*.

acoustics – (2) the acoustical principles that underlie a given device or phenomenon, as in *the ~ of the trombone*, or *the ~ of the Rijke tube*.

acoustics – (3) *used as a plural noun* the acoustical properties of a space. Depending on context, either subjective or objective properties may be implied. Compare ACOUSTIC (3).

acoustic scatterer – a passive obstacle (or an inhomogeneity in the medium) that produces a SCATTERED FIELD when irradiated by sound waves.

acoustic signature – a pressure time-history, usually of a transient type (e.g. *acoustic signature of a gunshot*), that is characteristic of a particular sound source. The meaning is sometimes broadened to include any distinctive feature by which an acoustic signal may be recognized, for example (in the case of a periodic signal) a particular harmonic amplitude spectrum.

acoustic sink – an object or region that acts as a net absorber of acoustic energy (under transient conditions), or of acoustic power (under steady-state conditions).

acoustic source – (1) anything that emits sound waves. Examples of acoustic sources in this sense are a radiating transducer, a fan, and a turbulent jet; however, the last two are more likely to be described as ***noise sources*** (= sources of unwanted sound). See SOUND, NOISE.

acoustic source – (2) *as opposed to an acoustic sink* an object or region that emits positive net acoustic energy (under transient conditions), or positive acoustic power (under steady-state conditions), into the surrounding medium.

acoustic source – (3) *as input to the wave equation* an ACOUSTIC SOURCE DENSITY defined over a specific limited region in space; a particular example is a POINT MONOPOLE. See also ACOUSTIC SOURCE STRENGTH.

acoustic source density – the quantity that appears on the right-hand side of the inhomogeneous wave equation that describes sound generation in a uniform fluid at rest. Specifically, if the wave equation for the acoustic pressure, p, is written as

$$\left(\frac{1}{c_0^2}\frac{\partial^2}{\partial t^2} - \nabla^2\right)p = \mathcal{F},$$

where t is time and c_0 is the sound speed in the undisturbed fluid, then $\mathcal{F}$ represents the acoustic source density for pressure. *Units* Pa m^{-2}.

Note: Alternative choices for the acoustic variable are the density variation or the velocity potential. Different source densities apply to pressure, density or velocity potential as acoustic variables; their values may be distinguished by writing $\mathcal{F}$, $\mathcal{F}_\rho$, $\mathcal{F}_\varphi$ for the three cases, respectively. See also SOURCE MOMENTS.

acoustic source distribution – (1) a distributed time-dependent VOLUME VELOCITY input applied to an acoustic medium. Expressed as volume velocity per unit volume; symbol Q. Compare ACOUSTIC SOURCE STRENGTH (1). *Units* s^{-1}.

Note: The *volume velocity distribution* Q is equivalent to $\mathcal{F}_\varphi$, the ACOUSTIC SOURCE DENSITY for the velocity potential φ, provided one adopts the sign convention that the fluid velocity is given by $-\nabla\varphi$.

acoustic source distribution – (2) an equivalent term for ACOUSTIC SOURCE DENSITY.

acoustic source strength – (1) the instantaneous unsteady VOLUME VELOCITY with which an acoustic source displaces the surrounding medium. *Units* $m^3 s^{-1}$.

acoustic source strength – (2) *with reference to an acoustic wave equation* the instantaneous volume integral of the corresponding ACOUSTIC SOURCE DENSITY. In mathematical form, the volume integral

$$S = \int \mathcal{F} dV \quad \text{(or a similar expression with subscript } \varphi \text{ on } S, \mathcal{F})$$

relates the source strength S to the source density $\mathcal{F}$. Equivalent terms are **monopole strength** and **simple source strength**. See also SOURCE MOMENTS.

Note: If the acoustic variable is taken to be the velocity potential φ, defined such that the particle velocity is $-\nabla\varphi$, then S_φ is equivalent to a localized volume velocity input U driving the medium; compare ACOUSTIC SOURCE STRENGTH (1). If the acoustic variable is taken to be the pressure p, then $S = \rho_0 \dot{U}$ where $\dot{U}$ is the equivalent local *volume acceleration* input and ρ_0 is the ambient fluid density.

acoustic source strength distribution – *in the context of far-field imaging of line sources* the mean square pressure radiated in a given direction per unit length of source region, multiplied by the square of the measurement radius and expressed as a distribution along a given axis. One way to measure this quantity is the POLAR CORRELATION TECHNIQUE. *Units* Pa^2 m.

acoustic streaming – a steady fluid flow set up by a sound source or oscillating solid boundary.

acoustic taxonomy – the identification and classification of animal species by their acoustic signature.

acoustic telescope – a device for imaging a distributed acoustic source from the far field. It may consist of an ellipsoidal reflector, with the remote focus placed near the source, or alternatively, of an array of microphones whose outputs can be processed to yield an image of the source distribution (see POLAR CORRELATION TECHNIQUE).

acoustic thermometry – the use of acoustic travel time data to infer the temperature along an underwater ray path, by inversion of standard relationships between temperature and sound speed. See ATOC, ACOUSTICAL OCEANOGRAPHY, and SOUND SPEED IN SEAWATER.

acoustic tomography – *in underwater acoustics* the analysis of coded signals exchanged between multiple pairs of transmitting and receiving systems, in order to study ocean or sea bed properties. As an example, travel time information over multiple paths can be inverted to construct a map of sound speed as a function of position in the ocean (see INVERSE PROBLEMS). Compare ACOUSTIC THERMOMETRY, ATOC.

acoustic trauma – instantaneous injury to or destruction of one or more components of the auditory system, caused by exposure to a very high transient sound pressure (e.g. from an explosion or weapons fire). The term is not to be confused with NOISE-INDUCED HEARING LOSS associated with chronic exposure, or with BAROTRAUMA.

acoustic turbulence – a term used to describe the finite-amplitude evolution of a broadband acoustic waveform, as it propagates under the combined effects of nonlinearity and dissipation. The term implies an analogy with hydrodynamic turbulence, where the same two effects operate (although in a more complicated manner, turbulence in fluid flow being three-dimensional). See also BURGERS EQUATION.

acoustic wallpaper – background sound deliberately introduced in order to mask other noise sources, thereby generating a more acceptable acoustic environment.

acoustic wavelength – *for sound at a specified frequency in a given medium* the quantity c/f, where f is the frequency and c is the sound speed in the medium. Equivalent terms are **wavelength of sound** and **sound wavelength**. *Units* m.

acoustic wavenumber – *for single-frequency sound in a lossless medium* the ratio $\omega/c = k_0$, where ω is the angular frequency and c is the sound speed in the medium. See also PROPAGATION WAVENUMBER. *Units* rad m^{-1}.

acoustoelastic effect – the phenomenon in which a static state of stress or strain applied to an elastic medium alters the speed of propagation of small-amplitude elastic waves (shear or compressional).

acoustoelasticity – the science of interactions between acoustic waves (stress waves) and steady strain fields, in a solid.

acoustooptic effect – the modulation of a light beam as it passes through an acoustically excited material, whose optical properties vary with the local state of strain.

ac power – see POWER (3).

action level – *in noise at work regulations* a critical value of some statistical indicator for noise – typically a NOISE EXPOSURE LEVEL but not necessarily a level in the logarithmic sense – at which either an employer is required to take remedial measures, or an employee is required to wear hearing protection.

Note: In Europe, action levels are also set by EU directives. See also PEAK ACTION LEVEL.

active control – *of sound or vibration* the use of secondary sources of excitation to cancel, or reduce, the response of a system to given primary sources; also to suppress self-excited oscillations of a system that is unstable.

active intensity – an equivalent term for ACOUSTIC INTENSITY, interpreted as a time-average quantity. The term is mainly applied to single-frequency sound fields, where it is useful to contrast active and reactive intensity components. See also REACTIVE INTENSITY, COMPLEX ACOUSTIC INTENSITY. *Units* W m^{-2}.

Note (1): If the acoustic pressure and particle velocity in a single-frequency 3D sound field are represented by

$$p = \text{Re}[P(\mathbf{x})\, e^{j\omega t}], \quad \mathbf{u} = \text{Re}[\mathbf{U}(\mathbf{x})\, e^{j\omega t}],$$

where $P(\mathbf{x})$ and $\mathbf{U}(\mathbf{x})$ are the complex pressure and velocity amplitudes at vector position $\mathbf{x}$, then the *active intensity vector* $\mathbf{I}$ is

$$\mathbf{I} = \frac{1}{2}\text{Re}(P^*\mathbf{U}), \qquad (* = \text{complex conjugate}).$$

An equivalent expression based on the rms pressure p_{rms} and the phase gradient $\nabla\varphi$ is

$$\mathbf{I} = -\frac{p_{\text{rms}}^2}{\omega\rho}\nabla\varphi \qquad (\varphi = \arg P),$$

where ρ is the fluid density.

Note (2): These relations apply to lossless fluids with no mean flow. In real situations, they are useful approximations provided the attenuation per wavelength is small, i.e. $\alpha\lambda \ll 1$, and also $M \ll 1$ where M is the mean-flow Mach number. For finite M, see note (2) under INSTANTANEOUS ACOUSTIC INTENSITY.

active noise control – see ACTIVE CONTROL.

active sonar – use of a sound-emitting transducer to project sound waves at objects to be detected underwater; the scattered or reflected signal is then detected either at the emitting location (MONOSTATIC) or at another location (BISTATIC). Subsequent signal processing yields the direction, range, and closing speed of the reflecting object (with simultaneous determination of position and speed subject to the uncertainty principle).

active transducer – a transducer whose operation relies, at least in part, on energy from an external source other than the input signal. Examples are a strain gauge and a condenser microphone. An alternative term is ***modulator***. See TRANSDUCER.

active vibration control – see ACTIVE CONTROL.

acuity – *of hearing* ability to detect faint sounds, especially in the presence of masking sounds or noise.

A-D, A/D – abbreviations for ANALOGUE-TO-DIGITAL, as in ∼ *converter*.

adaptation – a reversible reduction in a particular neurophysiological response to a given stimulus, associated with continued exposure to the stimulus.

adaptive beamforming – *for the detection of incoming acoustic waves in the presence of noise* automatic adjustment of an array's BEAM PATTERN so as to minimize the gain in the noise source direction(s), while maximizing the gain in the direction of the source to be detected. The array weighting factors are calculated adaptively in real time, in response to variations in the signals at each array element.

ADC – abbreviation for ANALOGUE-TO-DIGITAL CONVERTER.

added mass – alternative term for VIRTUAL MASS. *Units* kg.

adiabat – a curve representing the relationship between any pair of state variables at the beginning and end of a thermodynamic process, when the process is ADIABATIC (but not necessarily REVERSIBLE). The compression process across a shock wave, for example, is described by the ***shock adiabat*** in P–V coordinates (P = pressure, V = specific volume).

adiabatic – (1) *in thermodynamics* without heat transfer (~ *compression*); impenetrable to heat (~ *boundary*). During an ***adiabatic process***, a system can exchange energy with its surroundings, but only via WORK.

adiabatic – (2) *in wave propagation* without conversion of wave energy between different modes of propagation, as in a slowly-varying environment (see ~ APPROXIMATION, ~ MODES). The underlying idea is that no energy transfer occurs between modes.

adiabatic approximation – *for sound propagation in a waveguide whose properties vary slowly in the propagation coordinate direction* the assumption that waveguide modes are uncoupled, each mode propagating with the wavenumber it would have in a uniform waveguide with the same local geometry.

adiabatic exponent – alternative term for ISENTROPIC EXPONENT. Compare POLYTROPIC EXPONENT. *Units* none.

adiabatic invariant – a quantity that is conserved under ADIABATIC (2) changes.

adiabatic modes – *in a waveguide whose properties vary slowly in the propagation direction* transverse modes determined by regarding the waveguide as locally uniform. In the limit of slow axial variations in either waveguide geometry or boundary impedance, the mode shapes and amplitudes evolve gradually during propagation, without exchange of energy between modes.

adjoint matrix – see HERMITIAN TRANSPOSE.

admittance – *in acoustics* a generic term for the reciprocal of an IMPEDANCE. Equivalently, it is a FREQUENCY RESPONSE FUNCTION in which velocity u (in a given direction) or volume velocity U (across a specified surface) is regarded as the output, and pressure p or force F (in a given direction) is regarded as the input. See ACOUSTIC ~, MECHANICAL ~, SPECIFIC ACOUSTIC ~.

admittance matrix – the inverse of an IMPEDANCE MATRIX.

admittance ratio – the SPECIFIC ACOUSTIC ADMITTANCE multiplied by ρc, where ρ is the density of the acoustic medium and c is the sound speed. *Units* none.

advection – *in fluid dynamics* convective transport of a passive component in a fluid mixture. The term can also be used for momentum or energy, as in *advective transport of momentum by turbulent eddies*.

aeolian tones – sound with a tonal character, generated by regular vortex shedding from a circular cylinder. Such vortex shedding occurs spontaneously in a certain range of Reynolds numbers (see below), when the cylinder is placed in a steady fluid flow with its axis mounted transversely to the oncoming flow direction. The sound is related to unsteady aerodynamic forces exerted by the cylinder on the fluid; it has an intensity maximum in the direction orthogonal to both the flow and the cylinder axis.

The frequency spectrum of the radiated sound depends on the REYNOLDS NUMBER $Re = ud/v$ (u = flow speed, d = cylinder diameter, v = kinematic viscosity). Over the range $400 < Re < 300\,000$, the spectrum contains a distinctive peak centred on frequency $f = 0.2\,u/d$, i.e. a STROUHAL NUMBER of $S = 0.2$. At higher Reynolds numbers, or with a roughened surface, or in the presence of incident sound, the flow approaches a fully-turbulent state and the Strouhal number shifts to around 0.27.

There is a transitional range of Reynolds numbers, roughly $300\,000 < Re < 3 \times 10^6$, over which the flow past a smooth circular cylinder lacks a regular coherent structure; aeolian tones are then absent, or much less prominent.

Note: A cylinder that can vibrate in the transverse direction will couple to the vortex shedding process when the aeolian tone frequency ($S \approx 0.2$) approaches that of a mechanical resonance. The coupled system produces modified tone frequencies that can "lock on" to successive resonances over a Strouhal number range of an octave or more.

aeroacoustics – the science of sound production by fluid flow, or by the interaction of flows with solid bodies. The term is also used to describe flow–acoustic interaction phenomena generally.

aerodynamic force – the force on an object in a surrounding fluid, due to relative motion between them. The aerodynamic force vector is conventionally divided into two components: the *lift force* which acts transversely to the relative velocity, and the *drag force* which acts parallel to the relative velocity. *Units* N.

aerodynamic sound – sound that is excited by a region of turbulent or unstable flow, and radiates into the surrounding fluid. At low Mach numbers, the radiation can be greatly amplified through scattering of the hydrodynamic near field, either by solid objects in the flow, or by local regions of different density or compressibility.

A distinguishing feature of aerodynamic sound is that the radiated acoustic energy is derived from the flow itself, rather than being generated directly by motion of the boundaries: these remain fixed, or else act on the sound field as passive acoustic absorbers. Aeroelastic instabilities may nevertheless contribute to aerodynamic sound generation, by extracting energy from a steady flow to set up an unsteady flow which then in turn radiates sound. See AEOLIAN TONES.

aeroelasticity – the study of dynamic interactions between an elastic structure (e.g. an aircraft wing), and a surrounding mean flow; the flow may be steady, or have disturbances superimposed on it. Aeroelastic phenomena include *flutter*, which is flow-excited instability; *gust response*, which is structural vibration caused by flow unsteadiness; and *divergence*, which is static instability under steady aerodynamic loading.

aetiology – scientific study of the factors involved in disease causation; also, for a particular disease, the causative factors themselves. For example, the aetiology of noise-induced hearing loss covers the chain of causation from initial exposure to physiological damage, usually focusing on a particular population at risk.

age variable – *for transmission between two points on a ray path* an integral evaluated along an acoustic ray path, that allows the nonlinear distortion of a signal waveform to be quantified. Specifically, it gives the amplitude-dependent travel time of a wavelet as

$$\Delta t \approx (\Delta t)_{\text{lin}} - p_+ L,$$

where p_+ is the acoustic pressure of the wavelet at the starting point on the ray path, L is the age variable, and $(\Delta t)_{\text{lin}}$ is the travel time given by linear acoustics. *Units* s Pa^{-1}.

Note: The relation between the age variable and the REDUCED PATH LENGTH, $\tilde{x}$, takes the following form in a stationary medium:

$$L = \left(\frac{\beta}{\rho c^3}\right)_+ \tilde{x}.$$

Here β is the nonlinearity coefficient of the medium, and ρ, c are the density and sound speed; the $+$ subscript indicates that these are all evaluated at the start of the ray path.

agglomeration – see ACOUSTIC AGGLOMERATION.

airborne path – a sound transmission path in which energy is carried principally by the fluid (air), and only to a minor extent via structural or solid-borne waves.

airborne sound insulation index – an alternative term for WEIGHTED SOUND REDUCTION INDEX. *Units* dB.

air conduction – *in audiology* the transmission of sound as airborne pressure fluctuations through the external ear to the eardrum, and from there via the ossicles of the middle ear to the cochlea. Abbreviated as *a-c*.

air-conduction audiometry – see PURE-TONE AUDIOMETRY.

airframe noise – that part of the noise radiated from an aircraft in flight that is not associated with the engines.

aliasing – the apparent conversion of high-frequency signal energy into lower frequencies that results from discrete sampling at a finite rate. Components of the original signal at frequencies $|f| \geq f_s/2$, where f_s is the sampling frequency, are folded back into the range $|f| < f_s/2$ (**aliased**) by addition or subtraction of a multiple of f_s. Components of the original signal at frequencies $|f| < f_s/2$ are not affected. See ANTI-ALIASING FILTER.

ambient – *used as an adjective* prevailing, surrounding. Hence ∼ *conditions*, the temperature, pressure, etc. in the surrounding atmosphere (or other fluid environment); see also STANDARD AMBIENT CONDITIONS. In acoustics, the term ∼ *value* (of pressure, velocity, etc.) is commonly used to refer to a steady background value on which acoustic disturbances are superimposed.

ambient noise – (1) *in underwater acoustics* the naturally-occurring acoustic environment in the ocean, caused by wave breaking, marine life, etc. (but not ships or other human activity). For ambient noise in sonar, see definition (2).

ambient noise – (2) *in sonar detection* the noise from all unwanted sources of sound, apart from those directly associated with the sonar equipment and the platform on which it is mounted. Compare SONAR SELF-NOISE.

ambient noise – (3) a generic term for the ACOUSTIC ENVIRONMENT (3) that prevails under normal conditions, for example at a given outdoor location.

Note: In the context of noise measurement or environmental noise assessment, ambient noise is what remains after any noise source being investigated has either been turned off, or suppressed to the point where its contribution is insignificant.

ambient noise level – the SOUND PRESSURE LEVEL due to ambient noise. *Units* dB re p_{ref}^2.

Note: The ambient noise level is commonly expressed for environmental assessment purposes as an EQUIVALENT CONTINUOUS SOUND PRESSURE LEVEL, L_{eq}, with A-weighting applied. Compare RESIDUAL NOISE LEVEL, which is a technical term used in noise assessment standards.

amplification – (1) an increase in signal amplitude or rms value.

amplification – (2) the amount by which the peak or rms value of a signal is increased, expressed in decibels. *Units* dB.

amplitude – the peak value of a sinusoidal signal; more generally, the maximum departure from equilibrium in any oscillation. Mathematically, the amplitude is the positive real coefficient A in the expression $A \cos(\omega t + \alpha)$ for a signal of angular frequency ω; here t is time, and ω, α are real constants.

Note: If a sinusoidal CARRIER SIGNAL is amplitude-modulated, with instantaneous value $A(t) \cos(\omega t + \alpha)$, the time-varying coefficient $A(t)$ may still be called the amplitude, provided it varies slowly compared with the cosine term. For example, with $A(t) = A_0 e^{-\delta t}$ the expression above describes the exponentially decaying amplitude of a lightly-damped mode, during FREE OSCILLATION of a linear system. For a more general approach to defining instantaneous amplitude that applies to non-sinusoidal signals, see ANALYTIC SIGNAL.

amplitude attenuation coefficient – ◆ *of a single-frequency progressive wave system* an alternative term for ATTENUATION COEFFICIENT, by analogy with AMPLITUDE DECAY COEFFICIENT. *Units* Np m^{-1}.

amplitude decay coefficient – *of a linear system, for a single mode of free vibration* the coefficient δ in the temporal decay factor $e^{-\delta t}$, which describes the amplitude decay of damped free oscillations with time, t. Also known as **damping constant**. See also LOGARITHMIC DECREMENT, COMPLEX NATURAL FREQUENCY. *Units* Np s^{-1}.

amplitude distortion – *of a linear system* refers to distortion of the output waveform that is caused by the system GAIN FACTOR varying with frequency.

amplitude focal gain – *of an acoustic transducer that produces a single-frequency focused beam* the pressure amplitude at the focus divided by the pressure amplitude at the transducer face. Also known as **amplitude gain**, **focusing gain**, or **pressure gain factor**. *Units* none.

amplitude modulation – see MODULATION (1).

amplitude response function – *of a linear time-invariant system* the magnitude of the system FREQUENCY RESPONSE FUNCTION; also known as the **gain factor** of the system. Thus if $H(\omega)$ is the frequency response function at angular frequency ω, the amplitude response function is $|H(\omega)|$.

amplitude spectrum – *of a periodic continuous signal* the infinite set of discrete Fourier harmonic magnitudes, plotted against frequency. See FOURIER ANALYSIS, COMPLEX FOURIER AMPLITUDES. For a real periodic signal, the amplitude spectrum is even with respect to frequency.

Note: Compare MAGNITUDE SPECTRUM, which relates to the Fourier transform of a transient signal.

analog ∼ – see ANALOGUE ∼.

analogous circuit, analogous electrical circuit – see EQUIVALENT CIRCUIT.

analogous flow resistance – *of an acoustic lumped element* an equivalent term for FLOW RESISTANCE. *Units* Pa s m^{-3}.

analogue filter – a linear device that converts an analogue input signal $x(t)$ into an analogue output signal $y(t)$. In signal processing, analogue filters are used for pre-processing the input signal prior to A-D conversion; see ANTI-ALIASING FILTER.

analogue signal – a signal whose value over a specified time interval is defined continuously for all times, in contrast to a DIGITAL SIGNAL. Equivalent terms are *continuous signal* and *continuous-time signal*.

analogue system – (1) a hardware device that converts an analogue input signal $x(t)$ into an analogue output signal $y(t)$, or more generally a set of multiple input signals $\mathbf{x}(t)$ into a set of multiple output signals $\mathbf{y}(t)$.

analogue system – (2) a rule or transformation for mapping the set of inputs $\mathbf{x}(t)$ to the set of outputs $\mathbf{y}(t)$.

analogue-to-digital conversion – the process of converting an ANALOGUE SIGNAL into a DIGITAL SIGNAL. It consists of SAMPLING the analogue signal, usually at equal time intervals; quantizing the sampled values; and encoding the sequence thus produced into digital words of finite length. See also SAMPLING THEOREM, ALIASING.

analogue-to-digital converter – a device for converting an analogue signal into a digital signal.

analytic function – *of a complex variable* a function $f(z)$ that is **differentiable** with respect to the complex variable z, in the sense that the following limit exists:

$$\lim_{\delta z \to 0} \frac{f(z + \delta z) - f(z)}{\delta z} = f'(z).$$

Note this is more restrictive than saying that $\partial f/\partial x$, $\partial f/\partial y$ exist, where x and y are coordinates in the complex plane such that $z = x + jy$.

Note (1): A function is said to be analytic in domain D if it is differentiable over a limited region D of the complex plane. If $f(z)$ is analytic in D, it is infinitely differentiable (in the sense above) throughout D.

Note (2): The real and imaginary parts, $u(x, y)$ and $v(x, y)$, of any analytic function are related by the **Cauchy–Riemann equations**

$$\frac{\partial u}{\partial x} = \frac{\partial v}{\partial y}, \quad \frac{\partial u}{\partial y} = -\frac{\partial v}{\partial x}.$$

analytic signal – a complex time-domain signal, formed from a real signal $x(t)$ by adding an imaginary component $jx_H(t)$, where subscript H denotes the HILBERT TRANSFORM. Thus the analytic signal $\check{x}(t)$ based on $x(t)$ is

$\check{x}(t) = x(t) + jx_H(t)$.

As an example, if $x(t)$ equals $\cos \omega t$, then $\check{x}(t)$ equals $e^{j\omega t}$. Also known as the *pre-envelope signal*. Compare QUADRATURE FUNCTION.

Note: The analytic signal is obtained by suppressing the negative-frequency components of $x(t)$ and doubling the result. Therefore its energy is twice that of the original signal.

anechoic – non-reflecting with respect to sound waves; totally absorbing.

anechoic boundary condition – *in computational acoustics* an equivalent term for ABSORBING BOUNDARY CONDITION.

anechoic chamber, anechoic room – a room designed to simulate FREE-FIELD acoustic conditions; also known as a *free-field room*. The surfaces are covered with sound-absorbing material, often in the form of wedges pointing into the room, so that sound waves arriving at the room boundaries are almost entirely absorbed. See also SEMI-ANECHOIC ROOM, DIRECT FIELD.

anechoic tank – the equivalent of an ANECHOIC CHAMBER, but with water as the acoustic medium. The other difference is that the water in the tank usually has a free surface, which is almost perfectly reflecting.

anechoic termination – an ideal termination (e.g. in a duct or waveguide) that reflects none of the acoustic power incident on it. See also IMPEDANCE MATCHING. A true anechoic termination is not achievable in practice, but the term is often used for a real termination that is designed to be as nearly anechoic as possible.

angle – *of a complex number* the angle θ used to represent the complex number in POLAR FORM,

$$z = re^{j\theta} \quad (r, \theta \text{ real}).$$

Equivalent terms are *phase* or *argument*. In symbols, the statement that θ is the angle of z is written as

$$\theta = \angle z, \quad \theta = \sphericalangle z, \quad \text{or} \quad \theta = \arg z.$$

angle of incidence – *of a progressive plane wave arriving at a plane boundary* the angle between the surface normal $\mathbf{n}$ (pointing outward from the boundary into the incident medium) and the direction from which the arriving wave approaches. This is equivalent to saying that if the incident wave has WAVENORMAL $\mathbf{i}$, the angle of incidence θ_{inc} is given by

$$\cos \theta_{\text{inc}} = -\mathbf{i} \cdot \mathbf{n}.$$

See also GRAZING INCIDENCE, NORMAL INCIDENCE. *Units* rad (but commonly expressed in degrees).

angular frequency – *of a sinusoidal oscillation* 2π times the FREQUENCY; equivalently, the rate of change of INSTANTANEOUS PHASE with time. Also known as *circular frequency* or *radian frequency*. *Units* rad s^{-1}.

angular intensity distribution – ◆ *at a point in a reverberant sound field* the factor $I_\Omega(\theta, \phi)$ in the expression $dI = I_\Omega(\theta, \phi) \, d\Omega$, which gives the ACOUSTIC INTENSITY due to plane-wave components arriving at the given point within a narrow cone of propagation directions centred on (θ, ϕ). Here (θ, ϕ) are SPHERICAL POLAR COORDINATES, and $d\Omega = \sin \theta \, d\theta \, d\phi$ is the cone SOLID ANGLE.

Underlying this definition is the idea that the REVERBERANT FIELD is made up of infinitely many uncorrelated plane waves, whose contributions to the intensity in any given direction are additive. *Units* W m^{-2} sr^{-1}.

Note: The angular intensity distribution in acoustics is analogous to *radiance* in optics. An alternative term is *solid-angle distribution of intensity*.

angular momentum – *of a system about a specified point* a vector quantity obtained by taking the MOMENT, about the given point, of the linear MOMENTUM of each particle, and summing the result over the whole system.

In symbols, the contribution of each particle to the angular momentum about point A is $(\mathbf{r} - \mathbf{r}_A) \times \dot{\mathbf{r}}$ per unit mass, where $\mathbf{r}, \mathbf{r}_A$ are the positions of the element and point A respectively, and $\dot{\mathbf{r}}$ is the velocity of the element relative to an INERTIAL FRAME OF REFERENCE. If $\mathbf{s}$ denotes the separation vector $\mathbf{r} - \mathbf{r}_A$, the total angular momentum of the system about A may be written as

$$\mathbf{L} = \Sigma(\mathbf{s} \times \dot{\mathbf{s}})m_i + (\mathbf{s}_c \times \dot{\mathbf{r}}_A)m_{tot},$$

where m_i is the mass of the ith particle, m_{tot} is the total mass, and subscript c denotes the MASS CENTRE. *Units* kg m² s⁻¹ ≡ N m s.

Note: The conservation law for angular momentum states that the angular momentum, $\mathbf{L}$, of a closed system about point A changes with time according to

$$\frac{d\mathbf{L}}{dt} = \mathbf{G}$$

where $\mathbf{G}$ is the moment about A of the forces applied to the system.

angular power distribution – ◆ *of a sound source* the SOUND POWER per unit SOLID ANGLE in the far field. Suppose a particular radiation direction from the source is specified by (θ, ϕ) in SPHERICAL POLAR COORDINATES. If the angular power distribution is denoted by $W_\Omega(\theta, \phi)$, then the sound power radiated by the source within a narrow cone centred on (θ, ϕ) is

$$dW = W_\Omega(\theta, \phi)\, d\Omega,$$

where $d\Omega$ is the cone solid angle. It follows that the far-field radial intensity I_r at distance r from the source is given by

$$I_r(r, \theta, \phi) = \frac{1}{r^2} W_\Omega(\theta, \phi).$$

Units W sr⁻¹.

Note (1): For a related quantity in underwater acoustics, see SOURCE LEVEL.

Note (2): The angular power distribution in acoustics is analogous to *radiant intensity* in optics.

angular radiated-energy distribution – ◆ *of a transient sound source* the acoustic energy (equal to the time-integrated SOUND POWER) radiated from the source into the far field, per unit SOLID ANGLE. The definition is analogous to that for ANGULAR POWER DISTRIBUTION: the acoustic energy radiated by the source within a narrow cone centred on (θ, ϕ) is

$$dE = E_\Omega(\theta, \phi)\, d\Omega,$$

where $d\Omega$ is the cone solid angle. It follows that the radial TIME-INTEGRATED INTENSITY N_r at distance r from the source is given by

$$N_r(r, \theta, \phi) = \frac{1}{r^2} E_\Omega(\theta, \phi).$$

Units J sr⁻¹.

30 angular velocity

Note: For a related quantity in underwater acoustics, see ENERGY SOURCE LEVEL.

angular velocity – (1) a vector whose component in any direction defines a rotation rate, measured in radians per unit time, about an axis pointing in that direction. *Units* rad s^{-1}.

angular velocity – (2) *of one point in space with respect to another* the vector $(\mathbf{s} \times \dot{\mathbf{s}})/s^2$, where $\mathbf{s}$ is the separation vector of the two points and s is their separation distance. *Units* rad s^{-1}.

angular velocity – (3) *of a unit vector* if the direction of a UNIT VECTOR $\mathbf{e}$ changes with time, the angular velocity of the unit vector is $\boldsymbol{\omega} = \mathbf{e} \times \dot{\mathbf{e}}$. *Units* rad s^{-1}.

angular wavenumber – 2π times the number of cycles per unit distance, for a quantity that varies sinusoidally with position along a specified axis. Often referred to in acoustics simply as WAVENUMBER. Angular wavenumber may be considered as the spatial analogue of ANGULAR FREQUENCY. *Units* rad m^{-1}.

anharmonic component – *in a signal consisting of discrete-frequency components* an individual sinusoidal component whose period is not integrally related to the other periods present.

animal bioacoustics – the science that deals with sound production and reception by animals, birds, or fish. The sound may be airborne, ground-borne or waterborne. Compare BIOACOUSTICS.

anisotropic – having properties that vary according to direction. An *anisotropic medium* in acoustics implies a medium with different wave propagation properties in different directions (e.g. a crystal, or a fibre-reinforced composite with preferential alignment of the fibres). When plane waves propagate in such a medium, the direction of energy propagation is no longer normal to the wavefronts – see GROUP VELOCITY. Compare ISOTROPIC.

anti-aliasing filter – an analogue low-pass FILTER applied to a signal prior to sampling; its purpose is to remove high-frequency components that would otherwise be ALIASED, or folded down to lower frequencies.

anticausal response – having no CAUSAL part; an anticausal response occurs entirely before the input. Compare ACAUSAL RESPONSE.

antinode – a point of maximum amplitude in a one-dimensional standing wave field (e.g. *pressure* ∼ *in a standing-wave tube; velocity* ∼ *on a transversely* *vibrating string*); alternatively, a *line* of maximum amplitude in a 2D field, or a *surface* of maximum amplitude in a 3D field. Compare NODE.

antiphase – two sinusoidal signals at the same frequency are said to be in antiphase if the phase difference between them is π (or 180°). An equivalent statement is that the signals are of *opposite phase* or *opposite polarity*.

antiresonance – the occurrence of a minimum in the magnitude of a driving-point frequency response function, as the frequency is varied. In terms of a transmission line analogy, an antiresonance occurs when waves return to the driving point in antiphase with the original outgoing wave.

anti-sound – the use of secondary acoustic sources to achieve sound cancellation. Equivalent terms are *active control of sound*, and *active noise control*. Compare ACTIVE CONTROL.

aperiodic – non-periodic or non-repeating; the opposite of PERIODIC. A transient signal is aperiodic.

aperture shading – *of an acoustic transducer* introduction of a profile of surface vibration amplitude over the transducer face; also known as APODIZATION.

apodization – (1) application of amplitude weighting, or APERTURE SHADING, to the active radiating face of an acoustic transducer, especially in ultrasonics.

apodization – (2) an equivalent word for WINDOWING. In ultrasonics the term is used for spatial windowing, for example by an aperture. Applied to signals in the time domain, apodization can refer either to direct windowing, or to the truncation of a correlogram in order to limit it to finite time delays.

apparent mass – the reciprocal of the real part of the ACCELERANCE. *Units* kg.

apparent sound reduction index – *of a partition separating two rooms* the result of applying the standard level difference measurement procedure (as for LABORATORY TRANSMISSION LOSS), but in the field. This means that measurements are made between two rooms in a building where the sound fields may not be diffuse, and where there may be flanking transmission paths. Symbol R'. *Units* dB.

apparent source width (ASW) – *of a sound source in a room* the apparent angle subtended by the source at the listener's position, measured in degrees in the horizontal plane. The apparent source width provides a geometrical measure of SOURCE BROADENING that can be compared in test situations between listeners. *Units* deg.

architectural acoustics – the ACOUSTICS (1) of the built environment. Some of its elements are: room acoustics, acoustical design of auditoria and concert halls, control of noise transmission into and out of rooms or buildings, and relevant areas of sound perception and psychoacoustics.

arctangent – if $x = \tan \theta$, then the angle $\theta = \tan^{-1} x$ is called the arctangent of x, or alternatively the *inverse tangent*. See INVERSE CIRCULAR FUNCTIONS.

arg – abbreviation for the ARGUMENT function, arg z, that gives the phase of a complex number z. See also PHASE AND MAGNITUDE REPRESENTATION.

Argand diagram – the complex plane, with coordinates x and y denoting the real and imaginary parts of any complex number (**J R ARGAND 1806**).

argument – *of a complex number* the angle θ used to represent the complex number in POLAR FORM,

$$z = re^{j\theta} \qquad (r, \theta \text{ real}).$$

In symbols, the statement that θ is the argument of z is written as

$$\theta = \arg z, \quad \theta = \angle z, \quad \text{or} \quad \theta = \measuredangle z.$$

array – a composite transducer made up of a number of similar elements. Arrays are used both for receiving sound (*receiving array*) and for radiating sound (*transmitting array*, *source array*).

array gain – a decibel measure of the enhancement in signal-to-noise ratio (SNR) provided by an ARRAY of sensors, as compared with a single sensor element. Enhancement of the overall SNR beyond the capabilities of a single sensor relies on the wanted signal being COHERENT across array elements in a predictable manner. In general, the array gain depends on the properties of the signal and noise fields, as well as the properties of the array.

The simplest situation is where the noise signal has zero coherence between array elements. The array gain in this case, for an N-element array in which signal i is weighted by a factor w_i, is given by $10 \log_{10} \Gamma$ where Γ is defined by

$$\Gamma = \left| \sum_{i=1}^{N} w_i \right|^2 / \sum_{i=1}^{N} |w_i|^2.$$

For a uniformly weighted array, the array gain equals $10 \log_{10} N$. *Units* dB.

Note: The vector with elements w_i is called the *weighting function* of the array.

array impedance matrix – *of elements in a radiating array* see MUTUAL IMPEDANCE.

array processing – simultaneous combined processing of time-domain signals from separate elements of a sensing array, in order to cancel out noise plus interference and enhance a desired incoming signal.

array sensitivity – see BEAM PATTERN. *Units* V Pa^{-1}.

array shading – *of a receiving or transmitting array* the application of a sensitivity profile across the elements of a receiving array, or a voltage profile across a transmitting array. Compare APERTURE SHADING.

articulation index (AI) – a method for predicting the output INTELLIGIBILITY of a speech transmission channel, based on the signal-to-noise ratio at the listener. To some extent it has been replaced by the SPEECH TRANSMISSION INDEX (STI) which takes account of the channel impulse response as well as the signal-to-noise ratio. *Units* none.

artificial ear – *in audiology* an equivalent older term for EAR SIMULATOR; the latter is the preferred term in IEC Standards.

A-scan – *in diagnostic ultrasound* a technique based on detecting signals back-scattered from an ultrasound beam; it produces information in one spatial dimension only, namely depth measured along the beam axis. Its output shows the relative amplitude of the reflected signal plotted as a function of time delay, or equivalently as a function of depth into the sample. The use of single A-scans is now largely obsolete; compare B-SCAN.

asdic – obsolete term (UK) for SONAR, in use between the end of WWI and the end of WWII. According to the Longman Dictionary of the English Language, it originated from the acronym for Anti-Submarine Detection Investigation Committee. A similar obsolete term is ***asdics*** (*ASD-ics*, with the ending -ics analogous to that in *physics*).

associated Legendre functions – solutions of the second-order differential equation

$$(1 - x^2)g'' - 2xg' + \left(\lambda - \frac{\mu^2}{1 - x^2}\right)g = 0$$

for the unknown function $g(x)$. Here λ, μ are constants and derivatives of g are denoted by primes. The equation has two linearly independent solutions, $P_\nu^\mu(x)$ and $Q_\nu^\mu(x)$, known as associated Legendre functions of the first and second kind respectively. The upper index μ is sometimes called the *order*. The lower index ν is then called the *degree*; it is related to λ by $\nu(\nu + 1) = \lambda$.

Note: In the special case where the order $\mu = 0$, the $P_\nu^\mu(x)$ solutions are called **Legendre functions**. For positive integer values of ν, say $\nu = n$, these functions are polynomials of degree n in x, called LEGENDRE POLYNOMIALS. See also SPHERICAL HARMONICS.

asterisk (*) – the notation X^* means the COMPLEX CONJUGATE of X. The different notation $x(t) * w(t)$ or $x * w|_t$ means the CONVOLUTION of $x(t)$ and $w(t)$, defined by

$$x(t) * w(t) = x * w|_t = \int_{-\infty}^{\infty} x(\tau)w(t - \tau)d\tau.$$

Both notations are widely used in the fields of acoustics and signal processing.

ASW – (1) *in room acoustics* abbreviation for APPARENT SOURCE WIDTH. *Units* deg.

ASW – (2) *in underwater acoustics* abbreviation for Anti-Submarine Warfare.

asymptotic – refers to a value that is approached in some specified limit. Thus the difference between the total and static pressure in a fluid flow has the *asymptotic value* $\frac{1}{2}\rho u^2$, as $M \to 0$ (low Mach number or incompressible flow limit). A more precise mathematical statement of this result would be

$$\frac{P_{\text{tot}} - P}{\frac{1}{2}\rho u^2} = 1 + O(M^2)$$

which is an example of an *asymptotic equation*, i.e. one that becomes exact in the indicated limit. The *order symbol* $O(M^2)$ indicates that the error divided by M^2 remains finite (or tends to zero) as M is made arbitrarily small. See also ORDER (1).

asymptotically stable – a linear system is called asymptotically stable if its displacement tends to zero whenever the system is perturbed from equilibrium and released; equivalently, its free oscillations are damped. Such a system always has a bounded output for any bounded input.

Note: Lyapunov's definition of stability (broadly, that free oscillations do not diverge without limit) does not ensure asymptotic stability, since it allows the system transfer function in the S-PLANE to have poles on the imaginary axis (examples are a simple integrator, and a system with undamped natural modes). In both these examples, a finite input sustained for infinite time can lead to an infinite output. Compare STABLE SYSTEM.

ATOC – acronym for Acoustic Thermometry of Ocean Climate; the use of ACOUSTIC THERMOMETRY to measure, and monitor, changes in sea temperature on an ocean or global scale.

attached mass – *of an accelerating rigid body in an incompressible fluid* the mass of fluid that, if rigidly attached to the body, would provide the same inertial fluid loading. Equivalently, the coefficient of proportionality between the force required to overcome fluid inertia, and the acceleration of the rigid body. See also the alternative term VIRTUAL MASS, where further details are given. *Units* kg.

attenuation – (1) a generic term for a reduction in the amplitude or rms value of an acoustic field variable, such as sound pressure; the term can also refer to a reduction in a power-like variable, such as sound intensity. Sound waves suffer progressive attenuation with distance during propagation through real media; the principal attenuation mechanisms are ABSORPTION and SCATTERING.

attenuation – (2) the amount by which the LEVEL of a signal is reduced, based on either the mean square value (usually in a specified frequency band) or the squared peak value (for a transient signal). Compare ATTENUATION RATE. *Units* dB.

attenuation coefficient – *of a single-frequency progressive wave system* the coefficient α in the spatial attenuation factor $e^{-\alpha x}$, which describes the reduction in amplitude of a progressive wave with distance, x, in the propagation direction. The alternative term **amplitude attenuation coefficient** may be useful where there is risk of confusion with ENERGY ATTENUATION COEFFICIENT. Compare ATTENUATION RATE; see also PROPAGATION FACTOR, PROPAGATION COEFFICIENT. *Units* Np m^{-1}.

Note: The space and time dependence of a single-frequency wave propagating along the x axis may be expressed in phasor notation as $e^{j(\omega t - kx)}$, where k is the complex PROPAGATION WAVENUMBER; the amplitude attenuation coefficient is then given by $\alpha = -\operatorname{Im} k$. Equivalently, the attenuation coefficient is the real part of the PROPAGATION COEFFICIENT $\gamma = jk$.

attenuation constant – an alternative term for ATTENUATION COEFFICIENT. The "coefficient" version seems to be widely used, and has been preferred in the present work. *Units* Np m^{-1}.

Note: The 1994 ANSI standard on acoustical terminology recognizes *attenuation constant* (with the prefix "acoustic"), but not *attenuation coefficient*; the 1994 IEC standard recognizes attenuation coefficient but not attenuation constant.

attenuation length – *for linear waves* the distance over which the amplitude of a progressive wave drops by a factor $1/e$, in an absorbing or scattering medium. If α denotes the ATTENUATION COEFFICIENT, the attenuation length is given by $L_a = 1/\alpha$. An alternative term, appropriate in the absence of scattering, is *absorption length*. *Units* m.

attenuation rate – *at a given frequency* the rate at which the LEVEL of mean square pressure falls off with distance in a progressive sound wave, adjusted (where appropriate) for spreading, and expressed in decibels per unit distance. In a plane progressive wave whose amplitude decays spatially like $e^{-\alpha x}$, where α is the ATTENUATION COEFFICIENT of the signal and x is distance measured in the propagation direction, the attenuation rate a is given by

$$a = 8.686\,\alpha \qquad (8.686 = 20/\ln 10)\,.$$

Units dB m^{-1}.

Note: It is assumed for purposes of the definition above that measurements are made in the far field. In this situation the attenuation rate for mean square sound pressure is the same as that for acoustic intensity.

audibility – *of a given sound* detectability by ear, especially to human listeners. The audibility of sounds depends on their level and frequency content, and may be reduced by the presence of other sounds; see ABSOLUTE THRESHOLD, MASKING.

audio frequency range – (roughly) 15 Hz to 20 kHz; the frequency range over which the normal human ear is sensitive to sound.

audiogram – *in audiology* a chart or table of a person's HEARING THRESHOLD LEVELS for pure tones at different frequencies.

audiology – the science of hearing, especially human hearing, and its dysfunction. Because the INNER EAR contains both hearing and balance organs, audiology is often understood to include the sense of balance as well as hearing.

audiometer – *in audiology* an electroacoustic instrument, usually equipped with earphones, that presents a subject's ear with test signals at known sound pressure levels and is calibrated in a specified manner, using either an ACOUSTIC COUPLER or an EAR SIMULATOR. It is used to determine HEARING THRESHOLD LEVELS, one ear at a time.

audiometric zero – *in pure-tone air-conduction audiometry* a reference set of pure-tone sound pressure levels at specified frequencies, as measured in a specified ACOUSTIC COUPLER or EAR SIMULATOR during calibration of an audiometer. The audiometric zero is intended to typify the threshold of hearing of young otologically normal persons. See also HEARING THRESHOLD LEVEL.

audiometry – measurement of auditory function. The term is general, but is commonly understood to mean the determination of a person's pure-tone AUDIOGRAM. See also PURE-TONE AUDIOMETRY.

Note: Audiometric techniques can be either subjective – involving a voluntary response from the subject – or objective; examples of the latter are *cortical audiometry*, in which evoked electrical potentials from the cortex of the brain are measured, and *electrocochleographic audiometry*, which uses potentials measured in the middle ear or external ear canal.

audition – the sense of hearing.

auditorium – a building or part of a building designed to accommodate listeners for concerts, lectures, drama or other audio-visual performances.

auditory – related to hearing, or to the mechanism of hearing; e.g. *auditory filter*, *auditory nerve*. Compare AURAL.

auditory adaptation – *in psychoacoustics* the decline in apparent magnitude of a steady auditory stimulus over time, on a time scale of a few minutes. See ADAPTATION.

auditory critical band – one of a number of contiguous bands of frequency into which the audio-frequency range may be notionally divided, such that sounds in different frequency bands are heard independently of one another, without mutual interference. An auditory critical band can be defined for various measures of sound perception that involve frequency.

auditory critical bandwidth for loudness – *for a given centre frequency* the maximum bandwidth over which an acoustic signal can be spread, with its mean square pressure held constant, without affecting the LOUDNESS. Thus the loudness of a continuous sound that lies entirely within a critical bandwidth depends only on the signal level, and not on the bandwidth of the signal. The critical bandwidth for loudness is an increasing function of frequency. *Units* Hz.

auditory fatigue – *in psychoacoustics* the reduction in response to an auditory stimulus that occurs following exposure to high levels of the stimulus. Also known as *post-stimulatory auditory fatigue*. The related shift in absolute threshold is called TEMPORARY THRESHOLD SHIFT.

auditory nerve – the eighth cranial nerve, also known as nerve VIII. It consists of two branches, the **cochlear nerve** and the **vestibular nerve**, running respectively from the cochlea and the organs of balance to the brain stem.

aural – via the ear, or via the hearing mechanism or process; e.g. *aural detection*.

auralization – conversion of a digital waveform into audible form.

auricle – alternative medical term for the PINNA.

auscultation – listening to internal body sounds, particularly with the aid of an impedance-matching device such as a stethoscope.

autocorrelation – *in signal processing* abbreviation for AUTOCORRELATION FUNCTION.

autocorrelation coefficient, autocorrelation coefficient function – *of a real time-stationary random signal* the normalized AUTOCORRELATION FUNCTION (1). Thus if a signal $x(t)$ has autocorrelation function $R_{xx}(\tau)$, its autocorrelation coefficient as a function of time shift τ is

$$\rho_{xx}(\tau) = R_{xx}(\tau)/R_{xx}(0) \qquad (-1 \leq \rho_{xx} \leq 1).$$

Note: The autocorrelation coefficient is generally applied to signals whose mean value is zero. It is equivalent, in such cases, to a normalized AUTOCOVARIANCE FUNCTION.

autocorrelation function – (1) *of a real time-stationary continuous signal* the time-averaged product

$$R_{xx}(\tau) = \lim_{T \to \infty} \left\{ \frac{1}{T} \int_{t_0}^{t_0+T} x(t)x(t+\tau)dt \right\}$$
$$= \langle x(t)x(t+\tau) \rangle = \langle x(t-\tau)x(t) \rangle,$$

where angle brackets $\langle \ldots \rangle$ denote the time averaging operation, $x(t)$ is the signal as a function of time t, and τ is a time shift or delay. The autocorrelation function is an even function of the time delay, i.e. $R_{xx}(-\tau) = R_{xx}(\tau)$.

Note (1): Putting τ equal to zero gives $R_{xx}(0) = \langle x^2(t) \rangle$.

Note (2): The autocorrelation function defined above is the same as the AUTOCOVARIANCE FUNCTION, if $x(t)$ has zero mean.

autocorrelation function – (2) *of a complex non-stationary stochastic process* the ENSEMBLE-AVERAGED product

$$R_{xx}(t_1, t_2) = E\{(x(t_1) x^*(t_2)\},$$

where $x(t)$ is the process in question, * denotes the complex conjugate, and $E\{\ldots\}$ is the EXPECTATION OPERATOR.

autocorrelation function – (3) *of a real time-stationary discrete sequence* the time-averaged product

$$R_{xx}[m] = \langle x[n]\, x[n + m] \rangle = \langle x[n - m]\, x[n] \rangle,$$

where angle brackets $\langle \ldots \rangle$ denote an average with respect to the DISCRETE TIME variable n, $x[n]$ is the discrete sequence, and m is a shift or delay. The autocorrelation function is an even function of the delay, i.e. $R_{xx}[-m] = R_{xx}[m]$.

autocorrelation matrix – *of a vector of real time-stationary continuous signals* the $N \times N$ square matrix defined by

$$\mathbf{R}_{\mathbf{xx}}(\tau) = \langle \mathbf{x}(t + \tau)\, \mathbf{x}^T(t) \rangle = \langle \mathbf{x}(t)\, \mathbf{x}^T(t - \tau) \rangle$$

$$= [R_{ij}(\tau)],$$

where angle brackets $\langle \ldots \rangle$ denote the time averaging operation, $\mathbf{x}(t)$ is a column vector of signals $x_1, x_2, \ldots, x_N$ as a function of time t, superscript T denotes the vector TRANSPOSE, and τ is a time shift or delay. The autocorrelation matrix has the property $R_{ji}(\tau) = R_{ij}(-\tau)$; its diagonal elements are autocorrelation functions.

Note: The autocorrelation matrix is the special case of the CROSS-CORRELATION MATRIX that results when the two signal vectors are identical. Both matrices arise in connection with multiple-input multiple-output systems.

autocovariance function – *of a real time-stationary continuous signal* the time-averaged product

$$\langle x'(t)\, x'(t + \tau) \rangle = R_{xx}(\tau) - \bar{x}^2$$

where the brackets $\langle \ldots \rangle$ denote the time averaging operation; $\bar{x}$ is the mean value of the signal, and $x' = x - \bar{x}$ is the departure from the mean. Compare AUTOCORRELATION FUNCTION, which is the same except that the mean value is not subtracted out first.

autospectral density – *of a real time-stationary continuous signal* the Fourier transform of the AUTOCORRELATION FUNCTION of the signal. If the signal is denoted by $x(t)$, and its autocorrelation function for time shift τ is $R_{xx}(\tau)$, its autospectral density is given by

$$S_{xx}(f) = \int_{-\infty}^{\infty} R_{xx}(\tau)\, e^{-2\pi j f \tau}\, d\tau.$$

The autospectral density is an even real function of frequency, and it is continuous provided the signal contains no periodic components. Its integral over all frequencies gives the signal POWER:

$$\int_{-\infty}^{\infty} S_{xx}(f)\, df = R_{xx}(0) = \langle x^2(t)\rangle.$$

Alternative terms for the autospectral density are **autospectral density function**, **power spectrum**, and **power spectral density**. See also SINGLE-SIDED SPECTRAL DENSITY, AUTOSPECTRAL MATRIX.

Note (1): The omission of the prefix "power" from the term autospectral density (in contrast to ENERGY AUTOSPECTRAL DENSITY) is well established.

Note (2): It follows from the CONVOLUTION THEOREM that the autospectral density can also be expressed as

$$S_{xx}(f) = \lim_{T\to\infty} \frac{1}{T} E\{|X_T(f)|^2\},$$

where X_T is the Fourier transform of the finite RECORD x_T of length T (defined to equal the original signal over the range $t_0 < t < t_0 + T$, where t_0 is arbitrary, and zero otherwise).

autospectral density function – an equivalent term for AUTOSPECTRAL DENSITY. Compare DENSITY FUNCTION.

autospectral matrix – *of a vector of real time-stationary continuous signals* the $N \times N$ square matrix defined by

$$\mathbf{S_{xx}}(f) = \lim_{T\to\infty} \frac{1}{T} E\{\mathbf{X}_T(f)\mathbf{X}_T^H(f)\},$$

where $E\{\ldots\}$ denotes the expectation operator, $\mathbf{X}_T(f)$ is the Fourier transform of $\mathbf{x}_T(t)$, and $\mathbf{x}_T$ is a column vector of N finite RECORDS each of length T (with each record defined to equal the original signal over the range $t_0 < t < t_0 + T$, where t_0 is arbitrary, and zero otherwise). The superscript H denotes the HERMITIAN TRANSPOSE.

Note: The autospectral matrix is the Fourier transform of the AUTOCORRELATION MATRIX. Compare CROSS-SPECTRAL MATRIX.

auxetic material – *in engineering* a material whose Poisson's ratio, ν, lies in the range $-1 < \nu < 0$.

Note: A stable material cannot have ν less than -1 or greater than $\frac{1}{2}$. The limiting value $\nu = -1$ corresponds to a material with a finite bulk modulus but an infinite shear modulus; the ROD LONGITUDINAL-WAVE SPEED remains finite, but the speeds of bulk longitudinal waves, bulk shear waves, and Rayleigh waves are all infinite.

average – *with respect to time* various types of time average are used in acoustics. In the following examples, $q(t)$ refers to any time-dependent quantity. In the first two cases $q(t)$ is stationary, with time average $\langle q \rangle$; in the third, $q(t)$ is non-stationary, with running average $\bar{q}(t)$. See also MEAN.

(1) For *periodic* signals:

$$\langle q \rangle = \frac{1}{T} \int_0^T q(t) dt \qquad (\textit{single-period average})$$

where T is the period.

(2) For *time-stationary* signals:

$$\langle q \rangle = \lim_{T \to \infty} \frac{1}{T} \int_0^T q(t) dt \qquad (\textit{long-duration average}).$$

(3) For signals that are not stationary, one can define a running average over a moving window of finite duration:

$$\bar{q}(t) = \frac{1}{T} \int_{t-T}^t q(t') dt' \qquad (\textit{finite-duration running average}).$$

The expression above uses uniform weighting over a window of duration T; an alternative type of running average uses exponential weighting with a time constant τ, giving

$$\bar{q}(t) = \frac{1}{\tau} \int_{-\infty}^t q(t') e^{-(t-t')/\tau} dt' \qquad (\textit{exponentially-weighted running average}).$$

In many practical applications, the averaging time T or time constant τ is chosen to be short compared with the slowly-varying time scale of the non-stationary signal statistics, but long compared with the oscillatory time scale of the signal itself.

average modal energy – *over a specified frequency band that contains several resonant modes* the time-average energy per mode of a multimode vibrating system, obtained by summing the energy of all the modes resonant within the specified band (total number = N) and dividing by N. Units J.

average sound pressure level – *in a reverberant space* the LEVEL calculated from the spatially-averaged mean square sound pressure in the reverberant field. Regions close to a source where the direct field dominates, and regions within a quarter-wavelength of the room boundaries, are normally excluded from the spatial average. Compare TIME-AVERAGE SOUND PRESSURE LEVEL. *Units* dB re $(20 \, \mu\text{Pa})^2$.

Avogadro constant – the conversion factor from a number of MOLES to a number of particles:

$$N_A \approx 6.022 \, 136 \, 7 \times 10^{23} \, \text{mol}^{-1}$$

(A AVOGADRO 1811).

Note: The number above is determined by the number of atoms in 0.012 kg of neutral ^{12}C.

A-weighted sound pressure level – the LEVEL of a sound pressure signal to which A-WEIGHTING has been applied. See also SOUND LEVEL, FREQUENCY WEIGHTING. Symbol L_A (or L_{pA}). *Units* dB re $(20\ \mu\text{Pa})^2$.

A-weighting – a frequency-weighting procedure, in which the power or energy spectrum of a signal is progressively attenuated towards the high and low ends of the audio frequency range. Frequency components around 1–5 kHz are hardly affected, but the attenuation is large at low frequencies (e.g. 70 dB at 10 Hz).

Note: The A-weighting curve was originally based on the shape of the 30-phon equal loudness contour, with no attempt to allow for the ear canal resonance that enhances the free-field sensitivity of the ear by nearly 10 dB in the range 2–5 kHz. The intention was that A-weighting the incident pressure, followed by squaring and averaging with a suitable time constant, would simulate the sensitivity of the human ear to audio-frequency sound at sound pressure levels below about 50 dB re $(20\ \mu\text{Pa})^2$.

axially progressive wave – *in a waveguide* a single-frequency wave field that propagates along the waveguide with a space–time dependence of the form $e^{j(\omega t - kx)}\psi(y, z)$, where k is the axial wavenumber, x is the axial coordinate (i.e. parallel to the waveguide), and (y, z) are transverse coordinates. Compare BLOCH WAVE.

Note: Propagation of individual WAVEGUIDE MODES as axially progressive waves, with separation of the x and (y, z) dependence, strictly requires the waveguide to be uniform in the x direction. For slow axial dependence, asymptotic solutions of axially-progressive type can be obtained; see ADIABATIC MODES, ADIABATIC APPROXIMATION.

axial mode – a mode with no transverse dependence (i.e. the only spatial pressure variation is parallel to the axis), in a hard-walled room of cylindrical or prismatic shape with end walls normal to the room axis.

axial mode count – *in a hard-walled room of cylindrical or prismatic shape with end-walls normal to the room axis* the number of axial-mode EIGENVALUES less than a stated value. A statistical estimate for rooms of rectangular shape, with dimensions (L_x, L_y, L_z), is $N_{\text{ax}} \approx (k/\pi)(L_x + L_y + L_z)$; this gives the number of axial-mode eigenvalues less than k^2. For rooms with only one axis, $N_{\text{ax}} \approx kL_x/\pi$ where L_x is the distance between the end walls.

Note: To estimate the number of *axial-mode natural frequencies* below a given frequency in such a room, replace k above by ω/c, where ω is the angular frequency and c is the sound speed in the room.

axial phase speed – *of a time-harmonic wave field* the speed at which a point of constant phase moves, along a line parallel to a given axis. *Units* m s^{-1}.

axial propagation wavenumber – *in a waveguide* the coefficient k in the expression $\exp j(\omega t - kx)$, used to describe an AXIALLY PROGRESSIVE WAVE propagating along the waveguide in the axial (x) direction at angular frequency ω. *Units* (real part) rad m^{-1}; (imaginary part) Np m^{-1}.

axial quadrupole – a point quadrupole made up of two collinear DIPOLES of equal and opposite strength. An equivalent term is *longitudinal quadrupole*. Compare LATERAL QUADRUPOLE.

axial-quadrupole source distribution – an ACOUSTIC SOURCE DENSITY in the form of a double derivative with respect to one cartesian coordinate; e.g.

$$\mathcal{F} = \frac{\partial^2 F_{xx}}{\partial x^2},$$

(quadrupoles oriented in the x-axis direction).

axial source level – *in underwater acoustics* the SOURCE LEVEL of a radiating transducer measured along the beam axis. *Units* dB re 1 μPa^2 @ 1 m, or dB re 1 μPa^2 m^2.

axisymmetric – having rotational symmetry about an axis; for example, an *axisymmetric waveguide* has both its geometry and its acoustic properties independent of orientation about the waveguide axis. Also known as *rotationally symmetric*.

azimuthally symmetric – an equivalent term for AXISYMMETRIC.

azimuth angle – the angle that measures rotation about the coordinate axis, in a CYLINDRICAL or a SPHERICAL POLAR COORDINATE system. *Units* rad (but commonly expressed in degrees).

B

B/A – symbol for the NONLINEARITY PARAMETER of a fluid. *Units* none.

background noise – (1) *in a measured signal* refers to any type of NOISE in the sense of an unwanted signal, acoustic or electronic, that forms an unavoidable part of the measurement; (2) *for purposes of environmental noise assessment* see RESIDUAL NOISE.

background noise level – (1) *in a measured signal* the LEVEL of BACKGROUND NOISE (1). *Units* dB.

Note: The difference between the level of the desired signal and the background noise level is the ***signal-to-noise ratio*** (SNR) in decibels.

background noise level – (2) *for purposes of environmental noise assessment* the A-weighted sound pressure level of the RESIDUAL NOISE. In situations where the residual noise level varies over time, a suitable single-figure measure, or statistical indicator, is used to represent the background noise level. In the UK, for example, this has been chosen as the L_{90} value (see the L_N entry for further information). *Units* dB re $(20~\mu\text{Pa})^2$.

backscattering – *by a target located in the field of a sound source* scattering of incident sound back towards the source.

backscattering cross-section – a measure of the ability of an object to scatter sound back in the opposite direction, when irradiated by a plane incident wave of a given direction and frequency. If W_Ω is the backscattered power per unit solid angle corresponding to an incident intensity I_{inc}, the DIFFERENTIAL SCATTERING CROSS-SECTION in the backscattering direction is $S_{\text{back}} = W_\Omega/I_{\text{inc}}$ (expressed in $\text{m}^2~\text{sr}^{-1}$), and the backscattering cross-section is this quantity multiplied by 4π steradians:

$$\sigma_{\text{back}} = 4\pi S_{\text{back}} = 4\pi W_\Omega/I_{\text{inc}}.$$

An alternative term is ***acoustic cross-section***. Compare TARGET STRENGTH. *Units* m^2.

Note (1): For a finite scattering area that forms part of a boundary, the backscattering cross-section is defined with a factor 2π in place of the 4π appearing above.

Note (2): The target strength, TS, of a scattering object is related to the backscattering cross-section by $\text{TS} = 10\log_{10}(\sigma_{\text{back}}/4\pi r_{\text{ref}}^2)$, where r_{ref} is the reference distance of 1 m.

band-limited white noise 45

backscattering direction – the scattering direction that sends waves back towards the source of the incident sound.

backscattering strength – *in underwater acoustics* a generic term for the TARGET STRENGTH of a continuous distribution of scatterers, normalized to unit scattering area (or volume). See BOTTOM BACKSCATTERING STRENGTH, SURFACE BACKSCATTERING STRENGTH, VOLUME BACK-SCATTERING STRENGTH, COLUMN SCATTERING STRENGTH. All these are conventionally expressed in logarithmic form.

backward wave – *in a one-dimensional standing wave system* refers to a travelling-wave component whose propagation direction is towards a source, or away from a passive boundary; equivalent to a reflected wave. Compare FORWARD WAVE.

baffle – (1) a reflecting surface or structure used to modify the response of an acoustic transducer; for example a partition or panel, with the transducer mounted flush with its surface.

baffle – (2) a plane absorbent structure suspended in an enclosure, intended to increase the room absorption and also the diffuseness of the sound field. The term can also refer to a sound-absorbing splitter in a SPLITTER SILENCER. Compare SCREEN, BARRIER.

baffle – (3) *in duct silencers* an equivalent term for SPLITTER.

balance – the degree to which the sounds from the various instrument groups in an orchestra are balanced relative to one another, or the sound of soloists is balanced with that of the orchestra as a whole. Balance is influenced both by the design of the concert hall, especially around the stage platform, and by the players' performance.

band – *in acoustics* abbreviation for FREQUENCY BAND.

band-edge frequencies – the upper and lower frequency limits of the transmission range, for a BANDPASS FILTER. See also CUTOFF FREQUENCY (2).

band level – the LEVEL of a signal in a specified frequency band. See also under SPECTRUM LEVEL. *Units* dB.

band-limited function – a function of time is said to be band-limited if its Fourier transform is zero outside a finite frequency interval and its energy is finite.

band-limited white noise – a continuous random signal whose power spectrum is flat over a finite bandwidth. Compare WHITE NOISE.

bandpass filter – a filter whose INSERTION LOSS is almost zero over a certain frequency range defined by finite lower and upper limits (i.e. the PASSBAND), and is large outside that range. Some authors write *band-pass filter*.

Note: Two important specifications when designing a bandpass filter are the amount of ripple within the passband, and the cut-off rate for frequencies outside the passband.

band pressure level – the SOUND PRESSURE LEVEL that corresponds to components of a sound-pressure signal within a specified frequency band. The term is a special case of BAND LEVEL. *Units* dB re p_{ref}^2.

bandwidth – (1) the difference between the upper and lower limits of any frequency band; (2) *of a signal* the range of frequencies between lower and upper limits within which most of the signal energy is contained; (3) *of a linear system or transducer* the range of frequencies over which the system is designed to operate; (4) *of a bandpass filter* see FILTER BANDWIDTH; (5) *of a resonant response curve* see ENERGY BANDWIDTH (1), HALF-POWER BANDWIDTH (1). *Units* Hz.

bandwidth–time product – see BT PRODUCT. *Units* none.

barotrauma – an injury to the ear due to rapid changes in the ambient atmospheric pressure, e.g. decompression at high altitude. Most commonly, the middle ear is affected, resulting in a CONDUCTIVE HEARING LOSS. Rarely, the inner ear is involved with consequent SENSORINEURAL HEARING LOSS and disturbance of balance. Compare ACOUSTIC TRAUMA.

barrier – a sound barrier is any large structure that prevents line-of-sight sound transmission between a source and receiver, usually outdoors. Effective blockage of the transmission path requires the structure to be acoustically opaque, and also large compared with the sound wavelength.

basilar membrane – a membrane that acts as a lengthwise partition along the spiral duct of the COCHLEA. It is excited by pressure fluctuations in the fluid that fills the inner ear, and responds selectively to different frequencies at different positions along its length. The response is finally converted into electrical impulses by HAIR CELLS attached to the membrane.

Note: Each point along the basilar membrane, between the base and apex of the cochlea, has a preferred range of frequencies to which it responds (10 dB down relative bandwidth $\Delta f/f \approx 0.1$–0.3). This tuning mechanism allows the ear to act as a frequency analyser.

basis functions – *over a specified region* a set of FUNCTIONS, usually chosen to be ORTHOGONAL, that form a LINEARLY INDEPENDENT SET. An example is the set of functions used in FOURIER ANALYSIS, namely ($\cos nx$, $\sin nx$), defined over the interval $x = [-\pi, \pi]$ with $n = 0, 1, 2 \ldots$; any other function defined over $-\pi < x < \pi$ can be represented as a unique linear combination of the basis functions.

Note: Basis functions commonly incorporate a NORMALIZATION CONSTANT. See ORTHONORMAL FUNCTIONS.

basis vectors – a set of VECTORS (usually of unit magnitude and ORTHOGONAL) that forms a LINEARLY INDEPENDENT SET. An example is the set of unit vectors (**i, j, k**) in the cartesian coordinate directions (x, y, z). Any VECTOR (1) sharing the same dimensionality as the chosen basis vectors can then be expressed uniquely as a linear combination of those vectors.

bass ratio – *in a concert hall* the ratio between the reverberation time at low frequencies (typically at 125 Hz) and the reverberation time at mid-frequencies (typically the average of values at 500 Hz and 1000 Hz). The bass ratio is considered to relate to perceived warmth, though the relative sound level at bass to mid-frequencies is thought by some to be more relevant. *Units* none.

bass reflex – a type of loudspeaker design in which the low-frequency output is increased by allowing the volume velocity from the rear of the diaphragm to escape through a ***bass reflex port***. See also PORT.

bathythermograph – *in underwater acoustics* an instrument for measuring water temperature as a function of depth, usually for the purpose of determining sound speed (given the assumption of constant SALINITY). See SOUND SPEED IN SEAWATER.

BBS – *in underwater acoustics* abbreviation for the BOTTOM BACKSCATTERING STRENGTH of the sea bed, expressed in decibels. *Units* dB.

beam displacement – *for a beam of sound incident on an interface between two media* a lateral shift in the position of the reflected beam relative to the incident one, caused (usually) by a step change in sound speed. The shift is related to the derivative of the plane-wave REFLECTION PHASE with respect to the wavenumber component parallel to the interface. *Units* m.

beam pattern – *of an acoustic receiving or transmitting array* the normalized sensitivity of the array as a function of arrival direction (when functioning as a receiver), or the normalized output sensitivity as a function of radiation direction (when functioning as a transmitter). For a receiving array irradiated by plane waves at a given frequency, f, arriving from direction (θ, ϕ) in spherical polar coordinates, the ***array sensitivity***

$G(f, \theta, \phi)$ is defined as the ratio of the output voltage magnitude to the input pressure magnitude. The beam pattern in decibels is then given by

$$B(f, \theta, \phi) = 20 \log_{10} \frac{G(f, \theta, \phi)}{G(f, \theta_{max}, \phi_{max})}.$$

Here $(\theta_{max}, \phi_{max})$ is the arrival direction for which G is a maximum. *Units* dB.

Note: According to the RECIPROCITY PRINCIPLE, a given array will have the same beam pattern whether it operates as a receiver or transmitter.

beat frequency – see BEATS.

beats – slow amplitude modulation produced by INTERFERENCE between two single-frequency signals, when the frequency difference is a small fraction of either frequency. The frequency of the modulation, called the ***beat frequency***, is equal to the difference in frequency between the two original signals. See MODULATION (1).

Beats are observed in various situations in acoustics; three examples are given here. Beats occur in the response of a linear system when it is simultaneously excited by *two inputs at slightly different frequencies*. Secondly, when a lightly-damped system is driven at a frequency close to a NATURAL FREQUENCY of the system, beats occur for a while after the excitation is switched on: this is the result of *interference between the* TRANSIENT (1) *response and the* STEADY-STATE *response*. The last example occurs during the free decay of two lightly-damped natural modes with closely-spaced natural frequencies (e.g. in a room); in this case *interference between the two modal transients* causes beats to appear.

bel – a unit of measure for LEVEL or LEVEL DIFFERENCE (**A G BELL** *1847–1922*). If a quantity is increased by a factor 10^n, its level goes up by n bels.

Note: The bel is the preferred unit for the SOUND POWER output level of computers and peripheral equipment (with a sound power of 1 pW as the REFERENCE VALUE): thus an acoustic output of 1 μW is expressed as 6 B re 1 pW. Compare DECIBEL. *Unit symbol* **B**.

bending stiffness – *at a given point along a rod or beam* the ratio of the applied bending moment to the resulting increase in curvature of the beam axis. *Units* N m^2.

Note: An axisymmetric rod has the same bending stiffness in any azimuthal plane through the axis of the rod. For other beam cross-sections, the direction of bending must be specified. For plates, a ***bending stiffness per unit width*** is defined.

bending wave – *in a thin plate* a wave that propagates along the plate with particle displacements almost entirely normal to the surface; there is virtually no variation in displacement through the plate thickness. Bending waves are also called *flexural waves*; they belong to the general family of free waves in plates called LAMB WAVES.

Bending waves can also propagate along slender rods or beams; an additional feature in this case is the possibility of POLARIZATION in various directions.

Note: The plate or beam thickness is required to be small in comparison with the bending-wave wavelength, in order for the description above to be valid.

Bessel beam – a parallel or focused beam of single-frequency sound, driven by excitation over an infinite plane aperture. The axial extent of the beam, before it begins to diverge, is approximately equal to the RAYLEIGH DISTANCE. From a mathematical viewpoint, Bessel beams correspond to separable solutions of the scalar wave equation in cylindrical coordinates, with one-way propagation in the axial direction. Compare LOCALIZED WAVES.

Bessel equation – the differential equation

$$R'' + \frac{1}{r} R' + \left(\kappa^2 - \frac{\mu^2}{r^2} \right) R = 0$$

(F W BESSEL 1824). Here $R(r)$ is a function of the cylindrical radius r, and primes denote derivatives of the function; κ and μ are constants, restricted only by $\kappa \neq 0$.

Note (1): The standard form of the Bessel equation has $\kappa = 1$. The scaled version given above is useful for applications; it has the general solution

$$R(r) = A J_\mu(\kappa r) + B Y_\mu(\kappa r) = Z_\mu(\kappa r) \quad (A, B \text{ constants}).$$

For further information, see CYLINDER FUNCTIONS.

Note (2): The following generalized equation also has solutions that are expressible in terms of cylinder functions:

$$R'' + \frac{1 - 2n}{r} R' + \left[(\kappa \lambda r^{\lambda - 1})^2 - \frac{\lambda^2 \mu^2 - n^2}{r^2} \right] R = 0,$$

where n and λ are constants. Its general solution, using the Z_μ notation above, is

$$R(r) = r^n Z_\mu(\kappa r^\lambda).$$

Bessel function – the cylindrical Bessel function of the first kind, denoted by $J_\mu(z)$ where z is the argument of the function and μ is the order. See CYLINDER FUNCTIONS.

Bessel functions – a generic name for the CYLINDER FUNCTIONS of the first, second, or third kind.

Bethe–Zel'dovich–Thompson fluid – a fluid for which the isentropes $P(V|s = \text{const.})$ are concave downwards, as opposed to the normal situation (which includes the perfect-gas model) where they are concave upwards. Here P, V, and s denote the pressure, specific volume, and specific entropy of the fluid respectively. An equivalent definition is that the COEFFICIENT OF NONLINEARITY of such a fluid is negative (**H A BETHE 1942, YA B ZEL'DOVICH 1946, P A THOMPSON 1973**).

bias – *in spectral analysis* an alternative term for LEAKAGE.

BIBO stability – *of a linear time-invariant system* abbreviation for BOUNDED INPUT/BOUNDED OUTPUT STABILITY.

bicorrelation function – *of a real time-stationary random signal* the time-averaged product

$$R_{xxx}(\tau_1, \tau_2) = \lim_{T \to \infty} \left\{ \frac{1}{T} \int_{t_0}^{t_0+T} x(t - \tau_2) x(t - \tau_1) x(t) dt \right\}$$
$$= \langle x(t - \tau_2) x(t - \tau_1) x(t) \rangle,$$

where angle brackets $\langle \ldots \rangle$ denote the time averaging operation, $x(t)$ is the signal as a function of time t, and τ_1, τ_2 are time shifts or delays. The bicorrelation function for zero delay, $R_{xxx}(0, 0)$, is equal to $\langle x^3(t) \rangle$.

bilinear spring – a nonlinear spring whose force–deflection curve consists of two straight-line segments with a sudden change of slope.

binary representation – the representation of a number to base 2, using the binary digits 0 and 1. Integers are represented in closed form: e.g. 9 (decimal) becomes 1001. Non-integer real numbers require the binary equivalent of a decimal point, with digits after the point: e.g. 0.500 (decimal) becomes 0.100.

binaural – used with, or making use of, both ears; e.g. ~ *headphones*, ~ *hearing*.

bins – *in statistical data analysis* categories into which the data can be partitioned or subdivided. Data assigned to bins in this way is said to be *grouped*. Categories used as bins must be discrete, and also non-overlapping.

Note: In FFT analysis, discrete frequency components are often grouped in bins to provide smoothing.

bioacoustics – the science that deals with acoustical properties of biological materials; with methods of measuring them; and with biological effects of acoustic irradiation (including effects at ultrasonic frequencies). Compare ANIMAL BIOACOUSTICS.

bispectrum – *of a real time-stationary random signal* the 2-dimensional Fourier transform of the BICORRELATION FUNCTION of the signal. Thus if the signal is $x(t)$, with bicorrelation function $R_{xxx}(\tau_1, \tau_2)$, its bispectrum is

$$S_{xxx}(f_1, f_2) = \int_{-\infty}^{\infty} \int_{-\infty}^{\infty} R_{xxx}(\tau_1, \tau_2)\, e^{-2\pi j(f_1 \tau_1 + f_2 \tau_2)}\, d\tau_1\, d\tau_2.$$

The bispectrum is a complex quantity. Its 2-dimensional integral over all frequencies gives the mean cube of the signal:

$$\int_{-\infty}^{\infty} \int_{-\infty}^{\infty} S_{xxx}(f_1, f_2)\, df_1\, df_2 = R_{xxx}(0, 0) = \langle x^3(t) \rangle.$$

For Gaussian signals, the bispectrum is zero.

bistatic – (1) adjective used to describe an ACTIVE SONAR system in which the transmitter and receiver are at different locations. The relevant scattering direction from the target is therefore different from the backscattering direction.

bistatic – (2) adjective used in underwater acoustics to describe any source–scatterer–receiver geometry where the source and receiver are in different directions from the scatterer.

bistatic bottom scattering strength – *in underwater acoustics* a generalization of BOTTOM BACKSCATTERING STRENGTH to arbitrary scattering directions (i.e. not back towards the source). *Units* dB.

bistatic scattering strength – *in underwater acoustics* the SCATTERING STRENGTH of a target or scatterer, in any specified direction different from the backscattering direction (i.e. not back towards the source). See BISTATIC BOTTOM SCATTERING STRENGTH, BISTATIC SURFACE SCATTERING STRENGTH, BISTATIC VOLUME SCATTERING STRENGTH; all these are conventionally expressed in logarithmic form.

bistatic surface scattering strength – *in underwater acoustics* a generalization of SURFACE BACKSCATTERING STRENGTH to arbitrary scattering directions (i.e. not back towards the source). *Units* dB.

bistatic volume scattering strength – *in underwater acoustics* a generalization of VOLUME BACKSCATTERING STRENGTH to arbitrary scattering directions (i.e. not back towards the source). *Units* dB re 1 m^{-1}.

bivariate – statistics formed from two RANDOM VARIABLES, or from pairs of values taken from a RANDOM SIGNAL, are called bivariate, as opposed to the univariate statistics formed from a single variable. Bivariate statistics include the CROSS-CORRELATION FUNCTION and the CROSS-SPECTRAL DENSITY.

Note: One can also define MULTIVARIATE statistics, for any number of random variables in combination. See JOINT CUMULATIVE DISTRIBUTION FUNCTION, JOINT PROBABILITY DENSITY FUNCTION; and, for an example, GAUSSIAN DISTRIBUTION.

bivariate distribution – a statistical description of the frequencies with which particular pairs of values occur together, e.g. when two separate signals are sampled at the same instant; expressed mathematically as a JOINT PROBABILITY DENSITY FUNCTION of the two variables in question.

Bjerknes forces – ACOUSTIC RADIATION FORCES that cause migration of bubbles in an acoustic field; they arise from an interaction between bubble translational and compressional responses to the sound, and are a non-linear phenomenon (**C A BJERKNES, V F K BJERKNES 1880**). The *primary Bjerknes force* is responsible for migration of an individual bubble up or down a gradient of mean square pressure, in the absence of nearby boundaries; *secondary Bjerknes forces* (also known as *mutual Bjerknes forces*) arise between neighbouring bubbles, and cause bubble agglomeration or dispersal depending on the frequency of the sound field.

blade passage frequency – an alternative term for BLADE PASSING FREQUENCY. *Units* Hz.

blade passing frequency – *of a rotor with equally-spaced blades* the frequency at which successive blades pass a fixed point in the rotor plane. It is given by BN, where B is the number of blades and N is the rotational frequency of the rotor. *Units* Hz.

Note: If the blades are not identical and equally-spaced, the rotor will radiate sound at all multiples of the shaft frequency N, although the *average blade passing frequency* BN may still be dominant.

Blake threshold pressure – *for a bubble* the amount by which the liquid pressure must be reduced, under quasi-steady conditions, to cause a spherical bubble of fixed gas content to become unstable and grow explosively (**F G BLAKE 1949**).

blast wave – a transient pressure wave in a gas or liquid, caused by an explosion.

blend – the degree to which the sound from an orchestra performing in a concert hall appears to be integrated into a whole, rather than fragmented.

Bloch wave – *in a medium or waveguide whose structure is periodic in the propagation direction* a propagating single-frequency wave in which the same pattern is repeated within each cell or period of the structure, with a fixed phase shift from one cell to the next (**F BLOCH 1928**). Bloch waves are characterized by a dispersion relation in which the frequency axis is divided into alternating passbands and stopbands. Within each passband, waves propagate freely without attenuation (provided the waveguide is nondissipative); in between, they are exponentially attenuated. See also FOSTER'S REACTANCE THEOREM (2).

Bloch wavenumber – the ratio of the phase shift per cell, in a BLOCH WAVE, to the cell length or spatial period. In a lossless waveguide, the phase shift is either purely real (corresponding to a passband) or purely imaginary (corresponding to a stop band). *Units* (real part) rad m^{-1}; (imaginary part) Np m^{-1}.

blocked impedance – *of a linear two-port system* the INPUT IMPEDANCE of the system when the output is connected to a load of infinite impedance.

blocked pressure – the acoustic pressure that would be produced at the surface of a reflecting object, if the surface were rigid (infinite boundary impedance) and the incident field remained unchanged (corresponding to an infinite source impedance). *Units* Pa.

Note: In a one-dimensional acoustic system, the term refers to the pressure produced at a point in the system when the downstream section (the part further from the source) is replaced with a termination of infinite impedance. As before, the incident field is assumed to remain unchanged.

Blokhintsev invariant – *in geometrical acoustics* a quantity related to acoustic energy that is constant along a RAY TUBE; for time-stationary sound fields it is conveniently identified as the ray-tube sound power. In a moving medium, the latter is given by

$$W = \rho q^2 D v_{\text{ray}} A,$$

where A is the ray-tube area, and the remaining factors (defined below) together represent the acoustic intensity I directed along the ray tube (**D I BLOKHINTSEV 1946**). Invariance of W is equivalent to the vector intensity $\mathbf{I} = \rho q^2 D \mathbf{v}_{\text{ray}}$ having zero divergence. Here $\mathbf{v}_{\text{ray}}$ is the RAY VELOCITY (with magnitude v_{ray}); $D = 1 + w_n/c$ is a DOPPLER FACTOR relating frequencies in a fixed frame to frequencies moving with the fluid; ρ, c are the fluid density and sound speed; w_n is the wind velocity component in the local wave-normal direction; and $q = p_{rms}/\rho c$ is the rms acoustic particle velocity. *Units* **W**.

Bode diagram, Bode plot – *for a linear system* a pair of graphs that displays the FREQUENCY RESPONSE FUNCTION of the system in magnitude and phase as a function of frequency (**H BODE 1940**). The upper plot shows the AMPLITUDE RESPONSE FUNCTION, i.e. the magnitude of the frequency response function $H(\omega)$ as a function of angular frequency ω, plotted in dB as $20 \log_{10}|H(\omega)|$. The lower plot shows the PHASE-SHIFT FUNCTION, i.e. the phase arg $H(\omega)$ as a function of frequency. Both plots use the same logarithmic frequency scale on the horizontal axis.

body-diffraction effect – *for plane waves arriving from a given direction at a given frequency* refers to changes in amplitude or phase of the acoustic pressure at either eardrum, due to diffraction of an incident sound wave around the human body.

bond graph – a graphical description of a physical system that allows for reticulation of the system into its essential elements, and delineates the important bonds of interaction between these elements. Bond graphs focus on the power and energy flow in combined mechanical, electrical, thermo-fluid, and chemical systems. Conservation laws (energy, mass, momentum, charge, etc.) are naturally enforced by the structure of the graph, which provides a uniform description of mixed physical domains and permits effective modelling of such systems. Sign conventions and causality relations can be assigned in a systematic fashion, with the result that a bond graph can automatically produce standard analysis tools such as state differential equations, block diagrams, or frequency response functions. Bond graphs can also be used to generate digital computer code automatically for the simulation of complex and nonlinear dynamic systems.

bone conduction – transmission of sound to the inner ear via mechanical vibration of the skull, rather than via the ear canal. Compare AIR CONDUCTION.

bone-conduction audiometry – see PURE-TONE AUDIOMETRY.

boom – (1) *in automotive engineering* a coupled resonance of a car body and the enclosed air, typically occurring in the frequency range 70 to 120 Hz and perceived by the ear as a booming sound. Boom in 4-cylinder cars is usually caused by structural excitation at the second ENGINE ORDER, related to unbalanced engine forces at this frequency.

Note: The associated vibration of vehicle body components may be perceptible to the touch, and felt as a "buzz". This tactile aspect of boom may explain why some automotive engineers lump boom alongside audible harshness, under the general heading of HARSHNESS.

boom – (2) *in aeronautical engineering* see SONIC BOOM.

bottom backscattering strength – *of the sea bed, in underwater acoustics* the SOURCE LEVEL of the backscattered field radiated from unit bottom area, minus the sound pressure level of the incident plane wave in dB re 1 μPa^2. Compare BISTATIC BOTTOM SCATTERING STRENGTH. *Units* dB.

Note: The definition of bottom backscattering strength is closely analogous to that of SURFACE BACKSCATTERING STRENGTH, and further detail will be found under that entry.

boundary absorption – *in room acoustics* the contribution to the ROOM ABSORPTION made by the boundaries of an enclosed space, as opposed to the contribution made by propagation losses in the air (or other fluid) as sound waves travel between the boundary surfaces. *Units* m^2.

boundary condition – *on a field variable* a condition imposed on the value of the variable at a boundary, or – more generally – on some function of the variable and its spatial and temporal derivatives. For example, if φ is the field variable, then

$$f(\varphi, \partial\varphi/\partial t, \partial\varphi/\partial n) = 0 \quad \text{on } S$$

is a boundary condition involving φ, its time derivative, and its spatial derivative normal to the boundary S. See also ESSENTIAL ∼, NATURAL ∼, HOMOGENEOUS ∼, INHOMOGENEOUS ∼, IMPEDANCE ∼, DIRICHLET ∼, NEUMANN ∼, ORTHOGONALITY ∼.

boundary element method – a numerical algorithm for solving a boundary integral equation, in which the domain of integration is divided into subdomains referred to as *elements*. The spatial dependence of the field variables within each element is approximated by an algebraic function of position, with the unknown nodal values as parameters, and the integral equation is rewritten at a number of surface locations equal to the number of unknowns. This leads to a system of algebraic equations that can be solved to obtain the unknown nodal values. Functions used to approximate the geometry of each element are known as *shape functions*, and if the shape functions are identical to the functions used to interpolate the field variables, the elements are referred to as *isoparametric elements*.

An important advantage of the boundary element method is the reduction of dimensionality; i.e. the wave equation in a three-dimensional region can be rewritten as a boundary integral equation on a two-dimensional surface.

Note (1): The boundary element method differs from the FINITE ELEMENT METHOD in an interesting way. The finite element method uses interpolation functions that satisfy the essential boundary conditions exactly, and then solves for the unknown nodal values that approximately satisfy the governing differential equation over the domain. The boundary element method is based on the free-space Green's function, which by definition satisfies the governing differential equation exactly over the domain, and solves for the unknown nodal values that approximately satisfy the boundary conditions.

Note (2): The HELMHOLTZ EQUATION solution over the domain can be rewritten as a surface integral, with the assistance of the divergence theorem and the free-field Green's function. The discretized version of this formulation is called the direct boundary element method. For the expression used, see HELMHOLTZ INTEGRAL FORMULA.

Note (3): An acoustic field can also be represented by a (non-unique) combination of surface distributions of source and dipoles. A boundary integral equation can be written for the field represented by the source distributions, and solved for the strengths of the distributions; this is called the ***indirect boundary element method***.

boundary layer – (1) *in the mean flow sense* a thin region next to a solid boundary within which the relative mean velocity $u(y)$ varies with distance y from the wall, from zero (at $y = 0$) to 99% of its free-stream value (at $y = \delta$). The boundary layer qualifies as thin provided $\delta \ll L$, where L is the streamwise length scale (e.g. the distance upstream to the start of the boundary layer); this in turn requires the REYNOLDS NUMBER based on L to be large. The quantity δ is called the ***boundary layer thickness***.

boundary layer – (2) *in the acoustic sense* a thin region produced by a sound field next to a solid boundary, within which the oscillatory velocity parallel to the wall drops to zero (relative to the wall) as the wall is approached, as a result of viscosity. The unsteady viscous boundary layer qualifies as thin provided $\delta \ll \lambda$; here λ is the acoustic wavelength and $\delta = \sqrt{2\nu/\omega}$, where ν is the KINEMATIC VISCOSITY of the fluid. The quantity δ, sometimes called the ***viscous penetration depth*** or ***viscous skin depth***, is a measure of the boundary layer thickness.

Note: The velocity boundary layer associated with any acoustic field near a boundary is normally accompanied by a THERMAL UNSTEADY BOUNDARY LAYER, in which an adjustment takes place between the oscillatory temperature of the solid boundary and that in the sound field away from the boundary. The two boundary layers, of fluctuating fluid velocity and temperature, are referred to together as the ***viscothermal unsteady boundary layer***.

boundary layer noise – sound that is radiated from a turbulent boundary layer. The sound may be radiated into the moving fluid on the flow side of the boundary; alternatively, flow-excited vibration of the solid boundary may transmit sound into other regions. An example of the latter is **aircraft-cabin boundary layer noise**, caused by the action of the exterior turbulent boundary layer on the fuselage.

boundary value problem – a mathematical problem in which values of the unknown variable u, and/or its derivatives, are specified on the boundary of a domain, and a differential equation has to be solved to find u within the domain. In **single-frequency boundary value problems**, the time factor $e^{j\omega t}$ is suppressed and u is a function of position only. In **time-dependent boundary value problems**, boundary conditions (as above) are specified as a function of time t, for $t < t_{max}$; provided the problem is causal, this information is sufficient to determine the solution for any time up to t_{max}.

bounded input/bounded output stability – *of a linear time-invariant system* the property that for any bounded input, the output remains bounded for all times. Abbreviated as **BIBO stability**.

Note: Any ASYMPTOTICALLY STABLE system has this property.

breathing mode – (1) *of a sphere* a mode of vibration in which the surface of the sphere expands and contracts uniformly; thus the pattern of surface motion is spherically symmetric.

breathing mode – (2) *of a circular cylinder* a mode of vibration in which every cross-section of the cylinder expands and contracts uniformly. Thus the pattern of surface motion is axisymmetric, but not necessarily uniform in the axial direction.

Brewster angle – *at a plane interface between two lossless fluids* the ANGLE OF INCIDENCE for which the reflection coefficient of plane waves arriving at the interface is zero, and the incident-wave energy is totally transmitted (**D BREWSTER 1813**). Compare CRITICAL ANGLE FOR TRANSMISSION. *Units* rad (but commonly expressed in degrees).

Note: The acoustic Brewster angle, θ_B, is related to the characteristic impedance ratio $r = \rho_2 c_2 / \rho_1 c_1$ and the sound speed ratio $s = c_2/c_1$ by

$$\sin \theta_B = \sqrt{\frac{r^2 - 1}{r^2 - s^2}}.$$

Here c_1 and c_2 are the wave speeds in the incident and the transmitting medium respectively, and ρ_1 and ρ_2 are the densities. The existence of a real θ_B therefore requires either ($\rho_2 c_2 > \rho_1 c_1$, $c_2 < c_1$), or vice versa.

brig – a logarithmic measure of the amplitude ratio between two sinusoidal quantities, analogous to the NEPER but using base 10 rather than base *e* (**H BRIGGS** *1561–1630*). Not a recognized SI unit.

brilliance – the sensation for the listener created by a rich treble sound in concert halls.

broadband – refers to a signal or oscillatory quantity whose spectrum covers a wide range of frequencies (typically an octave or more, although the bandwidth implied may vary according to context). The descriptions ~ *noise* or ~ *random signal* imply a broadband continuous spectrum; the description ~ *multiple-harmonic signal* indicates a broadband line spectrum.

broadband shock-associated noise – *of an underexpanded jet* an equivalent term for SHOCK-CELL NOISE.

brown noise – a stationary random signal whose power spectrum falls off at a constant rate of 6 dB per octave, as the frequency increases. Compare WHITE NOISE, PINK NOISE.

Note: Brown noise is time-integrated white noise.

B-scan – *in diagnostic ultrasound* a two-dimensional image produced by combining a number of A-SCANS.

BT product – (1) *in spectral analysis* the dimensionless product of an analysis bandwidth (or frequency resolution), denoted by B, and a record length (or averaging time), denoted by T. See also DEGREES OF FREEDOM. *Units* none.

Note: The BT product determines the statistical variance of quantities such as power spectral density, when these are estimated from finite-length data samples. The expected variance is of order $(BT)^{-1}$ times the true value.

BT product – (2) *for a transient signal* the dimensionless product of the extent, or spread, of the signal in frequency and the extent or spread in time. For further detail, see UNCERTAINTY PRINCIPLE. *Units* none.

buffeting – *in aerodynamics* violent fluctuations in the aerodynamic loading on a solid surface, associated either with SEPARATION of the flow or with oncoming turbulence. At flight Mach numbers just below Mach 1, *transonic buffeting* occurs when local shocks interact with turbulence to cause unsteady regions of separated flow on flight surfaces.

bulk absorption coefficient – *of an acoustic medium* the ABSORPTION CROSS-SECTION per unit volume; also known as the ENERGY ATTENUATION COEFFI-

CIENT. A bulk absorption coefficient of m causes a plane progressive wave to decay in intensity with distance as e^{-mx}. *Units* m^{-1}.

bulk acoustic wave – *in a solid* a compressional wave that propagates through the bulk material, as opposed to a SURFACE ACOUSTIC WAVE.

bulk backscattering differential – *of a volume distribution of scatterers in an acoustic medium* an equivalent term for VOLUME BACKSCATTERING STRENGTH. Compare DIFFERENTIAL BULK SCATTERING COEFFICIENT, which has different units. *Units* dB re 1 m^{-1}.

bulk modulus – *of a fluid or solid* the slope of the curve of pressure plotted against the logarithm of the density, with either adiabatic or isothermal conditions imposed. The first of these defines the adiabatic or isentropic bulk modulus, B_s, and the second defines the isothermal bulk modulus, B_T. Thus

$$B_s = \rho \left(\frac{\partial P}{\partial \rho} \right)_s, \qquad B_T = \rho \left(\frac{\partial P}{\partial \rho} \right)_T,$$

where ρ is the density and P is the pressure. The subscripts denote which quantity is held constant during compression (s = specific entropy, T = temperature). *Units* Pa.

bulk-reacting – the opposite of LOCALLY-REACTING. A bulk-reacting duct liner transmits waves within the liner material.

bulk scattering coefficient – *of a volume distribution of scatterers in an acoustic medium* the SCATTERING CROSS-SECTION per unit volume, for plane waves incident in a specified direction. A bulk scattering coefficient of m_v implies that a plane progressive wave will decay in intensity with distance as $e^{-m_v x}$. Compare BULK ABSORPTION COEFFICIENT. *Units* m^{-1}.

Note: The equivalent term ***volume scattering coefficient*** is commonly used in underwater acoustics.

bulk viscosity – *in relaxing fluids* the coefficient μ_B in the relation

$$P_{\text{inst}} = P_{\text{eq}} - \mu_B \Delta.$$

Here P_{inst} is the instantaneous pressure in a fluid that is subjected to a finite DILATATION RATE Δ, and P_{eq} is the equilibrium pressure of the fluid (i.e. evaluated at the same density and specific internal energy, but with $\Delta = 0$). *Units* Pa s.

buoyancy frequency – *for a horizontally-stratified fluid acted on by gravity* the angular frequency N defined by

$$N^2 = -\frac{g}{\rho}\left(\frac{d\rho}{dz} + \frac{g\rho}{c^2}\right)$$

where z is the vertical coordinate (positive upwards), g is the gravitational acceleration, ρ is the density, and c is the speed of sound. Also known as the **Brunt–Väisälä frequency** or **Väisälä frequency** (V Väisälä 1925, D Brunt 1927). *Units* rad s^{-1}.

Note: Buoyancy frequencies in the ocean are highest in the upper-ocean thermocline, with corresponding periods of order 10 minutes. Similar oscillation periods are found in the atmosphere under stable conditions.

Burgers equation – the equation

$$\frac{\partial u}{\partial t} + u\frac{\partial u}{\partial x} = \nu\frac{\partial^2 u}{\partial x^2},$$

where $u(x, t)$ is a function of position x and time t, and ν is a diffusion coefficient. The Burgers equation was originally proposed as a one-dimensional model of turbulence (**J M Burgers 1939–40**). Ten years later, it was recognized that the same equation also modelled the effects of viscosity and heat conduction in plane progressive sound waves of finite amplitude. For the latter purpose, one introduces the retarded time variable $\tau = t - x/c$ (where x is the coordinate in the propagation direction and c is the sound speed); then the acoustic pressure, p, obeys the following Burgers-type equation in (τ, x) variables:

$$\frac{\partial p}{\partial x} - \left(\frac{\beta}{\rho c^3}\right)p\frac{\partial p}{\partial \tau} = \frac{\delta}{2c^3}\frac{\partial^2 p}{\partial \tau^2}.$$

Here ρ is the density of the fluid, β is the COEFFICIENT OF NONLINEARITY, and δ is the DIFFUSIVITY OF SOUND.

The **generalized Burgers equation** is an extension of the Burgers equation to cylindrical or spherical progressive waves, with x replaced by the radial coordinate r (the wavefront radius of curvature). It contains an additional term $(m/r)p$ on the left to allow for wavefront divergence:

$$\frac{\partial p}{\partial x} + \frac{m}{r}p - \left(\frac{\beta}{\rho c^3}\right)p\frac{\partial p}{\partial \tau} = \frac{\delta}{2c^3}\frac{\partial^2 p}{\partial \tau^2}.$$

The equation is asymptotically exact in the far-field limit $r \gg \lambda$. The coefficient m equals $1/2$ or 1, for cylindrical or spherical waves respectively.

Note: Converging waves, travelling in the $-r$ direction, are described by a similar equation with c replaced by $-c$. The effect is to change the sign of the δ and β terms.

buzz-saw noise – *of a supersonic fan rotor* a sequence of tones at multiples of the shaft rotation frequency, that can be observed both in the fan inlet duct and the far field. The envelope of the harmonic components in the far field typically peaks at around half the BLADE PASSING FREQUENCY. Also known as *multiple pure tone noise*.

Note: The shaft-order tones are produced by non-uniformities in shock spacing from one blade to the next in the rotor. The low-order tones make only a small contribution to the sound power leaving the rotor face; they become more prominent in the far field, however, because (provided they are above CUTOFF) they decay more slowly along the fan duct than the blade-passing tone.

B-weighting – a FREQUENCY-WEIGHTING procedure, in which the power or energy spectrum of a signal is progressively attenuated towards the high and low ends of the audio frequency range, but less so than with A-WEIGHTING.

Note: The B-weighting curve was originally based on the shape of the 70-phon equal loudness contour, with no attempt to allow for the ear canal resonance that enhances the free-field sensitivity of the ear by nearly 10 dB in the range 2–5 kHz. The intention was that B-weighting the incident pressure, followed by squaring and averaging with a suitable time constant, would simulate the sensitivity of the human ear to audio-frequency sound at sound pressure levels in the approximate range 50–90 dB re $(20 \, \mu\text{Pa})^2$. However, the B-weighting curve is no longer used in standards for noise assessment.

C

CA, CAA – abbreviations for COMPUTATIONAL AEROACOUSTICS. In the UK, *CAA* also stands for the Civil Aviation Authority.

calculated loudness level – *of a continuous sound* a numerical LOUDNESS LEVEL that has been calculated by a specified procedure from the measured pressure signal, rather than found from subjective experiments. *Units* phon.

calculus of variations – the branch of mathematics concerned with minimizing or maximizing the values of given FUNCTIONALS, through an appropriate choice of the FUNCTIONS on which they depend. An example that is called the first problem of the calculus of variations is typical. Let $y = y(x)$ be an arbitrary function defined on a closed interval $[x_1, x_2]$; the end-point values $y(x_1)$ and $y(x_2)$ are given. The problem is to choose the function $y(x)$ so as to minimize the integral J, defined by

$$J = \int_{x_1}^{x_2} F(x, y, y') dx,$$

where the function F is given and $y' = dy/dx$.

The solution proceeds by supposing that $y(x)$ represents the function being sought, and defining a COMPARISON FUNCTION Y by $Y = y + \varepsilon \eta$, where ε is a parameter and $\eta = \eta(x)$ is an arbitrary function that vanishes at x_1 and x_2. Expressed in terms of Y, the functional J depends on ε:

$$J(\varepsilon) = \int_{x_1}^{x_2} F(x, Y, Y') dx.$$

By definition, $J(\varepsilon)$ is stationary with respect to ε when $\varepsilon = 0$, yielding the necessary condition

$$\delta J = \frac{dJ(\varepsilon)}{d\varepsilon}\bigg|_{\varepsilon=0} = 0.$$

The symbol δJ defined by this operation is called the VARIATION of J. Evaluating δJ for this example yields

$$\delta J = \int_{x_1}^{x_2} \left(\frac{\partial F}{\partial y}\eta + \frac{\partial F}{\partial y'}\eta'\right) dx = 0.$$

Using integration by parts, the equation can be written as

$$\delta J = \int_{x_1}^{x_2} \left[\frac{\partial F}{\partial y} - \frac{d}{dx}\left(\frac{\partial F}{\partial y'}\right)\right] \eta \, dx = 0.$$

The only way the integral can vanish for arbitrary $\eta(x)$ is for the factor in brackets to be zero:

$$\frac{\partial F}{\partial y} - \frac{d}{dx}\left(\frac{\partial F}{\partial y'}\right) = 0, \quad x \in [x_1, x_2].$$

This result, called the **Euler–Lagrange equation**, provides a differential equation with which to determine the function y.

Statements such as those made above, that J is required to be a maximum or minimum or that the variation δJ vanishes, are examples of VARIATIONAL PRINCIPLES. They are equivalent to the Euler–Lagrange equation, in the sense that a function $y(x)$ that satisfies the variational principles also satisfies the equation.

calibration curve – a plot of CALIBRATION FACTOR versus frequency.

calibration factor – *of a transducer* the numerical value of the transducer SENSITIVITY, at any specific frequency. May be given as a single average value for a range of frequencies, or else as a table or graph showing how the sensitivity varies with frequency.

calorimeter – *in ultrasonics* see ULTRASONIC CALORIMETER.

carrier signal – a high-frequency signal that is modulated by a signal of lower frequency. See also MODULATION.

Carson's theorem – if a transient deterministic signal is repeated at random intervals with mean frequency f_0, then the power spectrum $G(f)$ of the resultant waveform is related to the energy spectrum $E(f)$ of a single transient event by

$$G(f) = f_0 E(f) + \text{dc term},$$

where the dc term arises if the transient has a non-zero mean (**J R CARSON 1931**).

cartesian coordinates – an equivalent term for RECTANGULAR COORDINATES (**R DESCARTES 1637**). Both capital (Cartesian) and lower-case spellings are used.

cartesian tensor – a TENSOR whose components are expressed in rectangular coordinates.

Cauchy principal value – *of a definite integral* a procedure for assigning a definite value to an integral, when the range of integration passes through a singularity of the integrand. The Cauchy principal value of the improper integral

$$I = \int_a^b f(x)dx, \quad \text{with} \quad |f(x)| \to \infty \text{ as } x \to x_0 \, (a < x_0 < b),$$

is defined as

$$\wp \int_a^b f(x)dx = \lim_{\epsilon \downarrow 0}\left\{\int_a^{x_0-\epsilon} f(x)dx + \int_{x_0+\epsilon}^b f(x)dx\right\}.$$

Provided that in the neighbourhood of $x = x_0$ the integrand behaves like an ODD FUNCTION of $x - x_0$, the contributions from either side of the singularity can cancel to produce a finite limit, even though the two separate integrals may tend to infinity.

Cauchy–Riemann equations – equations that relate the real and imaginary parts of an ANALYTIC FUNCTION in the complex plane (**A L CAUCHY 1814; G F B RIEMANN 1851**). Also known as *Cauchy–Riemann conditions*.

causal – obeying the principle that events follow causes rather than preceding them. A *causal system* is one whose behaviour is uninfluenced by future events. See also KRAMERS–KRONIG RELATIONS.

caustic – an envelope formed by a family of intersecting RAYS. Equivalently, a caustic is the locus of wavefront singularities caused by the intersection of adjacent rays. Each point on the caustic surface corresponds to a cylindrical focus, as ray tubes shrink to zero width in one dimension while remaining unchanged in the other dimension. On passing the focus, the signal associated with a given ray is advanced in time by one-quarter period, equivalent to taking the negative HILBERT TRANSFORM of the original signal.

Note: A caustic is touched by approaching rays, but is not penetrated by them, since each ray leaves on the same side of the caustic that it arrived from. The name *caustic* (which means *burning*) indicates that it is a region of relatively high intensity.

cavitation – the growth and collapse of gas bubbles in a liquid in response to pressure fluctuations in the liquid, e.g. as imposed by a sound field.

Note: *Stable cavitation* refers to acoustically-driven bubble oscillations that persist over many complete cycles. The term *transient cavitation* refers to a form of INERTIAL CAVITATION where the bubble oscillations typically last for only a few cycles (possibly less than one), ending in violent collapse. See also BLAKE THRESHOLD PRESSURE.

cavitation noise – the sound field radiated from cavitation bubbles, especially during the collapse phase.

cdf – abbreviation for CUMULATIVE DISTRIBUTION FUNCTION.

cent – a unit of LOGARITHMIC FREQUENCY INTERVAL corresponding to a frequency ratio of $2^{1/1200}$. The interval, in cents, between a lower frequency f_1 and a higher frequency f_2 is therefore $1200 \log_2(f_2/f_1)$. *Unit symbol* cent.

Note: A trained singer can be expected to stay in tune to within 1/50 oct, or ±24 cents. Musical instruments are typically tuned to an accuracy of ±2 cents at 440 Hz.

central limit theorem – states that the sum of N independent random numbers, denoted by

$$X = x_1 + x_2 + \ldots + x_i + \ldots + x_N,$$

has a probability density function $f(X)$ that approximates a Gaussian distribution in the limit $N \to \infty$, regardless of the individual probability density functions of the components x_i. (It is necessary to assume that the moments of N are not dominated by a small number of the component variables x_i.)

centre frequency – (1) *of a filter for fixed-bandwidth frequency analysis* the arithmetic mean of the upper and lower band limits, or filter cutoff frequencies. *Units* Hz.

Note: The *geometric centre frequency* of a third-octave filter is the midpoint of the band on a logarithmic frequency scale. See also MIDBAND FREQUENCY.

centre frequency – (2) *of a narrowband noise signal* the frequency that divides the energy or power of the signal into equal parts. For example, half the power of a stationary noise signal lies above the centre frequency, and half below. *Units* Hz.

Note: If the narrowband signal results from random MODULATION of a sinusoidal carrier signal, the centre frequency is the carrier frequency.

centre time – *at a point in a room* the difference in arrival time between the direct sound and the centre of gravity of the squared impulse response. The centre time is closely related to the CLARITY INDEX or ratio of early-to-late sound energy, but avoids the discrete division of the impulse response into early and late periods. *Units* s.

change of phase – (1) *of a signal* equivalent term for PHASE CHANGE (1). *Units* rad.

change of phase – (2) a transition between any two of the three states of matter (gas, liquid, or solid). See PHASE (4).

channel axis – *in underwater acoustics* see SOFAR CHANNEL.

chaotic signal – an unpredictable signal generated by a nonlinear process, such as turbulence. Such signals are for practical purposes non-deterministic; see also DETERMINISTIC SIGNAL.

characteristic decay time – an alternative term for ENERGY DECAY TIME. *Units* s.

characteristic function – an alternative term for EIGENFUNCTION.

characteristic impedance – *for a specified type of wave* the impedance associated with one-dimensional propagation in a given direction. In acoustics, two types of characteristic impedance are commonly used: the first is the complex ratio of the pressure to the particle velocity component in the direction of propagation (giving a specific impedance), and the second is the complex ratio of the pressure to the volume velocity (giving an acoustic impedance). Examples are given below.

(1) The *characteristic specific impedance of plane progressive waves* in an ideal fluid, commonly referred to as the *characteristic impedance* of the fluid, has the value ρc where ρ is the fluid density and c is the sound speed. *Units* Pa s m^{-1}.

(2) The *characteristic acoustic impedance* of a uniform fluid-filled waveguide with rigid walls is $\rho c/S$, where S is the cross-sectional area. *Units* Pa s m^{-3}.

characteristic length for thermal effects – *in a fluid-saturated open porous material* twice the ratio of the pore volume to the pore surface area, evaluated over a representative sample of the material. In symbols, the thermal characteristic length is given by

$$\Lambda_t = \frac{2V}{S} \quad (V = \text{total pore volume}; S = \text{pore area}).$$

See also CHARACTERISTIC LENGTH FOR VISCOUS EFFECTS. *Units* m.

characteristic length for viscous effects – *in a fluid-saturated open porous material* a measure of the volume to surface area of the pores, weighted according to the square of the local fluid velocity corresponding to ideal incompressible potential flow through the pores in a specified direction:

$$\Lambda_v = \frac{2\int_V u^2(\mathbf{x})dV(\mathbf{x})}{\int_S u^2(\mathbf{x}')dS(\mathbf{x}')}.$$

Here V is the region occupied by fluid, with typical point $\mathbf{x}$; S represents the pore boundaries, with typical point $\mathbf{x}'$; and the weighting factor u^2 is the square of the velocity magnitude associated with the local potential flow. Unless the porous material is statistically isotropic, Λ_v is a function of the average flow direction. Compare the CHARACTERISTIC LENGTH FOR THERMAL EFFECTS, Λ_t, which is a non-directional property. *Units* m.

Note: The characteristic lengths Λ_v, Λ_t control the acoustic properties of most open-cell porous materials. However, fibre blankets of high porosity – e.g. glass wool of bulk density less than 10 kg m^{-3} – have acoustic properties that depend on the fibre diameter d, rather than on the much larger length scales Λ_v, Λ_t.

characteristic path length – *for sound in a room* an alternative term for the MEAN FREE PATH; commonly estimated as $4V/S$, where V is the room volume and S is the surface area of the room boundaries. *Units* m.

characteristic polynomial – *of a matrix* suppose $\mathbf{A}$ is an $n \times n$ matrix, and $\mathbf{I}$ is the IDENTITY MATRIX of the same size. The determinant $\det(\mathbf{A} - \lambda\mathbf{I})$ is a POLYNOMIAL in λ of DEGREE n, which is called the ***characteristic polynomial*** of matrix $\mathbf{A}$. Equating the characteristic polynomial to zero gives the ***characteristic equation*** of the matrix $\mathbf{A}$, whose roots are the ***characteristic values*** or EIGENVALUES of $\mathbf{A}$.

characteristics – *in one-dimensional wave motion* lines in the x–t plane that trace the progress of elementary propagating disturbances, or wavelets. Each set of waves (e.g. forward or backward travelling) has its own set of characteristics. For forward ($+x$) acoustic waves, for example, the characteristics are the lines $t - x/c_0 = $ const., and for forward waves in 1D unsteady compressible flow they are the lines $dt - dx/(u + c) = $ constant. Here c is the sound speed, c_0 is its undisturbed value, and u is the fluid velocity in the x direction.

Note: Along each set of characteristics, there is a particular variable that remains constant, called an ***invariant***. In the examples above, the acoustic forward-wave invariant is $u + (P - P_0)/\rho_0 c_0$, and the generalization of this to finite-amplitude waves is $u + \int dP/\rho c$. Here P, ρ are the fluid pressure and density, and subscript 0 indicates a value in the undisturbed fluid (i.e. where $u = 0$). See RIEMANN INVARIANTS.

characteristic value, characteristic vector – *of a square matrix* see EIGENVALUE, EIGENVECTOR.

characteristic wavenumber – *of single-frequency progressive waves propagating in a specified direction* the propagation of a progressive-wave field in the x direction can be described by the complex exponential factor $e^{j(\omega t - kx)}$. The characteristic wavenumber is the coefficient k that multiplies the PROPAGATION COORDINATE x in this expression. In general, k is a complex quantity; its imaginary part describes spatial attenuation (or amplification) of the waves as they propagate. See also PROPAGATION FACTOR. *Units* (real part) rad m^{-1}; (imaginary part) Np m^{-1}.

Note: The characteristic wavenumber is sometimes referred to as the ***free-wave complex wavenumber***. An equivalent term is ***propagation wavenumber***.

characteristic wavenumber vector – *in a lossless anisotropic medium* the 3-dimensional analogue of the CHARACTERISTIC WAVENUMBER; it describes free-field propagation of plane progressive waves, in a given direction at a given frequency. The corresponding propagation factor is $e^{j(\omega t - \mathbf{k} \cdot \mathbf{x})}$, where ω is the angular frequency, **k** is the characteristic wavenumber vector for propagation in the direction parallel to **k**, and vector **x** denotes position. *Units* (real part) rad m^{-1}; (imaginary part) Np m^{-1}.

chirp – see SWEPT SINEWAVE TESTING.

chi-square degrees of freedom – see the entry for CHI-SQUARE DISTRIBUTION. If X_i ($i = 1$ to N) are independent random variables with typical realization x_i, normally distributed with zero mean and unit variance, the quantity

$$\chi^2 = \sum_{i=1}^{N} x_i^2$$

follows a chi-square distribution with N degrees of freedom.

chi-square distribution – a family of continuous PROBABILITY DENSITY FUNCTIONS, with a parameter N called the ***degrees of freedom***. It is generated by the CHI-SQUARE RANDOM VARIABLE χ_N^2 whose mean value is N. The following additive property applies to two STATISTICALLY INDEPENDENT (1) chi-square variables, with chi-square degrees of freedom M and N:

$$\chi_N^2 + \chi_M^2 = \chi_{N+M}^2.$$

The sum of two such variables therefore also has a chi-square distribution. See also DEGREES OF FREEDOM (2).

Note: Chi-square is commonly written as χ^2.

chi-square random variable – name given to the random variable defined by the normalized sum

$$\chi_N^2 = \frac{1}{\sigma^2} \sum_{i=1}^{N} (x_i - \mu)^2.$$

Here the x_i are N values sampled from a continuous Gaussian random variable, with mean value μ and standard deviation σ.

The chi-square variable is itself continuously distributed, but in a non-Gaussian manner: its probability density function is called the CHI-SQUARE DISTRIBUTION with N degrees of freedom.

chi-square statistic – a statistic used to measure goodness of fit between measured data and a model of the data. It is defined as

$$\chi_N^2 = \sum_{i=1}^{N} \left(\frac{x_i - y_i}{\sigma_i} \right)^2,$$

where the x_i are N measured data points with individual standard deviations σ_i (due to random measurement error), and y_i are the corresponding model predictions. If the model has M parameters and these are adjusted to minimize χ_N^2, the resulting probability density function for χ_N^2 is the CHI-SQUARE DISTRIBUTION with $N - M$ degrees of freedom.

chord – *of an airfoil* the distance between the leading and trailing edges. *Units* m.

chromatic audition – *in human listeners* the ability to visualize colours in response to sound. The phenomenon is triggered in some people by hearing certain sounds; specific sounds are associated by each individual with specific colours. This is an example of SYNAESTHESIA.

circuit analogue – *for an acoustical or mechanical system* a representation of the linear equations governing the system by an equivalent electrical circuit, in which lumped elements are connected together by ideal conductors. Various versions of this basic idea have been used in acoustics: in the ***conventional or impedance analogy***, acoustic pressure is equated to voltage and acoustic volume velocity to current. In the ***acoustic-mobility analogy***, these equivalences are reversed.

circular convolution – (1) *of two periodic discrete sequences* the convolution sum evaluated modulo N, where N is the period of the two sequences. It is written in equation form as

$$x_1[n] \otimes x_2[n] = \sum_{i=0}^{N-1} x_1[i] x_2[n - i]_{\mathrm{mod}\, N},$$

where the index variables are defined to lie in the range 0 to $N - 1$. The notation mod N means that N is added or subtracted as required to bring the index into this range.

Note: An alternative notation for $x_1[n] \otimes x_2[n]$ is $x_1 \otimes x_2|_n$.

circular convolution – (2) an unwanted effect that can be encountered when the convolution of two finite sequences is calculated via the product of their DISCRETE FOURIER TRANSFORMS (DFTs). Unless precautions are taken, the inverse transform of this product will not be the same as the linear convolution; see CONVOLUTION THEOREM. The result "wraps around" owing to the periodicity of the DFT and its inverse.

The circular convolution obtained by this process can be made identical to the linear convolution by ZERO-PADDING before performing the DFTs; if the sequence lengths are N_1 and N_2, zeros are added to bring each sequence up to length $N_1 + N_2 + 1$.

circular frequency – an equivalent term for ANGULAR FREQUENCY. *Units* rad s^{-1}.

circular functions – a generic term for the trigonometric functions sin, cos, and tan.

circularly polarized – see POLARIZATION.

circulation – *in fluid dynamics* the line integral $\Gamma(C) = \oint \mathbf{u} \cdot d\mathbf{s}$, taken in a specified direction around a closed path C. Here $\mathbf{u}$ is the fluid velocity field, and $d\mathbf{s}$ is an elementary displacement along C. In a homogeneous IDEAL FLUID, **Kelvin's theorem** (W THOMSON, LORD KELVIN 1869) states that for any closed path formed by material elements, $\Gamma(C)$ remains constant with time. However if the fluid density and pressure are not one-to-one related, i.e. $\rho \neq \rho(P)$, pressure gradients in the fluid generate VORTICITY and the theorem does not apply. *Units* m^2 s^{-1}.

circumaural earphone – an EARPHONE that completely encloses the pinna, surrounding it like an earmuff.

clarity – the degree to which musical detail can be heard during a live performance or on a recording. Extremes of **muddy** and **clear** are often used. Clarity for music is equivalent to INTELLIGIBILITY for speech.

clarity index – *at a point in a room* see EARLY-TO-LATE SOUND INDEX. Symbol C_{80}. *Units* dB.

classical absorption – *of sound waves in a fluid* the ABSORPTION of sound due to viscosity and heat conduction (plus molecular diffusion of species, if the fluid is a mixture of different gases or different liquids).

Note: The corresponding plane-wave ATTENUATION COEFFICIENT, denoted by α_{cl}, varies as the square of the frequency in air and seawater over the audio frequency range. The same is true in any fluid at frequencies low enough that the product $\alpha_{cl}\lambda$ (λ = wavelength of sound) is much less than 1. Under these conditions the viscous, thermal, and diffusion contributions are additive.

classical attenuation coefficient – *for sound waves in a fluid* the ATTENUATION COEFFICIENT of plane acoustic waves due to CLASSICAL ABSORPTION mechanisms. Symbol α_{cl}. *Units* Np m^{-1}.

class intervals – a particular type of BIN, used for numerical data, and defined by dividing the available range into (usually equal) numerical intervals.

clear – *in concert-hall acoustics* see CLARITY.

closed system – a system with no transfer of matter across its boundaries.

coaxial – any two axisymmetric objects in three dimensions are said to be coaxial if they share the same axis. Examples: ***coaxial cylinders***, ***coaxial beams of sound***.

cochlea – the fluid-filled spiral-shaped organ of the inner ear, situated within the temporal bone of the skull, that accepts a mechanical input from the OSSICLES and produces an electrical output in the form of nerve impulses. See also HAIR CELL.

Note: The cochlea consists of a membraneous spiral duct of about $2\frac{1}{2}$ turns, tapering towards its far end, and attached as a kind of tail to the saccule of the vestibular system. The duct is partitioned lengthwise by two parallel membranes (the ***basilar membrane*** and ***Reissner's membrane***) into 3 spiral canals, the central one of which (scala media or cochlear duct) contains the cochlear sensory apparatus, i.e. the ***tectorial membrane*** and ***organ of Corti***.

Vibrations of the small bones in the MIDDLE EAR are transmitted via the oval window into the fluid (perilymph) that surrounds the cochlea, and hence into the basilar membrane. The latter responds selectively at different positions along its length to different frequencies. The response is finally converted into electrical impulses by hair cells attached to the surface of the membrane.

cochlear implant – a device that includes a microphone, signal processing hardware, and implanted electrodes. It takes over the function of the external and middle ear and the cochlea, and converts incoming sound into a train of electrical pulses that stimulate the auditory nerve. Some sensation of hearing can thereby be restored to a person with inner-ear deafness.

cocktail party effect – (1) *in room acoustics* the tendency for people talking in groups to raise their voices, and to move closer together, when many conversations are taking place simultaneously in a reverberant environment.

cocktail party effect – (2) *in psychoacoustics* the ability of listeners with normal hearing to focus on an individual speaker when many conversations are taking place simultaneously in a reverberant environment. Also known as the ***cocktail party phenomenon***. It appears to involve temporal comparison of the information coming from the two ears.

coefficient of nonlinearity – *for a fluid* the isentropic derivative $\beta = c^{-1}\partial(\rho c)/\partial\rho|_s$, evaluated at the ambient or undisturbed state. Here c is the sound speed and ρ is the density; subscript s indicates that the specific entropy is held constant. *Units* none.

Note: The coefficient of nonlinearity is so called because it occurs in the first-order approximation to the amplitude-dependent propagation speed, $v(u)$, of a point on a finite-amplitude plane progressive waveform:

$v(u) \approx c_0 + \beta u.$

Here c_0 is the ambient sound speed in the fluid at rest, and u is the particle velocity at the given point on the waveform.

coefficient of virtual inertia – *for a body immersed in fluid* the ratio of the virtual, or added, mass to the mass of fluid displaced by the body. See also VIRTUAL MASS. *Units* none.

coherence – (1) *applied in the general sense to two signals* see COHERENT.

coherence – (2) abbreviation for COHERENCE FUNCTION. *Units* none.

coherence function – *of two time-stationary random signals* a measure of the extent to which the two signals are linearly related at any given frequency. It is defined by

$$\gamma^2_{xy}(f) = \frac{|G_{xy}(f)|^2}{G_{xx}(f)G_{yy}(f)} \qquad (0 \leq \gamma^2_{xy} \leq 1),$$

where $G_{xy}(f)$ is the single-sided CROSS-SPECTRAL DENSITY of the signals $x(t)$, $y(t)$, and G_{xx}, G_{yy} are the respective autospectral densities (also single-sided in this definition). An equivalent term is ***ordinary coherence function***. *Units* none.

Note: If the coherence function is close to unity, this means that $y(t)$ may be linearly predicted from $x(t)$ with a relatively small mean square error. If two signals are linearly related through a FREQUENCY RESPONSE FUNCTION, their coherence is unity.

coherency squared function – the same as COHERENCE FUNCTION. *Units* none.

coherent – two signals are ***strictly coherent*** if their Fourier components, at each frequency, maintain a fixed phase relationship for all time. The term coherent is also used in a broader sense, to describe any two stationary signals whose COHERENCE FUNCTION is non-zero (usually over a specified frequency range); such signals are more precisely called PARTIALLY COHERENT.

coincidence – *at a plane boundary* a situation in which the TRACE VELOCITY of plane sound waves incident on the boundary coincides with the speed of matching free waves as they propagate along the boundary. "Matching" means that the wavelength of the waves on the boundary equals the TRACE

WAVELENGTH of the incident waves. The angle of incidence at which this occurs is known as the *coincidence angle*.

A similar definition may be applied to a cylindrical boundary, in terms of wavelengths and phase speeds in the axial direction; and likewise to a rod or beam. For coincidence in a waveguide, see PSEUDO-COINCIDENCE.

coincidence effect – *for a panel or partition* the dip in the curve of TRANSMISSION LOSS (2) against frequency that occurs when incident sound waves (at whatever angle of incidence) have the same trace wavelength as free bending waves in the panel.

coincidence frequency – the frequency at which COINCIDENCE occurs, when a plane boundary is excited with a specified trace velocity (e.g. plane acoustic waves incident from a given direction). *Units* Hz.

Note: The CRITICAL FREQUENCY is the lowest frequency at which coincidence can occur with incident sound. Coincidence at this frequency requires the exciting sound field to be at grazing incidence.

coincident spectral density – an expanded version of the term CO-SPECTRAL DENSITY.

collimated beam – a beam of sound or light that neither diverges nor converges.

coloured noise – alternative term for NON-WHITE NOISE.

column backscattering strength, column strength – *of a volume distribution of scatterers in shallow-water sound transmission* the TARGET STRENGTH of a column of water of unit cross-sectional area in the horizontal plane. Numerically, it is equal to the DIFFERENTIAL SCATTERING CROSS-SECTION of the water column in the backscattering direction, normalized per unit area and expressed in decibels. *Units* dB.

combination resonance – *of a forced system with quadratic nonlinearity* a resonance that occurs when the forcing frequency is equal to the sum, or difference, of two of the system natural frequencies.

combination tone – *in psychoacoustics* when the ear is stimulated simultaneously with two tones at frequencies f_1 and f_2, such that $\frac{1}{2}f_2 < f_1 < f_2$, a tone may be heard by some subjects with a pitch that corresponds to frequency $2f_1 - f_2$. This nonlinear phenomenon is called a combination tone, or alternatively a *combination product*. Although the $2f_1 - f_2$ combination tone is usually the most prominent, other combinations of the two primary frequencies can also be observed when the levels of the primary sounds are high enough.

combustion noise – noise radiated from turbulent or unstable combustion. See also CORE NOISE, RAYLEIGH'S CRITERION, RIJKE TUBE.

compact – for a given frequency, a radiating or scattering object is called compact if its maximum dimension is small compared with the acoustic wavelength. An equivalent term is *acoustically compact*.

comparison function – a function that satisfies all the boundary conditions of a given problem. See CALCULUS OF VARIATIONS.

complementary function – see HOMOGENEOUS SOLUTION.

complex acoustic intensity – (1) *in a single-frequency sound field* the complex vector quantity

$$\mathbf{I}_c = \mathbf{I} + j\mathbf{J},$$

where $\mathbf{I}$ is the ACTIVE INTENSITY and $\mathbf{J}$ is the REACTIVE INTENSITY. *Units* W m^{-2}.

complex acoustic intensity – (2) *in any time-stationary sound field* the complex vector quantity defined as the time-average product

$$\langle \check{p}\mathbf{u} \rangle = \mathbf{I}_c .$$

Here $\check{p}$ is the ANALYTIC SIGNAL corresponding to the local acoustic pressure p, and $\mathbf{u}$ is the acoustic particle velocity vector at the same point; angle brackets $\langle \ldots \rangle$ denote the time averaging operation. *Units* W m^{-2}.

complex amplitude – if a real sinusoidal signal $x(t)$ is represented by the real part of the complex phasor $\hat{x}e^{j\omega t}$, then the coefficient $\hat{x}$ is called the complex amplitude. The magnitude of the phasor, i.e. the modulus $|\hat{x}|$, is the AMPLITUDE of the signal.

Note: The time factor $e^{-i\omega t}$ is sometimes used in place of $e^{j\omega t}$; in that case the complex amplitude $\hat{x}$ is the complex conjugate of its value for $e^{j\omega t}$.

complex angular frequency – by assigning a complex value to the coefficient ω in the time factor $e^{j\omega t}$, an amplifying or decaying oscillation can be represented as a PHASOR. The quantity ω is then called the complex angular frequency of the oscillation; compare COMPLEX NATURAL FREQUENCY. *Units* (real part) rad s^{-1}; (imaginary part) Np s^{-1}.

complex attenuation coefficient – ◆ *of single-frequency plane progressive waves in a dispersive medium* the complex attenuation coefficient, α', is the complex number whose real part is the ATTENUATION COEFFICIENT, α, and whose imaginary part is the DISPERSION COEFFICIENT, β; thus $\alpha' = \alpha + j\beta$. *Units* (real part) Np m^{-1}; (imaginary part) rad m^{-1}.

complex coherence function – *of two time-stationary signals* the normalized CROSS-SPECTRAL DENSITY of the two signals, defined by

$$g_{xy}(f) = \frac{G_{xy}(f)}{\sqrt{G_{xx}(f)G_{yy}(f)}} = \gamma_{xy}(f)e^{j\varphi_{xy}(f)}$$

where $G_{xy}(f)$ is the single-sided CROSS-SPECTRAL DENSITY of the signals $x(t)$, $y(t)$, and G_{xx}, G_{yy} are the respective autospectral densities (also single-sided in this definition). In the final expression on the right, $\gamma_{xy}(f)$ is the square root of the COHERENCE FUNCTION, and φ_{xy} is the PHASE ANGLE OF THE CROSS-SPECTRAL DENSITY. *Units* none.

complex conjugate – *of a complex number* the complex conjugate of $x + jy$, where x and y are real, is $x - jy$. The complex conjugate of z is written as z^* or $\bar{z}$.

complex effective bulk modulus – *of a lossy fluid* the reciprocal of the COMPLEX EFFECTIVE COMPRESSIBILITY; denoted by $B(\omega) = 1/K(\omega)$. *Units* Pa.

complex effective compressibility – *of a lossy fluid* the complex ratio of the relative change in density (i.e. the acoustic condensation ζ) to the acoustic pressure p, for single-frequency excitation at small amplitudes. In symbols,

$$K(\omega) = \frac{\zeta}{p} \qquad \left(\zeta = \frac{\rho - \rho_0}{\rho_0}; |\zeta| \ll 1\right)$$

where $K(\omega)$ is the complex effective compressibility at angular frequency ω. In the expression for ζ, ρ is the fluid density and ρ_0 is the undisturbed density. It follows that the linearized continuity equation for the lossy fluid in the absence of mean flow is

$$\operatorname{div} \mathbf{u} = -K(\omega)j\omega p,$$

where $\mathbf{u}$ is the acoustic particle velocity vector. *Units* Pa^{-1}.

Note (1): The complex effective compressibility $K(\omega)$ is actually a function of wavenumber as well as frequency, although this dependence is often ignored. The compressibility that is generally implied is the *free-wave* complex compressibility. Similar remarks apply to other complex effective quantities (density, sound speed, bulk modulus), which despite their names are not true local properties.

Note (2): For rigid-frame porous materials, the linearized continuity equation for the lossy fluid in the absence of mean flow is

$$\operatorname{div}(\phi\bar{\mathbf{u}}) = -K(\omega)j\omega p,$$

where ϕ is the porosity of the material and $\bar{\mathbf{u}}$ is the acoustic seepage velocity vector, defined such that its component in any direction is the volume flowrate per unit transverse cross-sectional area, averaged over an area covering several pores.

See also note (2) under DYNAMIC TORTUOSITY.

complex effective density – *of a lossy fluid* the complex ratio of the negative pressure gradient to the fluid particle acceleration in the same direction, for single-frequency small-amplitude sound waves. In symbols, the complex density is the factor $\rho(\omega)$ in the lossless form of the linearized momentum equation,

$$\rho(\omega) j\omega \mathbf{u} = -\nabla p.$$

Here ω is the angular frequency, $\mathbf{u}$ is the particle velocity, and p is the acoustic pressure. Compare COMPLEX EFFECTIVE COMPRESSIBILITY. *Units* kg m^{-3}.

Note: For sound waves propagating through a rigid-frame porous material, the complex effective density definition above is generalized by writing the linearized momentum equation in terms of the macroscopic fluid velocity $\langle \mathbf{u} \rangle_V$, defined as an average over a small region that is large enough to enclose several pores. Thus the equation above is replaced by

$$\rho(\omega) j\omega \langle \mathbf{u} \rangle_V = -\nabla p \quad (\text{where } p = \langle p \rangle_V),$$

with angle brackets $\langle \ldots \rangle_V$ denoting a local volume average over the pore fluid.

Since the area average $\langle u_x \rangle_A$, over planes $x = \text{const.}$, is equivalent to the volume average $\langle u_x \rangle_V$, the volume flowrate per unit total cross-sectional area is given by the vector $\bar{\mathbf{u}} = \phi \langle \mathbf{u} \rangle_V$ where ϕ is the porosity. The linearized momentum equation may be expressed in terms of $\bar{\mathbf{u}}$ as

$$\frac{\rho(\omega)}{\phi} j\omega \bar{\mathbf{u}} = -\nabla p.$$

See also DYNAMIC TORTUOSITY.

complex effective sound speed – *of a lossy fluid* the complex quantity

$$c(\omega) = \omega/k(\omega) \quad (\text{at angular frequency } \omega),$$

where $k(\omega)$ is the PROPAGATION WAVENUMBER for freely-propagating plane acoustic waves. *Units* m s^{-1}.

complex flow resistivity – *of a bulk porous material at a specified frequency* see DYNAMIC FLOW RESISTIVITY. *Units* Pa s m^{-2}.

complex Fourier amplitudes – *of a periodic continuous signal* the infinite set of complex coefficients obtained by applying FOURIER ANALYSIS to a continuous signal. Their magnitudes define the *amplitude spectrum* of the signal, and their phases define the *phase spectrum*.

Note (1): For a periodic signal in discrete time, the corresponding Fourier coefficients are called the *spectral coefficients*; they form a periodic sequence whose repetition length is that of the original signal. (In terms of physical frequency, the period is equal to the sampling frequency.) See DISCRETE FOURIER SERIES.

Note (2): Compare COMPLEX FOURIER SPECTRUM, which relates to the Fourier transform of a transient signal.

complex Fourier spectrum – *of a transient signal* the FOURIER TRANSFORM of the signal, with its real and imaginary parts plotted as continuous functions of frequency. See also MAGNITUDE SPECTRUM, PHASE SPECTRUM (2).

Note: The spectrum of a signal sampled at equal intervals in discrete time is periodic; the spectrum of a continuous signal is nonperiodic.

complex modes – modes in which the phase varies continuously from point to point.

complex modulus – *of an isotropic linear viscoelastic material* a generic term for one of the three quantities YOUNG'S MODULUS, BULK MODULUS, SHEAR MODULUS when these are defined as the complex ratio of stress to strain under single-frequency excitation. *Units* Pa.

Note: Two of these three moduli – for example, the complex shear modulus, G, and the complex bulk modulus, B – are sufficient to describe any isotropic linear material whose properties do not vary with time. Other parameters such as the complex Young's modulus, E, and the complex Poisson's ratio, v, may then be deduced from these by using the relations

$$E = \frac{9GB}{G+3B}, \quad v = \frac{1}{2}\frac{3B-2G}{3B+G}.$$

complex natural frequency – *of a linear system oscillating in a single mode* a COMPLEX ANGULAR FREQUENCY, ω, whose real part corresponds to the DAMPED NATURAL FREQUENCY of the system and whose imaginary part gives the exponential decay coefficient of free oscillations. *Units* (real part) rad s^{-1}; (imaginary part) Np s^{-1}.

Note: The instantaneous displacement at any point of the system may be represented as Re$\{Ae^{j\omega t}\}$, where t is time and A is a constant. If the system is damped (i.e. asymptotically stable), the imaginary part of ω is positive.

complex number – any quantity of the form $x + jy$, where j stands for $\sqrt{-1}$ and (x, y) are real numbers. Complex numbers were introduced by Gauss in

1831; he also adopted the *i* notation proposed by Euler for $\sqrt{-1}$. See also ARGUMENT, MODULUS, POLAR FORM.

complex plane – the plane defined by coordinates x, y such that any complex number $x + jy$ is uniquely represented by a point in the plane. Its graphical representation is known as the ***Argand diagram***.

complex pressure – the PHASOR whose real part is the instantaneous acoustic pressure. *Units* Pa.

complex pressure reflection coefficient – *for single-frequency sound waves* the complex ratio of reflected to incident sound pressure in a standing wave, with both pressures measured at the same point. It is usually defined either for plane waves incident at a given angle on a uniform reflecting plane boundary, or for plane waves in a one-dimensional system (e.g. a pipe or waveguide) incident on a discontinuity in the system. *Units* none.

Note: The complex pressure reflection coefficient is sometimes called the ***sound pressure reflection coefficient***, the ***pressure reflection coefficient***, or (in acoustical contexts) simply the ***reflection coefficient***. See also REFLECTION FACTOR (1).

complex pressure transmission coefficient – *for single-frequency incident waves transmitted through a boundary or discontinuity* the complex ratio of transmitted to incident pressure, with the transmitted pressure measured immediately behind the boundary and the incident pressure measured immediately in front. For purposes of the definition, the transmitted field is assumed to be outgoing (i.e. no waves are reflected on the transmitted side). *Units* none.

Note (1): The complex pressure transmission coefficient is sometimes called the ***sound pressure transmission coefficient*** or the ***pressure transmission coefficient***, but preferably not (because of ambiguity) simply the ***transmission coefficient***.

Note (2): Compare the entry for TRANSMISSION COEFFICIENT. Also compare SOUND TRANSMISSION COEFFICIENT, which has a different meaning in architectural acoustics.

Note (3): The complex pressure transmission coefficient is usually defined either for plane waves incident at a given angle on a uniform plane boundary, or for plane waves in a one-dimensional system (e.g. a pipe or waveguide) being transmitted between specified points in the system. For more general situations (e.g. plane waves incident on a finite partition), the pressure transmission coefficient is a local quantity that varies from point to point over the boundary. For this reason the SOUND POWER TRANSMISSION COEFFICIENT (a global quantity) is often of greater practical interest.

complex specific flow impedance – ◆ *of a porous acoustic layer that is thin compared to an acoustic wavelength* the differential specific impedance given by

$$z_{\text{flow}} = \frac{\Delta p}{\bar{u}},$$

where Δp is the pressure drop through the layer, and $\bar{u}$ is the FACE VELOCITY of the flow relative to the layer (equivalent to the volume flowrate per unit surface area). Compare SPECIFIC FLOW RESISTANCE. *Units* Pa s m^{-1}.

complex stiffness – *of a mechanical lumped element* the complex ratio of transmitted force to relative displacement, at a given frequency. *Units* N m^{-1}.

complex time-domain signal – an equivalent term for ANALYTIC SIGNAL. Many electronic systems use an in-phase (I) and quadrature (Q) channel, corresponding to the real and imaginary parts of the analytic signal.

complex time-integrated intensity – ◆ *in a transient sound field* the complex vector quantity

$$\mathbf{N}_c = \int_0^T \check{p}\mathbf{u}\, dt \quad (t=\text{time}),$$

which represents the total acoustic energy transfer (both active and reactive components) per unit area at a fixed point, integrated over the duration T of the transient sound field. Here $\check{p}$ is the ANALYTIC SIGNAL corresponding to the local acoustic pressure p, and $\mathbf{u}$ is the vector particle velocity at the same point. Compare TIME-INTEGRATED INTENSITY, COMPLEX ACOUSTIC INTENSITY (2). *Units* J m^{-2}.

complex wavenumber – a WAVENUMBER (1) whose imaginary part is non-zero, corresponding to either attenuation or amplification of the wave as it propagates. See also CHARACTERISTIC WAVENUMBER. *Units* (real part) rad m^{-1}; (imaginary part) Np m^{-1}.

compliance – (1) see LUMPED ELEMENTS, VOLUME COMPLIANCE; (2) *for a viscoelastic material* see COMPLIANCE MATRIX.

compliance matrix – *of a linear viscoelastic material* the frequency-dependent complex matrix $\mathbf{C}$ in the following strain-stress relationship for a general anisotropic material under sinusoidal excitation:

$$\mathbf{e} = \mathbf{C}\,\mathbf{s}, \quad \text{or equivalently} \quad e_i = C_{ij} s_j.$$

Here $\mathbf{e} = \{e_i\}$ is a column vector of the 6 strain components, expressed as complex amplitudes, and $\mathbf{s} = \{s_i\}$ is the column vector of corresponding stress components. Compare STIFFNESS MATRIX. *Units* Pa^{-1}.

compliant – deformable, usually in respect of volume or length.

compressibility – *of a fluid or solid* the reciprocal of the BULK MODULUS. In materials subject to thermal expansion, it is necessary to distinguish between the ISENTROPIC (or ADIABATIC) COMPRESSIBILITY, K_S, and the ISOTHERMAL COMPRESSIBILITY, K_T. For gases in particular the difference is significant; an ideal gas has $K_T = \gamma K_s$, where γ is the SPECIFIC-HEAT RATIO.

In acoustics, the term compressibility commonly implies the isentropic value, which (for any fluid) is related to the fluid density ρ and sound speed c by $K_s = 1/\rho c^2$. *Units* Pa^{-1}.

compressional waves – waves characterized by local expansion and compression of the medium; also known as DILATATIONAL WAVES. Sound waves are compressional waves.

Note: In a uniform isotropic elastic solid, two types of small-amplitude disturbance propagate independently: these are shear (S) and compressional (P) waves. The latter have two distinguishing features: they involve local changes of volume, and they are IRROTATIONAL. For further information, see LONGITUDINAL WAVES.

compressional wave speed – *in a uniform isotropic elastic medium* in a bulk medium, plane compressional waves travel at speed c given by $c = [(B + \frac{4}{3}G)/\rho]^{1/2}$. Here ρ is the density, B is the isentropic BULK MODULUS, and G is the SHEAR MODULUS. *Units* m s^{-1}.

compression wave – *in a fluid or solid* a progressive wave, or wavefront, that causes compression of the medium rather than expansion. Compare RAREFACTION WAVE.

computational acoustics – the use of numerical simulation to model sound radiation and scattering by boundaries of arbitrary shape.

Note: Computational acoustics usually involves solving the Helmholtz equation by finite element or boundary element methods. Compare COMPUTATIONAL AEROACOUSTICS.

computational aeroacoustics – the use of numerical simulation to model sound production by unsteady flows, or to model the transmission of sound through flows. Abbreviated as *CA* or *CAA*.

Note: Computational aeroacoustics, like computational fluid dynamics, usually involves solving the equations of fluid motion by finite difference methods. Compare COMPUTATIONAL ACOUSTICS.

computational fluid dynamics – the use of numerical simulation to describe fluid flows, by employing a suitable system of algebraic equations to approximate the continuous equations that govern the flow variables. Abbreviated as **CFD**.

concha – the deep depression in the PINNA that leads directly to the EAR CANAL.

condensation – *in acoustics* the increase in density of a material under stress, divided by the original density; i.e. the negative of the VOLUME STRAIN or DILATATION. *Units* none.

Note: The use of condensation in this sense is becoming less common, and the term does not appear at all in Pierce's *Acoustics* (first published 1981).

conditional probability – the PROBABILITY that some specified event will occur, given that another specified event has already occurred. The conditional probability of event A, given event B, is written $P(A|B)$. *Units* none.

Note: The conditional probability $P(A|B)$ is related to the probability of *both* events occurring, $P(A, B)$, by

$$P(A|B) = \frac{P(A, B)}{P(B)}.$$

In the particular case where A and B are STATISTICALLY INDEPENDENT, the relation above reduces simply to

$$P(A|B) = P(A);$$

i.e. the fact that B has occurred has no influence on the likelihood of A.

conductance – the real part of an ADMITTANCE.

conductive hearing loss – HEARING LOSS caused either by blockage of the external ear (e.g. wax in the ear canal) or by disease or damage in the middle ear, so that the signal amplitude reaching the inner ear is reduced.

conductivity – (1) *for oscillatory irrotational inviscid flow between two constant-pressure surfaces or ports* the ratio $C = \rho \dot{U}/\Delta p$, where ρ is the fluid density (assumed constant and uniform), $U(t)$ is the volume velocity across either surface (directed from surface 1 to surface 2), and Δp is the unsteady pressure difference $p_1(t) - p_2(t)$ between the two surfaces. Also known as the **Rayleigh conductivity**. *Units* m.

Note: The ratio $\Delta p/\dot{U} = \rho/C$ is sometimes called the ACOUSTIC INERTANCE, m_{ac} (see LUMPED ELEMENTS); its physical significance is that $m_{ac}U^2/2$ is the instantaneous kinetic energy of the flow between surfaces 1 and 2.

conductivity – (2) see THERMAL CONDUCTIVITY.

cone of silence – *in jet noise* a region of relative silence near the downstream jet axis. According to geometrical acoustics, high-frequency acoustic radiation from within the jet shear layer cannot penetrate into a cone defined by a maximum angle

$$\theta_{\text{crit}} = \cos^{-1}\frac{c_0}{u_s + c_s}$$

measured from the jet axis (c_0 = ambient sound speed; u_s, c_s = mean axial velocity and sound speed in the source region).

Note: One can imagine sound reaching the cone of silence via EVANESCENT WAVES in the jet flow; however, more accurate models show that this simplified picture greatly overestimates the attenuation.

conservation law – *for a finite control volume* a statement of the form

$$\frac{dB}{dt} = P + X,$$

where B is the amount of the "conserved quantity", e.g. mass, in a specified region; X is the inflow rate of the conserved quantity across the boundaries of the region; and P is the rate of production of the "conserved quantity" within the region, or the rate of supply from external sources.

By applying the statement above to a small fixed region of space, the conservation law may also be written in differential form: for example,

$$\frac{\partial \rho}{\partial t} = -\text{div}\rho\mathbf{u}$$

expresses the local conservation of mass in a fluid flow with no mass sources. See CONTINUITY EQUATION.

conservative force – a force **F** applied at a moving point *P* is said to be conservative if its magnitude and direction depend only on the position of *P*, and in addition the work done around any closed path is zero. An example is the restoring force applied by one end of a lossless spring, when it is displaced with the other end held fixed.

Note: When a conservative force acts on a system in periodic motion, the net energy added to the system over one cycle is zero.

conservative force field – *acting on a solid or fluid medium* a FIELD that acts on the medium with force **f** per unit mass, where **f** is derived from a *potential* Φ: thus $\mathbf{f} = \nabla\Phi$. The work done by such a field on a material element when it moves round a closed path is zero.

conservative system – a system for which one can introduce the notion of a total energy, made up of a potential energy that depends only on the displacements of the system, and a kinetic energy that depends only on the time derivatives of the displacements and the displacements. The motion of the system is then governed by energy conservation: the sum of the potential and kinetic energies remains constant.

Note: In such a system, work is done only by CONSERVATIVE FORCES. This means that all external forces that do work on the system, as well as internal forces that exchange energy between different parts of the system, are conservative and do no net work in a cyclic process; it is usually convenient to account for conservative external forces by means of a POTENTIAL ENERGY (e.g. gravitational potential energy).

constant-bandwidth frequency analysis – the partitioning of signal energy or power into CONTIGUOUS BANDS of fixed width.

constitutive relation – *for a given material* a relation that connects stresses at a point in the material with the time-history of the local strains. For examples, see STRESS-RELAXATION FUNCTION, RELAXATION SPECTRUM.

constructive interference – see INTERFERENCE.

contiguous bands – a sequence of adjacent frequency bands of either constant or proportional bandwidth, used as passbands for spectral analysis of noise and vibration measurements: for example, the standard 1/3-octave bands centred on $10^{n/10}$ Hz. By choosing a suitable combination of bandwidth and record length, the statistical variance of the relevant spectral estimates can be made as small as required (see BT PRODUCT).

continuity equation – *in fluid or solid mechanics* a differential equation that expresses the fact that mass is conserved locally. If the fluid or solid has density $\rho(\mathbf{x}, t)$ and vector velocity field $\mathbf{u}(\mathbf{x}, t)$, one version of the continuity equation is

$$\frac{\partial \rho}{\partial t} + \nabla \cdot (\rho \mathbf{u}) = 0.$$

An equivalent version in terms of the MATERIAL DERIVATIVE operator D/Dt is

$$\frac{1}{\rho}\frac{D\rho}{Dt} + \nabla \cdot \mathbf{u} = 0.$$

For the special case of an IDEAL FLUID, the first term in this version of the equation may be rewritten in terms of the pressure, P, to give

$$\frac{1}{\rho c^2}\frac{DP}{Dt} + \nabla \cdot \mathbf{u} = 0.$$

Here c is the sound speed of the fluid. See also EULER EQUATIONS.

continuous random variable – a RANDOM VARIABLE whose possible values form a continuous range or interval. The range of values may be bounded or infinite.

continuous signal – a SIGNAL can either be a continuous function of time, or a sequence of discrete values. The former type is called a continuous signal. Equivalent terms are *analogue signal* and *continuous-time signal*.

The process of SAMPLING a continuous signal produces a discrete sequence: see DIGITAL SIGNAL.

continuous spectrum – *in underwater acoustics* the component of the sound field in a horizontally-stratified waveguide that corresponds to VIRTUAL MODES. The word *spectrum* here refers to the horizontal wavenumber spectrum of the depth-dependent Green's function, at a given frequency.

continuous system – (1) *in dynamics* a mechanical system whose kinetic and potential energy and overall energy dissipation rate are continuously distributed, rather than being modelled by LUMPED ELEMENTS. Continuous systems have an infinite number of degrees of freedom, and are governed by partial differential equations. Compare DISCRETE SYSTEM.

continuous system – (2) *in a signal processing context* the same as an ANALOGUE SYSTEM.

continuous-wave (CW) – *used as an adjective* refers to a steady-state oscillation, as opposed to a transient.

contrast agent – *in diagnostic ultrasound* an acoustic medium, or additive, that is introduced to provide increased backscatter in the target area. In order to improve cardiac imaging, for example, microbubbles in a soluble coating are introduced through a vein and are carried to the heart by the blood flow.

control group – *in an experiment that involves human subjects* a group of people, statistically identical to the subject group in all relevant characteristics, who are not exposed to the stimuli whose effect is being investigated. Post-exposure comparison of the control and subject groups can provide evidence that the effect being investigated is real, and can be used to quantify the effect in relation to the stimulus. Control groups are also used in retrospective studies, e.g. in epidemiology.

convection velocity – *of a travelling-wave space–time pattern* the phase velocity of each frequency component. If the pattern is a stationary random function of space and time, rather than a travelling wave, a convection velocity can be defined statistically in terms of contours of the two-point wavenumber–frequency cross power spectrum. See also FROZEN PATTERN. *Units* m s^{-1}.

convective amplification – *for a moving source in a stationary medium* the increase in amplitude of the radiated pressure caused by source motion towards the observer. The factor by which the square of the pressure increases, for a given direction of emission and observer distance, is called the ***convective amplification factor***.

Note (1): Convective amplification is fundamentally an effect of source motion relative to the surrounding medium; the same effect occurs when the source is stationary, and the medium (plus observer) are in uniform motion in the opposite direction. In either case, however, it is important that the observer has the same EMISSION TIME COORDINATES relative to the source, in both the moving and non-moving cases.

Note (2): For various types of elementary source the convective amplification factor is expressible as D^{-n}, where n is a positive integer that depends on source type, and

$$D = 1 - [v_s]/c \qquad (v_s = -\dot{r})$$

is the DOPPLER FACTOR. Here v_s is the approach velocity of the source, r is the source–observer distance with time derivative $\dot{r}$, c is the sound speed, and square brackets denote the value at the time the sound is emitted. The table below gives n for some basic types of point source:

Volume velocity	Monopole	$n = 4$
	Dipole	$n = 6$
Volume displacement	Monopole	$n = 6$
	Dipole	$n = 8$

The general relation for compact sources is $n = 2(\mu + \nu + 1)$, where μ is the spatial order (e.g. dipole = 1, quadrupole = 2) and ν is the temporal order of the source (e.g. volume velocity = 1, volume displacement = 2). Note that ν depends on the choice of wave variable (pressure in this case), as well as on the source mechanism.

convergence zone – *in underwater acoustics* a region where rays reconverge after being emitted from the same source in different directions. In the ocean, convergence tends to occur at ranges of tens of km; it is generally associated with the upward refraction of rays once they reach a certain depth in the ocean (due to the increase of sound speed with hydrostatic pressure).

converging wave – another term for INCOMING WAVE.

convolution – (1) *for continuous signals* the convolution of two continuous signals $x(t)$ and $w(t)$, evaluated for argument t, is defined as

$$x(t) * w(t) = x * w|_t = \int_{-\infty}^{\infty} x(\tau)w(t-\tau)d\tau$$
$$= \int_{-\infty}^{\infty} x(t-u)w(u)du.$$

As indicated above, the operation is commutative: $x * w = w * x$. Note that the arguments of the two functions add up to t for all values of the dummy variable.

convolution – (2) *for discrete sequences* the convolution of two discrete sequences $x[n]$ and $w[n]$ is defined as

$$x[n] * w[n] = x * w|_n = \sum_{i=-\infty}^{\infty} = x[i]w[n-i]$$
$$= \sum_{q=-\infty}^{\infty} x[n-q]w[q].$$

As in the continuous case, the arguments of the two functions in either summation add up to a constant value; see CONVOLUTION (1).

convolution integral – *for the response of a system in continuous time* to obtain the response of a linear time-invariant system to an input $x(t)$, one forms the convolution $x(t) * h(t)$ where $h(t)$ is the response to a unit impulse $\delta(t)$; see IMPULSE RESPONSE FUNCTION. The response is accordingly given by

$$y(t) = \int_{-\infty}^{\infty} x(\tau)h(t-\tau)d\tau = \int_{-\infty}^{\infty} x(t-u)h(u)du,$$

and is called the convolution integral.

Note (1): The result above follows directly from the CONVOLUTION (1) definition, with w replaced by h. The second integral represents the summed output at time t, due to the superposition of elementary impulses $x(t-u)\,du$ applied to the system; u in this integral is the time delay by which the output lags the elementary input.

Note (2): Integration from $-\infty$ to ∞ is not required for systems that are CAUSAL. If the input to a causal system begins at time $t = t_0$, the output for any $t > t_0$ is given by

$$y(t) = \int_{t_0}^{t} x(\tau)h(t-\tau)d\tau = \int_{0}^{t-t_0} x(t-u)h(u)du.$$

convolution sum – *for the response of a system in discrete time* to obtain the response of a linear time-invariant system to an input $x[n]$, one forms the convolution $x*h|_n$, where $h[n]$ is the response to a unit impulse $\delta[n]$. The output $y[n]$ is given by either of the equivalent expressions

$$y[n] = \sum_{i=-\infty}^{\infty} x[i]h[n-i] = \sum_{q=-\infty}^{\infty} x[n-q]h[q],$$

referred to as the convolution sum or **superposition sum**.

Note (1): The result above follows directly from the CONVOLUTION (2) definition, with w replaced by h.

Note (2): Summation from $-\infty$ to ∞ is not required for systems that are CAUSAL. If the input sequence $x[n]$ to a causal system begins at $n = n_0$, the output for any $n > n_0$ is given by

$$y[n] = \sum_{i=n_0}^{n} x[i]h[n-i] = \sum_{q=0}^{n-n_0} x[n-q]h[q].$$

Compare CONVOLUTION INTEGRAL.

convolution theorem – (1) *for continuous signals* states that the FOURIER TRANSFORM of the product of two continuous functions equals the CONVOLUTION (1) of their Fourier transforms. In mathematical form, if $x_1(t) \leftrightarrow X_1(f)$ and $x_2(t) \leftrightarrow X_2(f)$, then

$$x_1(t)\, x_2(t) \leftrightarrow X_1(f) * X_2(f)$$

and also

$$X_1(f)\, X_2(f) \leftrightarrow x_1(t) * x_2(t).$$

Here $\leftrightarrow$ denotes a Fourier transform pair, and $*$ denotes the CONVOLUTION operation. The convolution theorem is also known as **Plancherel's theorem**.

convolution theorem – (2) *for transient discrete-time signals* states that the DISCRETE FOURIER TRANSFORM of the term-by-term product of two sequences equals the CONVOLUTION (2) of their DFTs. This means that if $x_1[n]$, $x_2[n]$ are discrete sequences of respective lengths N_1, N_2, and both sequences are extended to length N by zero-padding to yield N-point DFTs $X_1[k]$, $X_2[k]$, then

$$\sum_{n=0}^{N-1} x_1[n]x_2[n] W_N^{-kn} = \sum_{l=0}^{N-1} X_1[l]X_2[k-l]_{\mathrm{mod}N};$$

here the augmented period $N \geq N_1 + N_2 + 1$, and symbol W_N denotes the Nth root of unity, i.e. $e^{j(2\pi/N)}$. The notation mod N is explained in the note below.

By interchanging the time and frequency domains, the following related result is obtained:

$$\frac{1}{N}\sum_{k=0}^{N-1} X_1[k]X_2[k]W_N^{kn} = \sum_{i=0}^{N-1} x_1[i]x_2[n-i]_{\text{mod}N}.$$

It states that multiplying the DFTs of the extended signals and taking the inverse DFT (over one cycle) yields the convolution of the original signals.

Note: The time and frequency indices n and k in the results above are defined to lie in the range $0 \leq (n,k) \leq N-1$. The notation mod N means that N is added or subtracted as required to bring the index into this range; see CIRCULAR CONVOLUTION (2) for further discussion.

coordinates – see RECTANGULAR ∼, CYLINDRICAL ∼, SPHERICAL POLAR ∼.

core noise – *in the context of aircraft engine noise* a component of the broadband noise that escapes from the core of a turbofan engine through the hot exhaust nozzle. It is typically associated with the combustion system and the low-pressure turbine.

correction – *to a numerical value* the amount that has to be added to the original value to give the correct result. It is the negative of the ERROR.

correlated signals – (1) *for time-stationary analogue signals* two signals whose time-averaged product is non-zero.

Note: The definition above allows two signals to be uncorrelated yet perfectly COHERENT: an example is $\cos\omega t$ and $\sin\omega t$. For a definition that allows for a time delay between signals, see CROSS-CORRELATION FUNCTION, CROSS-CORRELATION COEFFICIENT FUNCTION.

correlated signals – (2) *for transient analogue signals* two signals whose product integrated over time is non-zero.

correlated signals – (3) *for digital data* two sampled signals $x[n]$, $y[n]$, of the same length are correlated if the sum $\Sigma x[n]y[n]$ is non-zero. See also CORRELATION.

Note: The term ***partially correlated*** is used when the CORRELATION COEFFICIENT of the data is between 0 and 1.

correlation – *between two variables* a measure of the extent to which their values are connected. This can be visualized graphically by constructing a ***scatter diagram***: N pairs of values (x_i, y_i) of the 2 variables are plotted in the x–y plane. If the points form a diffuse cloud, this indicates lack of correlation between x and y values. On the other hand if the points cluster into a line, this indicates close correlation. See also CORRELATION COEFFICIENT.

correlation coefficient – (1) *of two real signals that are jointly stationary with respect to time* the value of the COVARIANCE FUNCTION at zero time delay, normalized by the standard deviations of the two signals. In mathematical form, the correlation coefficient is

$$\rho_{xy} = \frac{C_{xy}(0)}{\sigma_x \sigma_y}.$$

Here $C_{xy}(\tau)$ is the covariance function of signals $x(t)$ and $y(t)$, and σ_x, σ_y are their respective STANDARD DEVIATIONS $\sqrt{C_{xx}(0)}$, $\sqrt{C_{yy}(0)}$. *Units* none.

Note: The definition above does not involve any delay or shift between the two variables. Compare the CROSS-CORRELATION COEFFICIENT FUNCTION, which is defined as a function of time delay for two stationary random functions of time.

correlation coefficient – (2) *of two random variables* the normalized COVARIANCE, defined by

$$\rho_{XY} = \frac{\text{cov}(X, Y)}{\sigma_X \sigma_Y}.$$

Here X and Y denote the two random variables; cov(X, Y) is their COVARIANCE; and σ_X, σ_Y are their individual STANDARD DEVIATIONS. *Units* none.

correlation function – abbreviation for CROSS-CORRELATION FUNCTION.

correlogram – a plot of either auto- or cross-correlation coefficient against time delay.

co-spectral density – *of two real signals that are jointly stationary with respect to time* the real part of the CROSS-SPECTRAL DENSITY. If $x(t)$ and $y(t)$ represent the two signals, their co-spectral density at frequency f is related to the CROSS-CORRELATION FUNCTION, $R_{xy}(\tau)$, by

$$C_{xy}(f) = \int_0^\infty [R_{xy}(-\tau) + R_{xy}(\tau)] \cos 2\pi f\tau \, d\tau.$$

Note: The co-spectral density is normally given as a SINGLE-SIDED spectrum,

$$\bar{C}_{xy}(f) = 2C_{xy}(f) \qquad (f > 0).$$

co-spectrum – an equivalent term for CO-SPECTRAL DENSITY.

Coulomb damping – an equivalent term for FRICTION DAMPING, as distinct from STRUCTURAL DAMPING or VISCOUS DAMPING.

coupled modes – modes of vibration that cannot vibrate independently, but exchange energy between themselves; thus excitation of one mode implies excitation of the others.

coupler – see ACOUSTIC COUPLER.

coupling factor – *of a passive transducer* a measure of the efficiency with which the input energy is transferred to the output; given by the ratio of the peak energy stored in the secondary segment to the total peak energy stored, during single-frequency operation. *Units* none.

coupling impedance – *at a given cross-section in a multimode waveguide* an off-diagonal term in a MODAL IMPEDANCE MATRIX or TERMINATION IMPEDANCE MATRIX. *Units* Pa s m^{-1}.

coupling loss factor – *in statistical energy analysis* if two subsystems are coupled, and each has many resonant modes excited within a specified narrow band of frequencies, the energy flow from one subsystem to the other is proportional to the coupling loss factor between the subsystems, multiplied by their difference in average modal energy. In equation form, the time-average energy flow per unit frequency bandwidth is written as

$$\Pi_{\alpha\beta} = \eta_{\alpha\beta}\omega n_\alpha(\varepsilon_\alpha - \varepsilon_\beta) \quad \text{(from subsystem } \alpha \text{ to } \beta\text{)},$$

where $\eta_{\alpha\beta}$ is the coupling loss factor, n_α is the average MODAL DENSITY of subsystem α, and ε_α, ε_β are the subsystem AVERAGE MODAL ENERGIES (i.e. the energy per resonant mode within the band). The angular frequency ω corresponds to the geometric centre frequency of the band considered. *Units* none.

Note: The corresponding expression for energy flow from β to α is

$$\Pi_{\beta\alpha} = \eta_{\beta\alpha}\omega n_\beta(\varepsilon_\beta - \varepsilon_\alpha).$$

Since this must equal the negative of $\Pi_{\alpha\beta}$, it follows that the coupling loss factors in the two directions are related:

$$\eta_{\alpha\beta} n_\alpha = \eta_{\beta\alpha} n_\beta.$$

coupling material – *in non-destructive testing and medical ultrasonics* a substance, usually in liquid or gel form, that is introduced between an ultrasonic transducer and a test object (or the patient's skin). By excluding air interfaces, the coupling material minimizes unwanted reflections and provides IMPEDANCE MATCHING between the transducer and sample.

coupling reflection coefficient – *for a termination in a multimode waveguide* an off-diagonal term in a REFLECTION COEFFICIENT MATRIX. *Units* none.

covariance – *of two random variables* the expected value of the product of the variables, formed after subtracting their mean values. The covariance is defined mathematically by

$$\text{cov}(X, Y) = E\{(X - \mu_X)(Y - \mu_Y)\},$$

where X and Y denote the two variables and μ_X, μ_Y their individual expected values. Symbol E stands for the EXPECTATION OPERATOR.

Note (1): If X and Y are INDEPENDENT RANDOM VARIABLES, their covariance is zero.

Note (2): For *sample covariance*, see the note under SAMPLE VARIANCE.

covariance function – *of two real signals that are jointly stationary with respect to time* the time-averaged product of one signal with a time-shifted version of the other, formed after subtracting the mean value of each signal:

$$C_{xy}(\tau) = \langle x'(t) y'(t + \tau) \rangle = R_{xy}(\tau) - \bar{x}\bar{y}.$$

Here angle brackets $\langle \ldots \rangle$ denote the time-averaging operation; $\bar{x}$, $\bar{y}$ are the mean values of each signal, and the primes denote signal variations about the mean. Thus $x' = x - \bar{x}$, and $y' = y - \bar{y}$. See also CROSS-CORRELATION FUNCTION.

CPSD – abbreviation for *cross-power spectral density*. For a definition, see CROSS-SPECTRAL DENSITY.

creep function – *for a linear viscoelastic material with time-independent properties* the function $\Psi(t)$ that appears in the integral

$$\varepsilon(t) = \int_{-\infty}^{t} \Psi(t - \tau)\dot{\sigma}(\tau)d\tau,$$

where the dot denotes a time derivative and $\varepsilon(t)$ and $\sigma(t)$ are the time-dependent strain and stress at a point in the material. This expression gives the strain at time t that is produced by a stress history $\sigma(\tau)$; compare STRESS-RELAXATION FUNCTION. *Units* Pa^{-1}.

creeping rays – see GEOMETRICAL THEORY OF DIFFRACTION.

creeping waves – diffracting acoustic waves that "cling" to a smooth convex boundary, and thus carry energy into the SHADOW ZONE.

critical angle for transmission – *at a plane interface between two lossless media* the plane-wave ANGLE OF INCIDENCE for which the transmission angle is 90 degrees; often called the *critical angle*. Its value is

$$\theta_{\text{crit}} = \sin^{-1}\frac{c_1}{c_2}.$$

Here c_1 and c_2 are the wave speeds in the incident and the transmitting medium respectively, and c_1 must be less than c_2. Both wave speeds are real, since the media are treated as lossless in this definition.

For angles beyond the critical angle, i.e. $\theta > \theta_{\text{crit}}$, the incident wave energy is totally reflected at the interface: see TOTAL INTERNAL REFLECTION. *Units* rad (but commonly expressed in degrees).

Note (1): At a fluid–solid interface, incident sound waves in the fluid (speed c_1) are transmitted into the solid partly as shear waves (speed b_2), and partly as longitudinal waves (speed c_2). There is then a second critical angle, given by $\theta_{\text{crit}} = \sin^{-1}(c_1/b_2)$, that relates to shear-wave transmission.

Note (2): In underwater acoustics, the term refers to the ***critical grazing angle***, $\theta_{\text{crit}} = \cos^{-1}(c_1/c_2)$.

critical band – *in psychoacoustics* abbreviation for AUDITORY CRITICAL BAND.

critical bandwidth – *in psychoacoustics* see AUDITORY CRITICAL BAND, AUDITORY CRITICAL BANDWIDTH FOR LOUDNESS. *Units* Hz.

critical coincidence frequency – *for a panel or partition* an equivalent term for CRITICAL FREQUENCY. *Units* Hz.

critical damping – the minimum value of the VISCOUS DAMPING COEFFICIENT that is required to prevent overshoot, when a linear system is displaced and released. A system that has just sufficient damping to avoid overshoot is said to be ***critically damped***.

A more detailed definition requires mathematics. Suppose FREE OSCILLATIONS of a linear system are described by the equation

$$m\ddot{x} + R\dot{x} + kx = 0,$$

where $x \equiv x(t)$ is the displacement at time t, with first and second derivatives ($\dot{x}$, $\ddot{x}$), and m, R, k are constants. There is a critical value of the viscous damping coefficient R, given by $R_{\text{crit}} = 2\sqrt{mk}$, beyond which the solutions of the free-oscillation equation are non-oscillatory, i.e. they do not pass through zero. The term ***critical damping*** describes the situation where $R = R_{\text{crit}}$. See also DAMPING RATIO.

critical damping coefficient – *for a single-degree-of-freedom system with viscous damping* see VISCOUS DAMPING COEFFICIENT. *Units* N s m^{-1}.

critical frequency – *for a panel or partition* the frequency at which the speed of free bending waves in the panel equals the speed of sound in the adjacent fluid. If the panel is anisotropic (different bending stiffness in different directions) the critical frequency will vary with direction along the panel. *Units* Hz.

Note (1): The critical frequency is the lowest frequency at which COINCIDENCE can occur with incident sound waves. Coincidence at this frequency requires the exciting sound field to be at grazing incidence.

Note (2): The concept of critical frequency is inappropriate in situations of heavy fluid loading, e.g. steel plates in water, where the presence of the fluid alters the bending wave speed significantly from its *in vacuo* value. Under such conditions it is essential to regard the propagation of waves in the fluid-loaded plate as a fully-coupled problem.

critical state – *of a fluid* the unique thermodynamic state in which the saturated liquid and saturated vapour coincide in density. The fluid is then said to be at its **critical temperature**, **critical pressure**, and **critical density**.

Note: Above the critical temperature, a gas cannot be liquefied by increasing the pressure.

cross admittance – see the definition for CROSS IMPEDANCE, which can be applied (with the appropriate changes of wording) to admittances as well as impedances.

cross-axis sensitivity – *for a rectilinear transducer, at a given frequency* the ratio of output magnitudes, $|y_2/y_1|$, when inputs of the same magnitude are applied in turn along the sensitive axis (output y_1), and transversely to the sensitive axis (output y_2). The cross-axis sensitivity is generally expressed logarithmically, as $20 \log_{10} |y_2/y_1|$. *Units* dB.

cross-bicorrelation function – *of two real signals that are jointly stationary with respect to time* the time-averaged product

$$R_{xxy}(\tau_1, \tau_2) = \lim_{T \to \infty} \left\{ \frac{1}{T} \int_{t_0}^{t_0+T} x(t-\tau_2)x(t-\tau_1)y(t)dt \right\}$$
$$= \langle x(t-\tau_2)x(t-\tau_1)y(t) \rangle,$$

where angle brackets $\langle \ldots \rangle$ denote the time averaging operation, $x(t)$ and $y(t)$ are the two signals as functions of time t, and τ_1, τ_2 are time shifts or delays.

cross-bispectrum – *of two real signals that are jointly stationary with respect to time* the 2-dimensional Fourier transform of the CROSS-BICORRELATION FUNCTION of the signals. Thus if the signals are $x(t)$ and $y(t)$, with cross-bicorrelation function $R_{xxy}(\tau_1, \tau_2)$, their cross-bispectrum is

$$S_{xxy}(f_1, f_2) = \int_{-\infty}^{\infty} \int_{-\infty}^{\infty} R_{xxy}(\tau_1, \tau_2) e^{-2\pi j(f_1 \tau_1 + f_2 \tau_2)} d\tau_1 d\tau_2.$$

For two signals that are jointly Gaussian, the cross-bispectrum is zero.

cross-correlation – *of two real signals that are jointly stationary with respect to time* an abbreviation (but not recommended where ambiguity might arise) for CROSS-CORRELATION FUNCTION.

cross-correlation coefficient function – *of two real signals that are jointly stationary with respect to time* the normalized COVARIANCE FUNCTION defined by

$$\rho_{xy}(\tau) = \frac{C_{xy}(\tau)}{\sigma_x \sigma_y} \qquad (-1 \leq \rho_{xy} \leq 1).$$

Here $C_{xy}(\tau)$ is the covariance function of the two signals $x(t)$ and $y(t)$, evaluated at time delay τ, and σ_x, σ_y are the respective STANDARD DEVIATIONS $\sqrt{C_{xx}(0)}$, $\sqrt{C_{yy}(0)}$. *Units* none.

Note (1): The shorter term *cross-correlation coefficient* is often used for the quantity defined above.

Note (2): Use of the cross-correlation coefficient is normally limited to signals whose mean value is zero. The coefficient $\rho_{xy}(\tau)$ defined above may then be regarded as a normalized CROSS-CORRELATION FUNCTION.

cross-correlation function – (1) *of two real signals that are jointly stationary with respect to time* the time-averaged product of one signal with a time-shifted version of the other. Conventionally, the cross-correlation function of signals $x(t)$ and $y(t)$ is defined as

$$R_{xy}(\tau) = \lim_{T \to \infty} \left\{ \frac{1}{2T} \int_{-T}^{T} x(t) y(t + \tau) dt \right\}$$
$$= \langle x(t) y(t + \tau) \rangle = \langle x(t - \tau) y(t) \rangle,$$

where the brackets $\langle \ldots \rangle$ denote the time averaging operation. It is the same as the COVARIANCE FUNCTION, if either $x(t)$ or $y(t)$ has zero mean.

Note: In this dictionary the notation $R_{xy}(\tau)$ always indicates that signal x is delayed by τ relative to y prior to averaging (rather than the other way round, with signal y delayed by τ relative to x). Although not universal, this convention is widely followed in the literature.

cross-correlation function – (2) *of two complex non-stationary stochastic processes* the ENSEMBLE-AVERAGED product

$$R_{xy}(t_1, t_2) = E\{x(t_1) y^*(t_2)\},$$

where $x(t)$, $y(t)$ are the processes in question, * denotes the complex conjugate, and E is the EXPECTATION OPERATOR.

cross-correlation matrix – *of two sets of real time-stationary continuous signals* the $N \times N$ square matrix defined by

$$\mathbf{R}_{\mathbf{xy}}(\tau) = \langle \mathbf{y}(t+\tau)\,\mathbf{x}^T(t) \rangle = \langle \mathbf{y}(t)\,\mathbf{x}^T(t-\tau) \rangle,$$

where angle brackets $\langle ... \rangle$ denote the time averaging operation, $\mathbf{x}(t)$ is a column vector of signals $x_1, x_2, ..., x_N$ as a function of time t, $\mathbf{y}(t)$ is a similar column vector of signals $y_1, y_2, ..., y_N$, superscript T denotes the vector TRANSPOSE, and τ is a time shift or delay.

Note: The cross-correlation matrix arises in connection with multiple-input multiple-output systems.

cross impedance – refers to a MECHANICAL IMPEDANCE in which the output and input are both defined as components of vector quantities, but in different coordinate directions. For example, the cross impedance $Z_{xy} = F_x/v_y$ is the complex ratio of a force F_x in the x direction and a velocity v_y in the perpendicular y direction. Compare DIRECT IMPEDANCE. *Units* N s m^{-1}.

Note (1): Generalizing the example above, one can define a DRIVING-POINT IMPEDANCE MATRIX, denoted by $\mathbf{Z}$, such that $\mathbf{F} = \mathbf{Z}\mathbf{v}$. Then all the off-diagonal terms like Z_{xy} are called cross impedances, while the diagonal terms are the DIRECT IMPEDANCES in the x, y, z directions.

Note (2): The concept of cross impedances can also be extended to situations where a velocity input generates a moment output, or an angular velocity input generates a force output.

cross-power spectral density, cross-power spectrum – *of a pair of signals* alternative terms for the CROSS-SPECTRAL DENSITY of the two signals.

cross-spectral density – *of two real signals that are jointly stationary with respect to time* the Fourier transform of the CROSS-CORRELATION FUNCTION of the two signals. Thus if the signals are $x(t)$ and $y(t)$, their cross-spectral density (abbreviated *CSD*) is

$$S_{xy}(f) = \int_{-\infty}^{\infty} R_{xy}(\tau) e^{-2\pi j f \tau} d\tau$$
$$= C_{xy}(f) + j Q_{xy}(f).$$

The real part is called the *co-spectral density* or *co-spectrum*, and the imaginary part is called the *quad-spectral density* or *quad-spectrum*.

Note (1): Alternative terms for the quantity defined above are *cross-spectral density function*, *cross-power spectrum*, and *cross-power spectral density* (*CPSD*). The omission of the prefix "power" (in contrast to ENERGY CROSS-SPECTRAL DENSITY) is well established.

Note (2): Some authors define the quad-spectral density with the opposite sign: $S_{xy}(f) = C_{xy}(f) - j Q_{xy}(f)$.

Note (3): It follows from the CONVOLUTION THEOREM that the cross-spectral density can also be expressed as

$$S_{xy}(f) = \lim_{T\to\infty} \frac{1}{T} E\{X_T^*(f) Y_T(f)\},$$

where $E\{\ldots\}$ denotes the expected value, and X_T and Y_T are the Fourier transforms of the finite-length data segments x_T and y_T of length T. (These are equal to the original signal over the range $t_0 < t < t_0 + T$, where t_0 is arbitrary, and zero otherwise.)

cross-spectral density function – an expanded version of the term CROSS-SPECTRAL DENSITY. See DENSITY FUNCTION.

cross-spectral matrix – *of two sets of real time-stationary continuous signals* the $N \times N$ square matrix defined by

$$\mathbf{S_{xy}}(f) = \lim_{T\to\infty} \frac{1}{T} E\{\mathbf{Y}_T(f) \mathbf{X}_T^H(f)\},$$

where $E\{\ldots\}$ denotes the expected value, $\mathbf{X}_T(f)$ is the Fourier transform of $\mathbf{x}_T(t)$, and $\mathbf{x}_T$ is a column vector of N finite-length data segments each of length T taken from the vector signal $\mathbf{x}$. (Each record is defined to equal the original signal over the range $t_0 < t < t_0 + T$, where t_0 is arbitrary, and zero otherwise). The vector $\mathbf{Y}_T(f)$ is defined similarly to $\mathbf{X}_T(f)$ but for signal $\mathbf{y}$, and superscript H denotes the HERMITIAN TRANSPOSE.

Note: The cross-spectral matrix is the Fourier transform of the CROSS-CORRELATION MATRIX. Compare AUTOSPECTRAL MATRIX.

cross-spectrum – abbreviation for *cross-power spectrum*, i.e. the CROSS-SPECTRAL DENSITY of two signals.

CTD probe/sensor – *in underwater acoustics* abbreviation for *conductivity, temperature, and depth probe/sensor*. An instrument for measuring water electrical conductivity and temperature as a function of depth, usually for the purpose of determining sound speed; the conductivity determines the salinity of the water (compare BATHYTHERMOGRAPH). See SOUND SPEED IN SEAWATER.

cumulative distribution function (cdf) – *for a single random variable* the function $F(x)$ that gives the probability of the variable being less than or equal to a given value x. Thus the cumulative distribution function of a random variable X is given by

$$F(x) = P(X \leq x),$$

where the notation $P(A)$ means the probability that statement A is true. An equivalent name for the function $F(x)$ is the *probability distribution function*.

Note: The relationship between $F(x)$ and the PROBABILITY DENSITY FUNCTION $f(x)$ of the same random variable is

$$F(x) = \int_{-\infty}^{\infty} f(u)du.$$

cumulative frequency distribution – *of a finite data sample whose individual values are either quantized, or grouped into categories or bins* the set of paired integers (N_i, i), where integer i labels the bins in a defined sequence, and N_i is the number of samples in all the bins up to and including i. Compare FREQUENCY DISTRIBUTION.

cumulative nonlinearity – *during acoustic wave propagation* progressive distortion of a finite-amplitude waveform as it propagates. The distortion is caused by the dependence of propagation speed on the instantaneous local pressure (or on particle velocity); see COEFFICIENT OF NONLINEARITY. Compare LOCAL NONLINEARITY.

Curle's equation – *in aeroacoustics* an extension of KIRCHHOFF'S FORMULA in which the inhomogeneous wave equation of Lighthill's ACOUSTIC ANALOGY is solved for a region containing boundaries (S N CURLE **1955**). In Curle's equation, the boundaries of the solution domain are fixed; for the general case in which the boundaries are free to move, see FFOWCS WILLIAMS–HAWKINGS EQUATION.

cutoff – *of an acoustic mode in a waveguide with non-absorbing boundaries* refers to the situation where the modal propagation wavenumber is imaginary (i.e. the mode is EVANESCENT), and no sound power is carried along the waveguide by a mode travelling in only one direction.

cutoff frequency, cut-off frequency – (1) *of an acoustic mode in a waveguide with non-absorbing boundaries* the frequency below which the mode becomes EVANESCENT. If no mode is specified then the cutoff frequency refers to the lowest such frequency. *Units* Hz.

cutoff frequency – (2) *of a filter designed to give sharply-defined passbands and stop bands* a frequency to one side of which the filter attenuation increases sharply. Normally, this is defined as the value of frequency at which the attenuation reaches 3 dB; the term **nominal cutoff frequency** is sometimes used for this value. *Units* Hz.

cutoff ratio, cut-off ratio – *in a waveguide with non-absorbing or non-leaky boundaries* the ratio f/f_c for sound propagating at a given frequency f, in a given mode with CUTOFF FREQUENCY f_c. *Units* none.

cutoff wavenumber – *for an acoustic mode propagating in a waveguide with rigid walls* the transverse wavenumber associated with the given mode. If κ_N denotes the cutoff wavenumber for mode N, then the transverse eigenfunction $\psi_N(\mathbf{y})$ for that mode satisfies the two-dimensional Helmholtz equation over the waveguide cross-section A,

$$(\nabla_\perp^2 + \kappa_N^2)\psi_N(\mathbf{y}) = 0, \qquad \mathbf{y} \in A,$$

(where $\nabla_\perp^2$ is the LAPLACIAN in coordinates transverse to the waveguide axis), together with the rigid-wall boundary condition $\partial \psi_N(\mathbf{y})/\partial n = 0$ on the boundary of A. The corresponding cutoff frequency is $f_N = c\kappa_N/2\pi$, where c is the sound speed inside the waveguide. *Units* rad m^{-1}.

cut-on frequency – an equivalent term for CUTOFF FREQUENCY; successive modes are thought of as being "switched on" as the frequency increases. *Units* Hz.

CW – abbreviation for CONTINUOUS-WAVE.

C-weighting – a FREQUENCY-WEIGHTING procedure, in which the power or energy spectrum of a signal is progressively attenuated towards the high and low ends of the audio frequency range, but much less so at low frequencies than with A-weighting. The C curve is almost flat between 100 Hz and 2 kHz; at 10 Hz it is 14 dB down (compared with 70 dB for the A curve), and above 1 kHz it falls off slightly faster than the A curve (maximum 2 dB below).

Note: The C-weighting curve was originally based on the shape of the 100-phon EQUAL LOUDNESS CONTOUR, with no attempt to allow for the ear canal resonance that enhances the free-field sensitivity of the ear by nearly 10 dB in the range 2–5 kHz. The intention was that C-weighting the incident pressure, followed by squaring and averaging with a suitable time constant, would simulate the sensitivity of the human ear to audio-frequency sound at sound pressure levels above about 90 dB re $(20\ \mu\text{Pa})^2$.

cycle – *of a periodic signal or recurrent spatial pattern* a segment whose length equals one period.

cycle distance – *in underwater or atmospheric acoustics* the horizontal distance that a ray travels in a SOUND CHANNEL before returning to the corresponding point on its path (same vertical position, travelling in the same direction). Compare SKIP DISTANCE. *Units* m.

cyclostationary – describes a NONSTATIONARY random signal whose statistics vary periodically with time (or along the record).

cylinder – (1) a three-dimensional object formed by translating a plane curve (the cross-section) in a direction normal to the plane (the axial direction). The cross-section need not be circular.

Note: The object defined above is called by mathematicians a *right cylinder*, to distinguish it from the more general case where the end faces are not orthogonal to the axis.

cylinder – (2) a common abbreviation for a *circular cylinder*, i.e. one with a circular cross-section. Hence CYLINDRICAL COORDINATES.

cylinder functions – a generic name for solutions of the BESSEL EQUATION. The general solution is written as

$$R(r) = AJ_\mu(\kappa r) + BY_\mu(\kappa r), \qquad (A, B \text{ constants})$$

or alternatively as

$$R(r) = C_1 H^{(1)}_\mu(\kappa r) + C_2 H^{(2)}_\mu(\kappa r), \qquad (C_1, C_2 \text{ constants}).$$

The standard cylinder functions J_μ, Y_μ, $H^{(1)}_\mu$, and $H^{(2)}_\mu$ are listed below; these are known as *cylindrical Bessel functions* or simply *Bessel functions*. Note that the argument z and the *order* μ may be real or complex.

Function	Name	Order	Symbol
1st kind	Bessel function	μ	$J_\mu(z)$
2nd kind	Neumann function	μ	$Y_\mu(z)$ or $N_\mu(z)$
3rd kind	Hankel function	μ	$H^{(1)}_\mu(z)$, $H^{(2)}_\mu(z)$.

The Hankel functions are related to the Bessel and Neumann functions by

$$H^{(1)}_\mu(z) = J_\mu(z) + jY_\mu(z); \quad H^{(2)}_\mu(z) = J_\mu(z) - jY_\mu(z).$$

cylindrical Bessel functions – a generic name for the CYLINDER FUNCTIONS of the first, second, or third kind, otherwise collectively known as *Bessel functions*. Compare SPHERICAL BESSEL FUNCTIONS.

cylindrical coordinates – a set of coordinates (r, ϕ, z) based on radial distance r from a specified axis, angular rotation ϕ about the axis, and distance z measured along the axis.

cylindrical progressive wave – (1) *in two dimensions* a wave field whose space-time dependence in (r, θ) POLAR COORDINATES consists of a directional factor $e^{-jm\theta}$, multiplied by a function of time and radius that in the far field has the asymptotic form

$$\frac{1}{\sqrt{r}} f(t - r/c) \qquad \text{(for outgoing waves), or}$$

$$\frac{1}{\sqrt{r}} f(t + r/c) \qquad \text{(for incoming waves).}$$

Here t is time, c is the sound speed, and m is the azimuthal mode number ($m = 0, \pm 1, \pm 2$, etc.).

cylindrical progressive wave – (2) *in three dimensions* a single-frequency wave whose space-time variation in (r, ϕ, z) CYLINDRICAL COORDINATES consists of a factor $e^{j(\omega t - m\phi - k_z z)}$, multiplied by a function of r that in the far field has the asymptotic form

$$\frac{1}{\sqrt{r}} e^{-j\kappa r} \quad \text{(for outgoing waves), or}$$

$$\frac{1}{\sqrt{r}} e^{+j\kappa r} \quad \text{(for incoming waves).}$$

Here t is time, ω and k_z are the angular frequency and axial wavenumber respectively, and m is the azimuthal mode number ($m = 0, \pm 1, \pm 2, \ldots$). The radial wavenumber κ is defined by

$$\kappa = \sqrt{\left(\frac{\omega}{c}\right)^2 - k_z^2} \quad (\kappa \text{ real}),$$

$$\kappa = -j\sqrt{k_z^2 - \left(\frac{\omega}{c}\right)^2} \quad (\kappa \text{ imaginary}),$$

where c is the sound speed (here assumed real).

Note: More generally, at any radius r the mathematical form of a cylindrically progressive wave is

$$p = A e^{j\omega t} e^{-j(k_z z + m\phi)} H_m^{(2)}(\kappa r) \quad \text{(outgoing), or}$$

$$p = A e^{j\omega t} e^{-j(k_z z + m\phi)} H_m^{(1)}(\kappa r) \quad \text{(incoming).}$$

Here p represents a scalar variable such as acoustic pressure, A is a constant, and $H_m^{(1)}$, $H_m^{(2)}$ are HANKEL FUNCTIONS of order m (see CYLINDER FUNCTIONS for more information).

cylindrical radius – *in cylindrical coordinates* a distance measured from the coordinate axis. *Units* m.

cylindrical spreading – *in an outgoing wave field* a 1/r dependence of intensity on radius (r), as in the far field of a CYLINDRICAL PROGRESSIVE WAVE. It may be regarded as a consequence of energy conservation applied to wavefronts whose area increases in proportion to radius. Equivalently, in cylindrical spreading the level falls off at 3 dB per doubling of distance from the source.

cylindrical wave field – a wave field described in terms of a combination of outgoing and incoming CYLINDRICAL PROGRESSIVE WAVES. Such a description is particularly appropriate for sound fields with axisymmetric boundary conditions.

D

1D, 1-D – abbreviations for ONE-DIMENSIONAL. Likewise, 2D and 3D are abbreviations for two-dimensional and three-dimensional, as in *3D wave equation*.

D-A, D/A – abbreviations for DIGITAL-TO-ANALOGUE, as in ~ *converter*.

DAC – abbreviation for DIGITAL-TO-ANALOGUE CONVERTER.

daily personal noise exposure – a measure of an individual's occupational SOUND EXPOSURE over a 24-hour integration period, expressed as a NOISE EXPOSURE LEVEL. Symbols $L_{EP,d}$, $L_{EX,8h}$. *Units* dB re 1.15×10^{-5} Pa^2 s.

d'Alembertian operator – the wave equation differential operator defined by $\nabla^2 - c^{-2} \partial^2/\partial t^2$, sometimes written as $\Box$ or $\Box^2$. Also known as the *d'Alembertian* (J D'ALEMBERT 1747).

d'Alembert's solution – an initial-value solution of the unbounded one-dimensional wave equation

$$\frac{\partial^2 \varphi}{\partial x^2} - \frac{1}{c^2}\frac{\partial^2 \varphi}{\partial t^2} = 0 \quad (-\infty < x < \infty),$$

where c is the wave speed and x, t denote position and time. D'Alembert's solution allows the wave variable $\varphi(x, t)$ to be calculated everywhere from INITIAL CONDITIONS on φ and $\dot{\varphi}$: it is given by

$$2\varphi(x,t) = h(\alpha) + h(\beta) + \frac{1}{c}\int_\alpha^\beta k(s)ds.$$

Here $h(x) = \varphi(x, 0)$ and $k(x) = \dot{\varphi}(x, 0)$ represent the initial data for φ and its time derivative, and (α, β) are characteristic variables defined by

$$\alpha = x - ct, \beta = x + ct.$$

Dalton's law of additive pressures – *for a mixture of ideal gases* see GIBBS-DALTON LAW.

damped natural frequency – *for a given natural mode of a linear system* the frequency given by the number of zero crossings in the same direction per unit time, during damped free vibration; sometimes called the *pseudo-frequency*. *Units* Hz.

Note: The damped natural frequency is $(1/2\pi)$ times the real part of the (angular) COMPLEX NATURAL FREQUENCY.

damped oscillations – oscillations whose amplitude decreases with time (or with distance, in the case of a spatial waveform). See AMPLITUDE DECAY COEFFICIENT, ATTENUATION COEFFICIENT, CRITICAL DAMPING.

damper rate – an alternative term for VISCOUS DAMPING COEFFICIENT. *Units* N s m^{-1}.

damping – (1) the absorption of energy in a propagating wave, or the loss of energy from an oscillating system by dissipation or radiation.

damping – (2) *of a room* the addition of sound-absorbing material to the room surfaces, to reduce sound levels in the room or to shorten the reverberation time.

damping – (3) *of a structure* the addition of damping material to the vibrating parts of the structure, to reduce the vibration levels transmitted to the structure or to shorten the decay time of free vibration.

damping – (4) abbreviation for DAMPING TREATMENT.

damping coefficient – see VISCOUS DAMPING COEFFICIENT. *Units* N s m^{-1}.

damping constant, damping factor – *for damped sinusoidal oscillations* alternative terms for AMPLITUDE DECAY COEFFICIENT. *Units* Np s^{-1}.

Note: For a single-degree-of-freedom system with VISCOUS DAMPING, free oscillations of the system are described by the normalized equation below in which the coefficient δ is the damping constant:

$$\ddot{x} + 2\delta\dot{x} + \omega_0^2 x = 0.$$

Here $\omega_0/2\pi$ is the UNDAMPED NATURAL FREQUENCY, and x is the displacement from equilibrium. The damping constant for such a system is related to the DAMPING RATIO, ζ, and the LOGARITHMIC DECREMENT, ε, by

$$\delta = \omega_0 \zeta$$
$$\approx \omega_0 \frac{\varepsilon}{2\pi} \quad \text{(for } \zeta \ll 1\text{).}$$

If the system is UNDERDAMPED, $\delta = f_d \varepsilon$ where f_d is the DAMPED NATURAL FREQUENCY in Hz.

damping ratio – *for a single-degree-of-freedom system with viscous damping* the ratio of the VISCOUS DAMPING COEFFICIENT to its critical value; also known as the *fraction of critical damping*. For a system whose displacement $x(t)$ is described by the equation

$$m\ddot{x} + R\dot{x} + kx = 0,$$

the damping ratio is given by

$$\zeta = \frac{R}{2\sqrt{km}}.$$

Units none.

Note: If such a system has a damping ratio greater than unity, it returns smoothly to its equilibrium position after a displacement, without overshooting.

damping treatment – sound-absorbing material applied to room surfaces, or damping material applied to the vibrating parts of a structure.

Darcy's law – *for slow viscous flow through a porous solid* the statement that at low Reynolds numbers, the *seepage velocity* across a given plane (i.e. the volume flowrate per unit total cross-sectional area) is linearly proportional to the pressure gradient (**H DARCY 1856**). Darcy's law is written mathematically as

$$\bar{u}_i = -\frac{1}{\mu} K_{ij} \frac{\partial P}{\partial x_j} \quad (\bar{u}_i = \text{seepage velocity in the } x_i \text{ direction});$$

here P is the pressure, μ is the fluid viscosity, and the coefficient K_{ij} represents the *permeability* of the porous solid. Note that the permeability is anisotropic in general: a pressure gradient in one direction can produce a component of seepage velocity in a perpendicular direction. Compare FORCHHEIMER'S LAW.

dashpot – an idealized LUMPED ELEMENT that generates a resistive force F determined by the relative velocity Δv with which the two ends are moved apart. It may be visualized as a loose-fitting piston in a cylinder filled with viscous fluid. A *linear dashpot* has $F \propto \Delta v$: see VISCOUS DAMPING COEFFICIENT.

data – numerical values obtained from a series of measurements or simulations, or from sampling a RANDOM SIGNAL. A distinction is drawn between *analogue data*, which is continuous (e.g. a tape-recorded signal), and *digital data*, which may be produced by sampling analogue data. Digital data is discrete, in the sense that it contains a finite number of individual values.

Note: The word *data* is plural, because it refers to multiple values taken by an observed variable; on the other hand it is often used as a collective noun and treated as singular, as in the definition above.

data acquisition and analysis – see under SIGNAL PROCESSING.

dB – unit symbol for DECIBEL.

dc, DC – steady, non-oscillatory (by analogy with direct current).

dc power – see POWER (3).

dead room – a room in which the total ABSORBING AREA approaches the actual room surface area. Note that such a room cannot qualify as "live"; compare LIVE ROOM.

deafness – partial or complete hearing impairment.

decade – (1) a factor of 10.

decade – (2) a unit of LOGARITHMIC FREQUENCY INTERVAL. The interval in decades between a lower frequency f_1 and a higher frequency f_2 is $\log_{10}(f_2/f_1)$. *Unit symbol* dec.

decade – (3) a FREQUENCY BAND whose upper and lower limits are a factor of 10 apart.

decay coefficient – abbreviation for AMPLITUDE DECAY COEFFICIENT. *Units* Np s^{-1}.

decay rate – the rate at which the LEVEL of a mean square signal decreases with time, as a result of dissipation; compare ATTENUATION RATE, which is the spatial equivalent. An equivalent term is *rate of decay*. The decay rate equals $4.343/\tau$, for EXPONENTIAL DECAY with an energy decay time of τ. *Units* dB s^{-1}.

Note (1): For free vibration of a system in a single mode, the decay rate is given by $8.686\,\delta$ where δ is the AMPLITUDE DECAY COEFFICIENT of the mode ($8.686 = 20/\ln 10$).

Note (2): Applied to multimode systems, the *decay rate in a given frequency band* refers to the decay of energy in those modes whose damped natural frequencies lie within the band.

decay time – see ENERGY DECAY TIME, REVERBERATION TIME.

decibel – the standard unit of measure, in acoustics, for level or level difference. The decibel scale is based on the ratio $10^{1/10}$; multiplying a power-like quantity (such as sound power, or mean square sound pressure) by this factor increases its level by 1 decibel. If a power-like quantity is increased by a factor $10^{n/10}$, its level goes up by n decibels. *Unit symbol* dB.

Note: Separate units are commonly adopted in acoustics and signal processing for LEVELS, which relate to power-like quantities and are measured in bels or decibels, and LOGARITHMIC MAGNITUDE RATIOS, which relate to harmonically-varying quantities and are measured in nepers.

deconvolution – the process of finding the response of a linear system to one input $x_1(t)$, given the response to a different input $x_2(t)$. Typically x_1 is a delta function $\delta(t)$, and the aim in this case is to extract the IMPULSE RESPONSE of the system from the response to some realizable input pulse.

deformation rate – a measure of the local rate at which an element of material is expanding and changing its shape; also known as ***strain rate***. For three-dimensional deformations one can define a ***deformation rate tensor***, whose components d_{ij} are related to gradients of the velocity field as follows:

$$d_{ij} = \frac{1}{2}\left(\frac{\partial u_i}{\partial x_j} + \frac{\partial u_j}{\partial x_i}\right).$$

Here x_i are cartesian coordinates ($i = 1, 2, 3$), and u_i are components of the velocity vector $\mathbf{u}(\mathbf{x}, t)$ that represents the velocity of the material element passing position $\mathbf{x}$ at time t. *Units* s^{-1}.

degeneracy – the occurrence of modes that have identical EIGENVALUES, but different spatial distributions or EIGENFUNCTIONS.

degenerate modes – modes that have identical eigenvalues, but different spatial distributions or eigenfunctions. The term is also used for different modes of a system that have the same NATURAL FREQUENCY.

degree – (1) *of a polynomial* the largest exponent that appears in the polynomial. Thus $ax^2 + bx + c$ is a polynomial of degree 2 provided $a \neq 0$.

degree – (2) *of the associated Legendre functions* the upper index μ in the P_ν^μ and Q_ν^μ notation.

degrees of freedom – (1) *in mechanics* for any discrete or lumped-element system, the number of degrees of freedom is the number of independent coordinates necessary to define the configuration of the system with respect to a frame of reference. For example, an unconstrained point mass has 3 degrees of freedom. Two such masses joined by a massless spring have 6 degrees of freedom (3 to define the mass centre, 2 to define the orientation of the line joining the masses, and 1 to define their separation).

degrees of freedom – (2) *in statistical inference* the equivalent number of discrete independent variables in a SAMPLE. It normally equals the number of independent observations, minus the number of parameters to be estimated. See also BT PRODUCT, CHI-SQUARE DEGREES OF FREEDOM.

delta function – abbreviation for DIRAC DELTA FUNCTION.

density – (1) *in continuum mechanics* the mass per unit volume of a material; also known as *mass density*. Its reciprocal is the SPECIFIC VOLUME of the material. *Units* kg m^{-3}.

density – (2) *in physics generally* the amount of any specified quantity per unit volume: e.g. *charge* ∼, *mass* ∼, *energy* ∼.

density – (3) *of a distribution* an abbreviation for DENSITY FUNCTION. Examples are *power spectral* ∼ (with respect to frequency); **probability** ∼ (with respect to the numerical value of a random variable).

density function – *of a distribution with respect to a single variable* the amount of the distributed quantity per unit interval of the distribution variable. For example a string with mass $m(x)$ per unit length, where x is distance measured along the string, has a *mass density function* $m(x)$ with respect to x. An example from statistics is the PROBABILITY DENSITY FUNCTION $f(x)$ of a random variable: $f(x)\,dx$ is the probability that the value will lie in the range $(x, x + dx)$. In the second example the quantity being distributed is the probability of an observation, and the distribution variable x is the value taken by the measured variable. See also MOMENTS.

Note: The definition of a density function can be extended to two or more distribution variables. For example, the BIVARIATE statistical properties of a pair of random variables are described by a JOINT PROBABILITY DENSITY FUNCTION with respect to the values of the two variables.

depth-dependent Green's function – *for sound radiation in a horizontally stratified fluid in the absence of mean flow* the frequency response function that gives the sound field of a unit point source at a given depth, expressed in the radial wavenumber domain. Specifically, for a single-frequency source with time factor $e^{j\omega t}$, located at depth z_0, the complex acoustic pressure at horizontal range r and depth z is written as

$$\tilde{p}(\omega, r; z, z_0) = \int_0^\infty g(\omega, k_r; z, z_0) J_0(k_r r) \, k_r \, dk_r;$$

here $\tilde{p}$ is the unit-source field as a function of range, and g is the depth-dependent Green's function in the wavenumber domain. The Green's function is commonly written as $g(k_r; z, z_0)$, with the argument ω omitted. *Units* m.

Note (1): The depth-dependent Green's function is the solution of the following ordinary differential equation in z, where $\rho(z)$ is the local density, $\rho'(z)$ is its derivative with depth, and $k_0(z)$ is the local acoustic wavenumber $\omega/c(z)$:

$$\left\{\frac{d^2}{dz^2} - \frac{\rho'(z)}{\rho(z)}\frac{d}{dz} + k_0^2(z) - k_r^2\right\}g(k_r; z, z_0) = -2\delta(z - z_0).$$

The function $g(k_r; z, z_0)$ may be interpreted as the HANKEL TRANSFORM of the pressure field radiated by a point monopole of strength 4π.

Note (2): Such a point monopole, if placed in a uniform unbounded medium, would generate a pressure field $p = (1/R)e^{-jk_0 R}$, where R is the distance from the source; in other words, the "unit source" used to define the depth-dependent Green's function is one that produces unit pressure amplitude at unit radius.

destructive interference – see INTERFERENCE.

detection range – *of a sonar system* the target distance at which the SIGNAL EXCESS is zero. Also known as **sonar range**. Units m.

detection threshold – *in active sonar* the minimum signal-to-noise ratio (SNR) required, at the input to a sonar detector, in order to achieve target detection with a given probability (usually 50%) at a fixed false alarm rate. When the SNR equals the detection threshold, the SIGNAL EXCESS is zero. Units dB.

deterministic signal – a signal whose values can be predicted from current and past information. This definition excludes *chaotic signals*, whose unpredictability arises from extreme sensitivity to initial conditions (i.e. the underlying process generating the signal may be deterministic, but the chaotic signal itself is not).

detrending – alternative term for TREND REMOVAL.

Deutlichkeit – literally "distinctness" in German; defined as the fraction of sound energy arriving within 50 ms of the direct sound. Its value for the octave bands centred on 500, 1000, and 2000 Hz in particular is considered to be a measure of speech INTELLIGIBILITY. Deutlichkeit is uniquely related to C_{50}, the early-to-late sound index when 50 ms is used as the boundary between early and late sound (see CLARITY INDEX). Symbol D_{50}. Units none.

deviatoric strain – the deviatoric tensor formed by subtracting from the STRAIN tensor ε_{ij} the isotropic tensor having the same TRACE. The deviatoric strain has cartesian components

$$\varepsilon'_{ij} = \varepsilon_{ij} - \theta\delta_{ij} \quad (i, j = 1, 2, 3),$$

where $\theta = \frac{1}{3}\varepsilon_{kk}$ is the trace of ε_{ij}. Units none.

deviatoric stress – the deviatoric tensor formed by subtracting from the STRESS tensor σ_{ij} the isotropic tensor having the same TRACE. The deviatoric stress has cartesian components

$$\sigma'_{ij} = \sigma_{ij} - \sigma\delta_{ij} \quad (i,j = 1, 2, 3),$$

where $\sigma = \frac{1}{3}\sigma_{kk}$ is the trace of σ_{ij} (the negative of the PRESSURE). *Units* Pa.

deviatoric tensor – a tensor with zero TRACE.

DFS – *for periodic discrete-time signals* abbreviation for DISCRETE FOURIER SERIES.

DFT – *for transient discrete-time signals* abbreviation for DISCRETE FOURIER TRANSFORM.

diagnostic ultrasound – the use of ultrasound for diagnosis and monitoring in medicine and industry. An example of the latter is non-destructive testing.

diagonal elements – *of a square matrix* the elements, or entries, that form the main diagonal. Thus if a_{ij} is the element in row i and column j, and the matrix size is $n \times n$, the diagonal elements are $a_{11}, a_{22}, \ldots, a_{nn}$.

diagonal matrix – a SQUARE MATRIX whose only non-zero elements lie on the main diagonal.

diaphragm – (1) *with specific reference to acoustics* a thin membrane, usually disc- or cone-shaped, that radiates sound when vibrated, as in a ***loudspeaker diaphragm***, or alternatively responds to an incident sound field, as in a ***microphone diaphragm***.

diaphragm – (2) a thin partition or membrane, used to separate two regions.

Note: In optics the word has acquired a different meaning: it is used to describe an opaque partition with a central hole, that acts as an *aperture stop*.

difference equation – a relation connecting successive terms of a SEQUENCE; also known as a ***recurrence relation***. An equation that connects $k + 1$ successive terms is called a kth-order difference equation. For example, if the sequence is denoted by $x_0, x_1, x_2, \ldots, x_n, \ldots$, then the relation

$$x_{n+2} + a\,x_{n+1} + b\,x_n = 0$$

is of second order.

Note: The second-order difference equation above has a straightforward solution, if a and b are constants. Let the roots of the quadratic equation $m^2 + am + b = 0$ be $m = (\alpha, \beta)$. Then provided $\alpha \neq \beta$,

$x_n = A\alpha^n + B\beta^n$ (A, B constants).

Otherwise, if $\alpha = \beta$, $x_n = (A + Bn)\alpha^n$.

differentiable – *in the complex plane* see ANALYTIC FUNCTION.

differential admittance – ◆ *of a compliant lumped element* an alternative term for MOBILITY (2).

differential bulk scattering coefficient – *of a medium containing a distribution of scatterers* the DIFFERENTIAL SCATTERING CROSS-SECTION per unit volume of the medium. It is related to the BULK SCATTERING COEFFICIENT, m_v, by the integral

$m_v = \int m_{dv}(\theta, \phi)\, d\Omega;$

here $m_{dv}(\theta, \phi)$ is the differential bulk scattering coefficient for scattering in the direction defined by the spherical polar angles (θ, ϕ), and $\int (...)\, d\Omega$ represents a solid-angle integral over all possible scattering directions. Compare VOLUME BACKSCATTERING STRENGTH, which is a logarithmic version of $m_{dv}(\theta, \phi)$ measured in the BACKSCATTERING DIRECTION. *Units* $m^{-1}\,sr^{-1}$.

differential equation – see ORDINARY DIFFERENTIAL EQUATION, PARTIAL DIFFERENTIAL EQUATION.

differential impedance – *of a non-deforming lumped element* an impedance based on the pressure difference or net force applied to the lumped element. For example, the differential impedance of a thin panel is the complex ratio of the pressure difference across the panel, at a single frequency, to the local normal velocity of the vibrating panel at the same point. The essential feature of a differential impedance is that the "transmitted variable" is a velocity (or angular velocity, or volume velocity) that is invariant through the element. See also LUMPED ELEMENTS.

Note: Equivalent terms for the differential impedance of a panel are *transmission impedance* and *separation impedance*.

differential scattering cross-section – *of an object in a plane-wave incident field* the ratio of the scattered ANGULAR POWER DISTRIBUTION in a specified direction, W_Ω, to the incident intensity, I_{inc}, when the object is irradiated by plane progressive waves at a given frequency arriving from a given direction. In symbols, the differential scattering cross-section is given by

$$s_d(\theta, \phi) = \frac{W_\Omega(\theta, \phi)}{I_{inc}}$$

where the spherical polar angles (θ, ϕ) define the scattering direction. *Units* $m^2\,sr^{-1}$.

differential surface scattering coefficient – *of a surface distribution of scatterers* the DIFFERENTIAL SCATTERING CROSS-SECTION per unit area of surface. It is related to the SURFACE SCATTERING COEFFICIENT, m_s, by the integral

$$m_s = \int m_{ds}(\theta, \phi) d\Omega;$$

here $m_{ds}(\theta, \phi)$ is the differential surface scattering coefficient for scattering in the direction defined by the spherical polar angles (θ, ϕ), and $\int(\ldots) d\Omega$ represents a solid-angle integral over the hemisphere of all possible scattering directions. Compare SURFACE BACKSCATTERING STRENGTH. *Units* $m^{-1} sr^{-1}$.

diffracted rays – see GEOMETRICAL THEORY OF DIFFRACTION.

diffraction – non-specular reflection or SCATTERING of sound waves by an object or boundary, particularly into the SHADOW ZONE. Diffraction generally involves the distortion of incident wavefronts.

Note: It is not easy to draw a sharp distinction between diffraction and scattering. Roughly speaking, diffraction is limited to wave phenomena that cannot be accounted for, even approximately, by RAY ACOUSTICS. On the other hand scattering can refer to any reflection process, whether specular or not.

diffraction limit of resolution – *for acoustic sources viewed from the far field* the limitation that spatial source information on a scale smaller than $\lambda/2$ (half an acoustic wavelength) cannot be resolved from far-field acoustic measurements.

Note: The diffraction limit of resolution is explainable by noting that only part of the source wavenumber–frequency spectrum contributes to the far-field acoustic pressure. Specifically, the far field is determined by a band-limited version of the 3D spatial Fourier transform of the source, confined to wavenumbers of magnitude less than the acoustic wavenumber (i.e. $|\mathbf{k}| < k_0$). As a result, the fine spatial detail of the source distribution is lost. For ways to overcome the diffraction limit, see SUPERRESOLUTION.

diffuse field – an idealized sound field that consists of infinitely many uncorrelated plane progressive waves, with their intensity uniformly distributed with respect to direction. The resultant acoustic intensity is therefore zero.

Note (1): This definition implies that the single-point and two-point statistics of the pressure field are independent of both absolute position and orientation, so that the field is statistically homogeneous and isotropic.

Note (2): A *pure-tone diffuse sound field* consists of infinitely many uncorrelated single-frequency plane waves, all with the same infinitesimal amplitude but with randomly distributed phases; the waves arrive from all directions with equal probability. Ensemble averaging over multiple realizations is implied in this case. The resulting 2-point cross-correlation function between pressures p and p', separated in time by delay τ and spatially by distance r, has the normalized value

$$\rho_{pp'} = \frac{\sin kr}{kr} \cos \omega \tau,$$

where ω is the angular frequency and $k = \omega/c$ is the acoustic wavenumber.

diffuse-field intensity – *at a point in a diffuse sound field* the quantity $cw/4$, where w represents the mean ACOUSTIC ENERGY DENSITY and c is the sound speed. Physically, $cw/4$ is the mean rate, per unit surface area, at which energy in a DIFFUSE FIELD is incident on an imaginary plane surface from one side only. *Units* W m^{-2}.

diffusely-reflecting – see DIFFUSE REFLECTION.

diffuser – (1) a device placed in a REVERBERATION CHAMBER to bring the sound field closer to a diffuse field. See ROTATING DIFFUSER.

diffuser – (2) a reflecting structure or boundary treatment, used to reduce the strength of specular reflections (e.g. in a concert hall) without introducing absorption. See DIFFUSION, DIFFUSE REFLECTION, QUADRATIC RESIDUE DIFFUSER, SCHROEDER DIFFUSER.

diffuse reflection – an idealized model of sound reflection from a rough surface, in which the scattering directivity is independent of the angle of incidence. Each element of the surface acts as an independent source of scattered sound, with a fixed directional distribution of reflected energy. Thus incident plane waves lose all "memory" of their arrival direction.

Note: Diffuse reflection does not mean that each surface element radiates omnidirectionally; the actual directivity pattern of reflected energy depends on the absorption coefficient of the surface (see LAMBERT'S COSINE LAW), but always falls to zero at near-grazing angles. On the other hand, provided all surfaces in a room have a near-zero sound power absorption coefficient, diffuse reflection at the boundaries does lead to a DIFFUSE FIELD within the room, in the high-frequency limit when a large number of resonant modes is excited.

diffusion – *at specified points in a room* the randomization of sound-wave arrival directions. Diffusion may be enhanced by placing scatterers or diffusely-reflecting surfaces in a room; see DIFFUSER, SCHROEDER DIFFUSER, VOLUME DIFFUSERS.

diffusivity of sound – *in a thermoviscous fluid* a combination of viscous and thermal properties that determines the attenuation of progressive sound waves in the far field. It has the dimensions of a diffusion coefficient, and is given by

$$\delta = \nu\left(\frac{4}{3} + \frac{\mu_B}{\mu} + \frac{\gamma - 1}{\Pr}\right),$$

where ν is the kinematic viscosity μ/ρ (μ = viscosity, ρ = density); μ_B is the bulk viscosity (an approximate representation of relaxation effects, valid for frequencies much less than the relaxation frequency); γ is the specific-heat ratio, and Pr is the PRANDTL NUMBER. *Units* $m^2 \, s^{-1}$.

Note: The propagation wavenumber for plane acoustic waves in a thermoviscous fluid is

$$k \approx k_0\left(1 - j\frac{\omega\delta}{c^2}\right)^{1/2}, \qquad k_0 = \frac{\omega}{c};$$

here ω is the angular frequency and c is the sound speed.

digital filter – a DIGITAL SYSTEM, designed to produce a linear transfer function in the Z-PLANE with certain desired characteristics.

digital signal – a sequence of numbers $x[n]$, defined for every integer n over some specified range.

digital signal processing – see SIGNAL PROCESSING.

digital simulation – *of an analogue system* replication of a sampled version of the analogue system output, via a DIGITAL SYSTEM whose input is a sampled version of the analogue system input.

digital system – an algorithm, or hardware implementing an algorithm, that converts one or more digital input signals $x[n]$ into one or more digital output signals $y[n]$.

digital-to-analogue converter – a device for converting a digital signal into an analogue signal. The digital-to-analogue conversion is not unique; it is the inverse of analogue-to-digital conversion, in which information is lost. The analogue signal therefore has meaning only at frequencies below the FOLDING FREQUENCY.

digitization – conversion of analogue data into digital data; often applied to 2D data. See also SAMPLING.

dilatant fluid – *in viscoelasticity* equivalent term for SHEAR-THICKENING FLUID.

dilatation – an alternative term for VOLUME STRAIN; i.e. the increase in volume of an element of material, per unit initial volume:

$$\theta = \frac{V - V_0}{V_0}.$$

Here V is the SPECIFIC VOLUME of the material, and V_0 is its initial value. *Units* none.

dilatational waves – *in a continuous medium* an equivalent term for COMPRESSIONAL WAVES. Also known as DILATATION WAVES.

dilatation rate – the rate of increase in volume of an element of material, per unit *current* volume. Symbol Δ. *Units* s^{-1}.

Note: The local dilatation rate can be expressed mathematically either as the divergence of the velocity field **u**, or as the MATERIAL DERIVATIVE of the logarithmic volume strain, $\ln(V/V_0)$:

$$\Delta = \operatorname{div} \mathbf{u} = \frac{D}{Dt}\left(\ln \frac{V}{V_0}\right) \quad (V = \text{specific volume})$$

$$\approx \frac{D\theta}{Dt} \quad (\ln \frac{V}{V_0} \approx \theta \text{ for } \theta \ll 1).$$

Here subscript 0 denotes the initial or undisturbed value, and θ is the DILATATION. The dilatation rate approximates to the material derivative of θ, i.e. $D\theta/Dt$, in the limit of small dilatation (as in linear acoustics).

dimensions – generalized units attached to physical quantities. The basic dimensions are normally taken as mass (M), length (L) and time (T), with temperature (Θ) and current (I) added as appropriate. In the MLT system, a force F (for example) has dimensions MLT^{-2}, and this is written symbolically as $[F] = MLT^{-2}$.

dipole – *used as a noun* abbreviation for POINT DIPOLE.

dipole density – see ORDER (12).

dipole moment – the first spatial moment of an ACOUSTIC SOURCE DENSITY; also known as the ***dipole-moment vector***. For a dipole constructed from two equal and opposite POINT MONOPOLES of strength S and $-S$, with the $+S$ monopole displaced by **d** relative to $-S$, the dipole moment is the product *S***d**. See DIPOLE-ORDER SOURCE DISTRIBUTION.

dipole-order source distribution – one in which positive and negative elements cancel overall at every instant. It is describable by a SOURCE DENSITY $\mathcal{F}$ whose total strength defined by the volume integral S below is zero, but whose first spatial moment – with components D_i defined below – is non-zero (and is called the DIPOLE MOMENT or dipole strength). Mathematically, the dipole distribution is characterized by

$S = \int \mathcal{F} d^3 \mathbf{y} = 0$ (zero source strength);

$D_i = \int y_i \mathcal{F} d^3 \mathbf{y}$ (non-zero dipole moment).

Here y_i are cartesian components ($i = 1, 2, 3$) of the position vector $\mathbf{y}$, and $d^3 \mathbf{y}$ denotes a volume element. In the second expression, D_i are the cartesian components of the **dipole-moment vector**.

dipole source – a sound generation mechanism that drives the surrounding medium by a process equivalent to a time-varying applied force field. In the COMPACT limit, the sound power output from a dipole source varies as c^{-3}, where c is the sound speed in the radiating medium.

dipole source distribution – abbreviation for DIPOLE-ORDER SOURCE DISTRIBUTION.

Dirac delta function – a generalized function equivalent to zero everywhere, except at one point where the function is infinite (**P A M DIRAC 1927**). The *one-dimensional delta function*, written as $\delta(x)$, is defined by the properties

$$\int_{-\infty}^{\infty} \delta(x) f(x) \, dx = f(0); \qquad \delta(x) = 0 \text{ for } x \neq 0.$$

Here $f(x)$ is any function that is continuous at $x = 0$.

One can visualize $\delta(x)$ as an infinitely narrow symmetrical spike of unit area, located at $x = 0$; thus for arbitrarily small positive numbers ε,

$$\int_{-\varepsilon}^{\varepsilon} \delta(x) dx = \int_{-\varepsilon}^{\varepsilon} \delta(-x) dx = 1.$$

Likewise $\delta(x - x_0)$ is a spike of unit area located at $x = x_0$, and its product with any non-singular function $f(x)$ has the integral property

$$\int_{-\infty}^{\infty} \delta(x - x_0) f(x) \, dx = \int_{-\infty}^{\infty} \delta(x_0 - x) f(x) \, dx = f(x_0);$$

this result is known as the **sifting relation**.

In three dimensions, the delta function $\delta(\mathbf{r} - \mathbf{r}_C)$ is a singularity located at vector position $\mathbf{r}_0$, and the sifting relation becomes a volume integral:

$$\int \delta(\mathbf{r} - \mathbf{r}_0) f(\mathbf{r}) \, dV(\mathbf{r}) = f(\mathbf{r}_0).$$

Here $f(\mathbf{r})$ is any function continuous at $\mathbf{r} = \mathbf{r}_0$, and the region of integration must include $\mathbf{r}_0$ but is otherwise arbitrary. This 3D version of the sifting relation may be extended to any number of dimensions.

Note (1): The one-dimensional delta function $\delta(x)$ may be regarded as the derivative of $H(x)$, the HEAVISIDE UNIT STEP FUNCTION.

Note (2): The continuous delta function $\delta(x)$ has a discrete counterpart, sometimes called the **unit pulse function**. See KRONECKER DELTA.

direct field – *in a reverberant space* the field that would be measured if reflections were absent; the outgoing-wave radiation from a source.

direct impedance – refers to a MECHANICAL IMPEDANCE in which the output and input are vector components in the same direction. For example, the direct impedance $Z_{xx} = F_x/v_x$ is the complex ratio of a force F_x in the x direction and a velocity v_x in the same direction. Compare CROSS IMPEDANCE. *Units* N s m^{-1}.

directional response – *of a receiving or emitting acoustical transducer* a description of the transducer response as a function of incident or emitted wave direction in the far field (usually relative to the transducer axis).

direction cosines – *in rectangular coordinates* a name for the cartesian components of a UNIT VECTOR. So called because if a unit vector **n** makes angles α, β, γ with the coordinate axes Ox, Oy, Oz, then the components of **n** are

$$n_x = \cos\alpha, \quad n_y = \cos\beta, \quad n_z = \cos\gamma.$$

directivity – *of far-field radiation* the directional pattern of radiated mean square pressure (usually in a specified band of frequencies).

directivity factor – *of a sound source* the ratio of the far-field mean square pressure (at a given frequency, in a specified direction from the source) to the average mean square pressure over a sphere of the same radius, centred on the source. If the source has an axis of symmetry, the directivity factor is understood to refer to the on-axis direction. *Units* none.

Note (1): One can define a **receiver directivity factor** in a similar way, by replacing the far-field radiation pattern (i.e. the mean square pressure measured over a sphere centred on the transducer) with $|G(f, \theta, \phi)|^2$, where $G(f, \theta, \phi)$ is the SENSITIVITY of the transducer to incident pressure waves of frequency f arriving from direction (θ, ϕ). Compare BEAM PATTERN.

Note (2): According to the RECIPROCITY PRINCIPLE, a reversible transducer has the same far-field directional response whether it operates as a source or a receiver. This means that the source diversity factor and the receiver directivity factor of the same transducer (or array of transducers) are identical.

directivity index – *of a sound source* $10 \log_{10} Q$, where Q is the DIRECTIVITY FACTOR. *Units* dB.

direct sound – *in auditorium acoustics* sound that travels directly from a source to a listener or receiver, as if in a free field. In a room, the sequence of arrivals is the direct sound first, followed by sound reflected from room surfaces.

Dirichlet boundary condition – *applied to solutions of a differential equation* a boundary condition that specifies the value of the solution at every point on the boundary (**P G L Dirichlet 1850**).

discontinuity distance – an alternative term for SHOCK-FORMATION DISTANCE. *Units* m.

discrete data – see QUANTIZED DATA.

discrete Fourier series (DFS) – *for periodic discrete-time signals* a representation of a periodic sequence $\tilde{x}[n]$, with period N, as a sum of Fourier components. The DFS for $\tilde{x}[n]$ is

$$\tilde{x}[n] = \sum_{k=\langle N \rangle} c_k e^{jk\Omega n} \qquad (\Omega = 2\pi/N).$$

The spectral coefficients c_k in the DFS form a periodic sequence of the same period as the original sequence. They are related to $\tilde{x}[n]$ by

$$c_k = \frac{1}{N} \sum_{n=\langle N \rangle} \tilde{x}[n] e^{-jk\Omega n}.$$

The notation $\sum_{k=\langle N \rangle}, \sum_{n=\langle N \rangle}$ denotes a summation with respect to k or n over one complete period (i.e. over N successive integers), starting at any point.

Note: The scaled coefficients Nc_k are identical with the Fourier components $X[k]$ obtained by applying the DISCRETE FOURIER TRANSFORM to one cycle of the periodic sequence.

discrete Fourier transform (DFT) – *for transient discrete-time signals* a discrete transformation defined by

$$X[k] = \sum_{n=0}^{N-1} x[n] e^{-j(2\pi/N)kn} = \sum_{n=0}^{N-1} x[n] W_N^{-kn},$$

$$x[n] = \frac{1}{N} \sum_{k=0}^{N-1} X[k] e^{j(2\pi/N)kn} = \frac{1}{N} \sum_{k=0}^{N-1} X[k] W_N^{kn}.$$

Here N is the length of the finite sequence $x[n]$, and W_N denotes the Nth root of unity, i.e. $e^{j(2\pi/N)}$. The first equation defines the N-point forward transform, and the second defines its inverse, called the ***inverse DFT***

(*IDFT*). Note that the IDFT allows the original sequence $x[n]$ to be recovered from the periodic DFT output $\tilde{X}[k]$.

Note (1): The procedure defined above results in an N-point DFT. If the sequence $x[n]$ contains fewer than N values, it may be extended by zero-padding, i.e. by adding zeros at either end to bring the length of the sequence up to N.

Note (2): The DFT of the finite sequence $x[n]$, with $n = 0, \ldots, N - 1$, is equivalent to taking one cycle from the DISCRETE FOURIER SERIES representation of the periodic sequence $\tilde{x}[n]$ based on $x[n]$. This is a sequence with repetition length N, defined by

$$\tilde{x}[n + iN] = x[n] \quad (i \text{ integer}; n = 0, \ldots, N - 1).$$

The spectral coefficients c_k of $\tilde{x}[n]$ form a sequence $\tilde{X}[k]$ of the same period,

$$c_{k + iN} = c_k \quad (i \text{ integer}; k = 0, \ldots, N - 1),$$

and the DFT consists of one period of c_k multiplied by N.

Note (3): The spatial DFT of a linear array's WEIGHTING FUNCTION gives the array far-field directivity.

discrete frequency components – *of a continuous signal* periodic components that appear as lines in the frequency spectrum.

discrete signal – a DIGITAL SIGNAL, produced for example by SAMPLING a continuous signal.

discrete spectrum – *in underwater acoustics* another name for the NORMAL-MODE FIELD in a horizontally-stratified waveguide, consisting of shallow-angle trapped rays. Also known as *trapped modes*. The word *spectrum* here refers to the horizontal wavenumber spectrum of the depth-dependent Green's function, at a given frequency. Compare VIRTUAL MODE, CONTINUOUS SPECTRUM.

discrete system – (1) the same as DIGITAL SYSTEM.

discrete system – (2) *in dynamics* an idealized mechanical system whose kinetic and potential energy, and overall energy dissipation rate, are discretized into LUMPED ELEMENTS rather than being modelled as continuously distributed. Discrete systems have a finite number of degrees of freedom, and are governed by ordinary differential equations. Compare CONTINUOUS SYSTEM.

discrete time – the sequence of real integers n used to label successive terms in a DIGITAL SIGNAL sequence. In situations where the digital signal is obtained by sampling an ANALOGUE SIGNAL at intervals Δ in time, discrete time n and physical time t are related by

$$t = n\Delta.$$

The constant Δ is called the *sampling interval*. *Units* none.

discrete-time Fourier transform – *for transient discrete-time signals* a transformation of the finite sequence $x[n]$ into a continuous periodic function of period 2π in the digital angular frequency domain, denoted by $X(\Omega)$. The discrete-time Fourier transform is defined by

$$X(\Omega) = \sum_{n=-\infty}^{\infty} x[n] e^{-j\Omega n}.$$

Its inverse represents the original signal as a Fourier integral,

$$x[n] = \frac{1}{2\pi} \int_{(2\pi)} X(\Omega) e^{j\Omega n} d\Omega,$$

where the notation $\int_{(2\pi)}$ denotes integration over one complete period (an interval of 2π), starting at any point. The transform pair is represented schematically by

$$x[n] \leftrightarrow X(\Omega) \qquad \text{(discrete-time Fourier transform pair)}.$$

Note: The discrete-time Fourier transform is equivalent to the Z-TRANSFORM with z replaced by $e^{j\Omega}$. For this reason, some authors prefer to write $X(\Omega)$ as $X(e^{j\Omega})$.

discrete variable – see QUANTIZED VARIABLE.

dispersion – *as a property of free wave propagation* the propagation of different frequency components at different speeds. Hence *dispersive*, as in *dispersive waves*.

dispersion coefficient – *in a dispersive medium* a measure of the extent to which the PHASE SPEED v_{ph} varies with frequency, defined as

$$\beta = \omega \left(\frac{1}{v_{ph}(\omega)} - \frac{1}{v_{ph}(0)} \right)$$

where ω is the angular frequency. *Units* rad m^{-1}.

dispersion relation – *for an acoustic medium or waveguide* a relation that gives the real and imaginary parts of the PROPAGATION WAVENUMBER as functions of frequency, or vice versa. Plane waves propagating in an ANISOTROPIC MEDIUM are described by the generalized dispersion relation $\omega = \Phi(k_1, k_2, k_3)$; here ω is the angular frequency and (k_1, k_2, k_3) are components of the wavenumber vector, for a plane wave propagating in an arbitrary direction.

dispersive medium – a medium in which freely-propagating waves exhibit DISPERSION.

displacement – a vector that measures a change of position. For small-amplitude mechanical oscillations, the displacement of any point in a system is usually measured from the rest position. *Units* m.

displacement transducer – a transducer that produces a response directly proportional to the input displacement (in some specified direction), over a given frequency range.

dissipation – (1) the irreversible conversion of energy from mechanical or electrical (i.e. organized) forms into thermal (disorganized) form, with an associated increase in ENTROPY.

dissipation – (2) abbreviation for DISSIPATION RATE. *Units* W.

dissipation rate – the rate at which mechanical or other organized forms of energy in a lossy system are converted into thermal energy, as a result of DAMPING or other irreversible processes. See DISSIPATION. *Units* W.

distortion – a change in the waveform of a signal, usually undesired, that is produced during transmission, transduction, or other processing of the signal. See also TOTAL HARMONIC DISTORTION, AMPLITUDE DISTORTION, PHASE DISTORTION. For nonlinear distortion of acoustic waveforms, see FINITE-AMPLITUDE WAVE, CUMULATIVE NONLINEARITY.

distortion level – *of a periodic waveform* see TOTAL HARMONIC DISTORTION. *Units* dB.

distributed source – (1) a radiating surface or volume whose maximum dimension is large in terms of acoustic wavelengths.

Note: An example is an array of ACOUSTIC SOURCES (1) that extends over several wavelengths in one direction (even though the individual elements of the array may be compact).

distributed source – (2) an ACOUSTIC SOURCE DENSITY that cannot be modelled as being COMPACT.

distribution – see DENSITY FUNCTION, FREQUENCY DISTRIBUTION, CUMULATIVE FREQUENCY DISTRIBUTION, GAUSSIAN DISTRIBUTION.

distribution function – an abbreviation used in statistics for the CUMULATIVE DISTRIBUTION FUNCTION of a random variable.

divergence loss – (1) *for free-field sound propagation between two points* the decrease in sound pressure level attributable to geometrical spreading of wavefronts; thus spherically spreading waves incur a far-field divergence loss of 6 dB per doubling of distance from the source. Also known as *spreading loss*. *Units* dB.

divergence loss – (2) *in underwater acoustics, for sound propagation between a source and receiver* the component of the TRANSMISSION LOSS (3) that is due to geometrical spreading of wavefronts. Thus at 1000 m from a source in a homogeneous ocean, the divergence loss is 60 dB re 1 m^2. See also PROPAGATION LOSS. *Units* dB re 1 m^2.

divergence theorem – another name for GAUSS'S DIVERGENCE THEOREM.

diverging wave – an equivalent term for OUTGOING WAVE.

Doppler effect – a shift in acoustic frequency (between source and observer reference frames) caused by relative motion between source and observer. The observer hears a higher frequency when the source is approaching than when it is receding, as a result of the source–observer travel time changing at a finite rate (**J C DOPPLER 1842**).

Doppler factor – (1) *for a moving constant-frequency source in a stationary uniform medium* the ratio f_s/f_0, where f_s is the source frequency and f_0 is the frequency heard by a stationary observer. If the source velocity in the observer direction was $[v_s]$ at the time the sound was emitted,

$f_s/f_0 = 1 - [v_s]/c$;

here c denotes the sound speed in the uniform medium. See the note under DOPPLER FACTOR (2) for a restriction on this equation. *Units* none.

Doppler factor – (2) *for a moving observer in a stationary uniform medium* the ratio f_m/f_0, where f_m is the frequency heard by the moving observer, and frequency f_0 is what a stationary observer would hear at the same position and the same instant. The Doppler factor in this case is given by

$f_m/f_0 = 1 - v_m/c$,

where the component of the observer velocity in the propagation direction of the arriving sound is v_m. *Units* none.

Note: In this and the previous definition, the assumption is made that the time scales on which v_s or v_m vary are much longer than the period of the source.

Doppler frequency relation – *for both source and observer moving through a uniform medium* the frequency of the sound heard by a moving observer, f_m, is related to the frequency emitted earlier by the source, $[f_s]$, by

$$\frac{f_m}{(1 - v_m/c)} = \frac{[f_s]}{(1 - [v_s]/c)}.$$

Here square brackets denote values at the time of emission of the signal, and unbracketed symbols denote values at the time of reception. For the meaning of v_s and v_m, see DOPPLER FACTOR (1), (2).

Note (1): The source and observer frequencies f_s and f_m are defined in different frames of reference, moving with the source and observer respectively.

Note (2): The source and observer velocities v_s and v_m are defined in relation to both the position of the observer at the time the sound is received, and the arrival direction of that sound; compare EMISSION TIME COORDINATES. Specifically, $[v_s]$ is the source velocity in the direction of the eventual observation point, at the time the sound was emitted; v_m is the observer velocity parallel to the propagation direction of the arriving sound, at the time the sound is received; and both velocities are measured relative to the medium.

Note (3): The time scales on which (f_s, v_s, v_m) vary are assumed to be long in comparison with the source period.

dot (˙) – symbol used over a quantity to denote a time derivative, as in $\dot{x}$.

double-wall resonance – see MASS–AIR–MASS RESONANCE.

drag – see AERODYNAMIC FORCE. *Units* N.

drag coefficient – the normalized drag on an object, given by

$$C_D = D/Aq$$

where A is an appropriate area (usually the projected area of the object in the relative flow direction), q is the free-stream DYNAMIC PRESSURE of the relative flow, and D is the drag force. *Units* none.

drift – *in signals produced by analogue instrumentation* an unwanted additive component, usually a monotonic function of time, caused by limitations in the instrumentation. Its effect is equivalent to a slow change in the short-duration AVERAGE of the signal.

drive-point admittance/impedance – alternative terms for DRIVING-POINT ADMITTANCE/IMPEDANCE.

driving-point admittance – see definitions (1) and (2) for DRIVING-POINT IMPEDANCE. These apply also to admittance-type quantities, with the appropriate changes of input and output variables.

Note: The real and imaginary parts of a driving-point admittance are called the ***driving-point conductance*** and ***driving-point susceptance*** respectively.

driving-point impedance – (1) *for a point-excited mechanical system* the complex ratio of applied force to velocity at a single frequency, when the velocity is measured at the point of application of the force. See also DRIVING-POINT IMPEDANCE MATRIX. *Units* $N \; s \; m^{-1}$.

driving-point impedance – (2) *for any linear system* a general term for an impedance-type frequency response function, where the input and output are measured at the same point in the system. Equivalent terms are ***point impedance*** or ***point input impedance***. The type of point impedance can be indicated by adding a prefix such as ***mechanical***, ***electrical***, or ***acoustic***.

For a mechanical system with two or more discrete excitation points, or a network with two or more ports, the driving-point impedances are the diagonal terms in the MULTIPORT IMPEDANCE MATRIX. Compare TRANSFER IMPEDANCE.

Note: The real and imaginary parts of a driving-point impedance are called the ***driving-point resistance*** and ***driving-point reactance*** respectively.

driving-point impedance matrix – ◆ *for a point-excited mechanical system* a square matrix of complex coefficients that relates the applied force vector **F** to the driving-point velocity vector **v**, as in the relations

$$\mathbf{F} = \mathbf{Z}\mathbf{v} \quad \text{or} \quad F_j = Z_{ji}v_i.$$

Here **Z** is a MECHANICAL IMPEDANCE, expressed in matrix form; the elements of **v** are the input velocity components v_i, and the elements of **F** are the force components F_j at the driving point that are implied by the given velocity vector **v**. See also CROSS IMPEDANCE, DIRECT IMPEDANCE. *Units* $N \; s \; m^{-1}$.

Note (1): It is conventional with mechanical systems to consider forces as inputs and velocities as responses (hence the common use of such response functions as RECEPTANCE and ACCELERANCE); but provided the system is invertible, there is no reason why one should not reverse these roles.

Note (2): If **F** and **v** *were* interchanged in the equation above, **Z** would be called a ***driving-point admittance matrix***.

driving-point receptance – *of a point-excited mechanical system* the complex ratio of displacement to applied force at a single frequency, when the displacement is measured at the point of application of the force. See also RECEPTANCE. *Units* m N^{-1}.

dropout – inadvertent momentary reduction in output level during analogue signal reproduction, or accidental disappearance of a data bit during digital signal transfer.

DSP – abbreviation for digital signal processing. See SIGNAL PROCESSING.

duct attenuator – an equivalent term for DUCT SILENCER. See also SPLITTER SILENCER.

duct liner – layer of material placed on the inner surfaces of a duct, in order to attenuate fluid-borne sound waves propagating along the duct.

duct lining – alternative term for DUCT LINER.

duct silencer – a device for reducing the sound power transmitted along a duct system. Installed between the source and the part of the system to be protected, it is designed to give a specified INSERTION LOSS, which may be achieved using reactive or resistive elements or both.

Duhamel integral – *for a linear time-invariant system* the convolution integral that gives the time-domain response of the system to an arbitrary input, in terms of the impulse response function (**J-M-C DUHAMEL 1833**). If $x(t)$ and $y(t)$ denote the system input and output at time t, and $h(t)$ is the impulse response (assumed causal), the Duhamel integral is written as

$$y(t) = \int_{-\infty}^{t} x(\tau)h(t-\tau)d\tau$$
$$= \int_{0}^{\infty} x(t-u)h(u)du.$$

Note: An alternative version of the Duhamel integral uses the system *step response* $\bar{h}(t)$, defined by

$$\bar{h}(t) = \int_{-\infty}^{t} h(\tau)d\tau.$$

The output and input are then related by

$$y(t) = \int_{-\infty}^{t} \dot{x}(\tau)\bar{h}(t-\tau)d\tau$$
$$= \int_{0}^{\infty} \dot{x}(t-u)\bar{h}(u)du,$$

where $\dot{x}(t)$ is the time derivative of the input $x(t)$.

D-weighting – a FREQUENCY WEIGHTING procedure that provides a practical approximation to the sensitivity of the human ear, at levels likely to be associated with moderate annoyance [specifically, audio-frequency sound pressure levels in the approximate range 50–90 dB re $(20\ \mu\text{Pa})^2$]. It improves on the otherwise-similar B-WEIGHTING curve by including an allowance for ear canal resonance in the range 2–5 kHz. However, unlike A- and C-weighting, D-weighting is not widely recognized apart from its use in assessing aircraft flyover noise.

dynamic compliance – an alternative term for RECEPTANCE. *Units* m N^{-1}.

dynamic flow resistance – *of a porous acoustic lumped element whose dimensions are small compared to an acoustic wavelength* the real part of the differential acoustic impedance (complex ratio of the pressure drop across the element to the volume velocity flowing through the element), at a specified frequency. Compare FLOW RESISTANCE. *Units* Pa s m^{-3}.

dynamic flow resistivity – *of a bulk porous material at a single frequency* the real quantity $\sigma(\omega)$ in the equation

$$[\sigma(\omega) + j\tau(\omega)]\bar{u}_x = -\frac{\partial p}{\partial x} \quad (\text{where } p = \langle p \rangle_V),$$

with angle brackets $\langle \ldots \rangle_V$ denoting a local volume average over the pore fluid. Here the volume flowrate per unit total area, across planes $x = $ const., is denoted by $\bar{u}_x$. Compare COMPLEX EFFECTIVE DENSITY, which generalizes the concept of FLOW RESISTIVITY to dynamic situations in a slightly different way. *Units* Pa s m^{-2}.

Note: The quantity $\sigma(\omega) + j\tau(\omega)$ that appears above is the ***complex flow resistivity*** of the material.

dynamic magnification factor – *for a linear system with a finite response at zero frequency* the ratio of the amplitude response function at a given frequency to its value at zero frequency:

$$G = \frac{|H(f)|}{|H(f \to 0)|}.$$

Here G is the dynamic magnification factor, and $H(f)$ is the frequency response function of the system at frequency f. *Units* none.

dynamic pressure – *in a steady fluid flow* the quantity $q = \frac{1}{2}\rho u^2$, where ρ is the fluid density and u is the fluid velocity magnitude. *Units* Pa.

Note: In the special case of incompressible flow, the dynamic pressure q is equal to the difference between the STAGNATION PRESSURE and the STATIC PRESSURE.

dynamic range – (1) *of an acoustic variable* the ratio of the maximum and minimum mean square values, expressed in decibels. *Units* dB.

dynamic range – (2) *of an energy or power spectrum* the difference between the maximum and minimum spectrum levels (or band levels), across the frequency range. *Units* dB.

dynamic range – (3) *of a transducer* the available dynamic range of input signals that the transducer can handle, between the upper operating limit (e.g. as set by nonlinearity) and the lower limit (e.g. as set by noise). *Units* dB.

dynamic specific flow resistance – *of a porous acoustic layer that is thin compared to an acoustic wavelength* the real part of the complex specific flow impedance (complex ratio of the pressure drop across the layer to the relative FACE VELOCITY through the layer), at a specified frequency. Compare SPECIFIC FLOW RESISTANCE. *Units* Pa s m^{-1}.

dynamic stiffness – *of a point-excited mechanical system* the complex ratio of applied force to displacement, for single-frequency excitation. It is the inverse of the RECEPTANCE. *Units* N m^{-1}.

Note: The term dynamic stiffness is also used to mean the ***dynamic stiffness magnitude***, i.e. the modulus of the complex ratio defined above.

dynamic tortuosity – *of a rigid-frame porous material* the ratio of the COMPLEX EFFECTIVE DENSITY of the fluid in the pores to the equilibrium fluid density. In symbols,

$$\tilde{\alpha}(\omega) = \frac{\rho(\omega)}{\rho_0}$$

relates the dynamic tortuosity $\tilde{\alpha}(\omega)$ at angular frequency ω to the complex density, $\rho(\omega)$; here ρ_0 is the equilibrium density of the fluid. *Units* none.

Note (1): At high frequencies, the dynamic tortuosity may be estimated from

$$\tilde{\alpha}(\omega) \approx \alpha_\infty \left[1 + (1-j)\frac{\delta}{\Lambda_v}\right], \quad \delta \ll \Lambda_v;$$

here α_∞ is the TORTUOSITY of the rigid frame, δ is the viscous penetration depth, and Λ_v is the frame CHARACTERISTIC LENGTH FOR VISCOUS EFFECTS. For fibre blankets of low porosity, the condition $\delta \ll \Lambda_v$ is replaced by $\delta \ll d$, where d is the fibre diameter.

Note (2): A similar high-frequency asymptotic expression applies to the COMPLEX EFFECTIVE COMPRESSIBILITY, $K(\omega)$, for rigid-frame porous materials:

$$K(\omega) \approx \frac{1}{\rho_0 c_0^2}\left[1 - (1-j)\frac{(\gamma-1)\delta}{\Lambda_t\sqrt{\text{Pr}}}\right], \qquad \delta \ll \Lambda_t\sqrt{\text{Pr}}.$$

Here c_0 is the speed of sound in the pore fluid, γ is the specific-heat ratio, Pr is the PRANDTL NUMBER, and Λ_t is the CHARACTERISTIC LENGTH FOR THERMAL EFFECTS. For fibre blankets of low porosity, the condition $\delta \ll \Lambda_t\sqrt{\text{Pr}}$ is replaced by $\delta \ll d$.

dynamic vibration absorber – an equivalent term for VIBRATION ABSORBER; also known as *vibration neutralizer*.

dynamic viscosity – *of a fluid* equivalent term for VISCOSITY. *Units* Pa s.

E

e – the real number with the property that $(d/dx)e^x$ equals e^x. Its value is 2.718 281 83..., given by the sum of the series

$$1 + \frac{1}{1!} + \frac{1}{2!} + \cdots + \frac{1}{n!} + \cdots.$$

ear – the organ of hearing and balance. See also COCHLEA, INNER EAR, VESTIBULAR SYSTEM, MIDDLE EAR, EXTERNAL EAR.

ear canal – the air-filled duct between the concha and the eardrum, through which sound enters the MIDDLE EAR. It is more formally called the *external ear canal* or *external auditory meatus*, in order to distinguish it from the INTERNAL AUDITORY MEATUS.

eardrum – common term for the TYMPANIC MEMBRANE.

early decay time – *for specified source and receiver locations in a room* a measure of the reverberant decay rate over the first 10 dB of the decay process. The early decay time, abbreviated **EDT**, is expressed as a reverberation time (i.e. the time for 60 dB decay), with the slope of the decay curve computed by fitting a straight line between the 0 dB and -10 dB points. The measurement is conventionally performed in octave bands. The EDT is considered to be a measure of RUNNING REVERBERATION or subjective REVERBERANCE. *Units* s.

early lateral energy fraction – *at a point in a room* the proportion of acoustic signal energy within 80 ms of the direct sound that arrives from the side, expressed as a fraction of that which arrives from all directions during the same time period. The lateral energy is generally measured with a microphone having a figure-of-eight directivity with the null pointing towards the source, whereas the overall energy is measured with an omnidirectional microphone. Early lateral energy fraction, abbreviated as **LF**, is considered to be a measure of SOURCE BROADENING in concert halls. It is a frequency-dependent quantity, sometimes expressed in octave bands. *Units* none.

early-to-late sound index – *for sound arriving at a point in a room from a specified source* the level difference between the early and late arriving signal energy, conventionally measured in octave bands. Symbol C_{80} or C_{50} is used, depending on whether the boundary between "early" and "late" arrivals is placed 80 ms or 50 ms after the direct sound. The 80 ms early-to-late sound index (symbol C_{80}) is generally used for music, and is called the *clarity index*. For speech C_{50} is considered more appropriate. *Units* dB.

Note: The overall ratio of early-arriving to late-arriving signal energy is found by partitioning the squared impulse response of the room at a specified delay with respect to the direct sound arrival. The area under the early part of the curve is divided by the area under the later part to form the energy ratio; expressing this as a logarithmic ratio, in decibels, gives the early-to-late index. The same process can also be carried out in octave bands.

earmuff – a hearing protector designed to enclose the external ear.

earphone – an electroacoustic transducer that couples directly to the external ear, and is commonly used for hearing tests (as part of an AUDIOMETER). Earphones are calibrated using an EAR SIMULATOR. See also CIRCUM-AURAL ∼, INSERT ∼, SUPRA-AURAL ∼.

earplug – a hearing protector that is inserted in the ear canal.

ear simulator – *in audiology* a specific type of device used in the calibration of an EARPHONE; formerly known as an ARTIFICIAL EAR. It is designed to present the same acoustic impedance to an earphone as the average human ear would. A calibrated microphone in the ear simulator measures the sound pressure generated by a particular earphone under test. See also OCCLUDED-EAR SIMULATOR. Compare ACOUSTIC COUPLER (1).

echo – (1) *of a transient acoustic signal* a reflected version of the signal, usually received by the sender (or by another observer) with sufficient time separation from the original signal that the echo is distinct. See also FLUTTER ECHO.

echo – (2) a recurrence or repetition of any signal. The distinctive feature of an echo is that it arrives by an indirect path, either by a reflection process or through being returned by a receiver.

echogenic – capable of causing reflections in the backscattered direction, especially in diagnostic ultrasound or sonar.

echo level (EL) – (1) *in active sonar* the ENERGY LEVEL of the desired signal returned by scattering from a target. *Units* dB re 1 μPa^2 s.

echo level (EL) – (2) *in underwater acoustics* the SOUND PRESSURE LEVEL of the desired signal returned by scattering from a target. *Units* dB re 1 μPa^2.

echolocation, echo location – *by bats, whales, dolphins or other animals; also some cave-dwelling birds* the use of sound to navigate around obstacles, or to locate objects, by a principle similar to that of ACTIVE SONAR; also known as *biosonar*. It involves emitting a succession of tone bursts or other transient signals, and listening for reflections. Echolocation permits navigation or hunting under conditions of poor visibility.

echo sounder – a SONAR device for detecting objects under water, and for measuring their distance (from the time taken for an echo to be received). Most ships have echo sounders to measure the water depth beneath the ship; another name for these is ***depth sounders***.

edge diffraction – DIFFRACTION of incident sound by a sharp-edged boundary. See also GEOMETRICAL THEORY OF DIFFRACTION.

edge wave – *from a plane piston radiator* a component of the sound field that is radiated from a piston in a rigid baffle. When the piston is impulsively accelerated, the first edge-wave arrival comes from the point on the edge of the piston nearest the observer.

Note: The total field radiated from the piston may be regarded as the superposition of two fields, one due to motion of the whole plane in which the piston lies, the other due to a cancelling motion of the baffle in the opposite direction. The second field, i.e. that due to motion of the baffle, is the edge wave.

EDT – abbreviation for EARLY DECAY TIME. *Units* s.

eduction – *in signal processing* extraction of a nonperiodic organized component from a noisy signal by averaging over an ENSEMBLE of sample records, with the start of each record timed to coincide with some feature of the desired underlying signal (conditional sampling).

EFDL – *in underwater acoustics* abbreviation for ***energy flux density level***; a decibel measure of the TIME-INTEGRATED INTENSITY at a far-field measurement position. Compare ENERGY SOURCE LEVEL, which is a decibel measure of the angular radiated-energy distribution of a sound source. *Units* dB re $(1/\rho c)\mu Pa^2$ s.

effective bandwidth – *of a resonant response curve* alternative term for ENERGY BANDWIDTH (1). *Units* Hz.

effective length – *of a uniform duct with one or both ends open* the sum of the actual length and the END CORRECTION at the open end (or ends). *Units* m.

effective perceived noise level (EPNL) – *for aircraft noise* an adaptation of TONE-CORRECTED PERCEIVED NOISE LEVEL that contains an allowance for duration. If $L(t)$ denotes the tone-corrected perceived noise level as a function of time, the effective perceived noise level is defined as

$$L_{EPN} = 10 \log_{10} \left[\frac{1}{t_{ref}} \int 10^{L/10} dt \right].$$

The integral is in principle evaluated over the entire time-history of a noisy event (e.g. an aircraft flyover); in practice, portions of the time-history where $L(t)$ falls more than 10 dB below the peak are ignored. The reference time used for normalization is $t_{ref} = 10$ s. *Units* dB re $(20 \ \mu Pa)^2 \cdot 10$ s.

effective sound pressure – *at a point* alternative term for ROOT MEAN SQUARE (rms) sound pressure. For a periodic signal the average is understood to be over an integral number of cycles; for a time-stationary signal, the limiting long-time average is implied. In all other cases the time interval used for averaging should be stated; see AVERAGE. *Units* Pa.

effective value – *of a time-stationary quantity* the ROOT MEAN SQUARE value.

eigenfunction – (1) in general, the eigenfunctions of a particular mathematical operation are those functions that do not change when the operation is applied, other than being multiplied by a constant. Compare EIGENVECTOR; the meanings of the terms overlap. Two examples are given below; the usual acoustical meaning is example (2).

(1) *Matrix multiplication*: A (square) matrix multiplies a vector **x** to give another vector. If the resulting vector is a scalar multiple of **x**, then **x** is called an *eigenvector* of the matrix. Thus the eigenvectors of the square matrix **A** are the solutions of $\mathbf{A}\mathbf{x} - \lambda \mathbf{x} = 0$, where λ is a scalar to be determined.

(2) *Distributed linear systems in free oscillation*: The displacement of such a system from equilibrium is represented by a continuous function (of position **x** and time t) that can be separated into spatial and temporal parts as $\varphi(\mathbf{x}) e^{j\omega t}$ (see SEPARABLE SOLUTIONS). Of the functions $\varphi(\mathbf{x})$ that satisfy the spatial differential equation, the subset of functions $\varphi_n(\mathbf{x})$ that also satisfy the boundary conditions are called **spatial eigenfunctions**. In acoustics, where the spatial functions are often governed by the Helmholtz equation

$$\nabla^2 \varphi(\mathbf{x}) + \kappa^2 \varphi(\mathbf{x}) = 0,$$

one can say that the functions $\varphi_n(\mathbf{x})$ are eigenfunctions of the ∇^2 operator subject to the given boundary conditions, with EIGENVALUES $-\kappa_n^2$. (The minus sign is often omitted and the κ_n^2 are called eigenvalues.)

eigenfunction – (2) one of the discrete set of functions that are generated as solutions of an EIGENVALUE PROBLEM.

eigenmode – alternative name for one of the NORMAL MODES or EIGENFUNCTIONS of a linear system.

eigenray – *in an acoustic medium whose sound speed varies with position* a RAY that joins two specified points in the medium. It is possible to have **multiple eigenrays**, meaning that two or more ray paths in the same direction connect the same two points.

eigenvalue – (1) *of a square matrix* any scalar λ such that the determinant of $\mathbf{A} - \lambda \mathbf{I}$ vanishes. Here $\mathbf{A}$ is the matrix concerned, and $\mathbf{I}$ is the IDENTITY MATRIX of the same size.

eigenvalue – (2) *in an acoustic system* usually refers to one of the values κ_n^2 ($n = 1, 2, 3, \ldots$) for which the equation describing the acoustic pressure at a specified frequency, e.g. the Helmholtz equation $(\nabla^2 + \kappa^2)p = 0$, has a solution matching the boundary conditions. The resulting solution, in one, two, or three dimensions, is called a **spatial eigenfunction**. See example (2) under EIGENFUNCTION (1).

eigenvalue – (3) one of the discrete set of parameter values for which an EIGENVALUE PROBLEM has a solution meeting all the conditions of the problem.

eigenvalue problem – a mathematical problem with the property that solutions exist only when a parameter of the problem takes certain special values, called **eigenvalues** or **characteristic values**. For example, the equation

$$\frac{d^2 u}{dx^2} - \lambda u = 0,$$ with boundary conditions $u = 0$ at $x = (0, 1)$,

has solutions $u = A \sin n\pi x$ if and only if the parameter $\lambda = -n^2$ (where n is a real integer); no solutions exist unless this condition is met. The corresponding solutions, in this case $\sin n\pi x$, are called **eigenfunctions** of the problem.

eigenvector – *of a square matrix* the eigenvector of a matrix $\mathbf{A}$ is any vector $\mathbf{x}$ that has its direction unaltered when the vector is multiplied by $\mathbf{A}$. In symbols,

$$\mathbf{A}\mathbf{x} = \lambda \mathbf{x} \qquad (\lambda = \text{scalar})$$

is the condition for $\mathbf{x}$ to be an eigenvector of $\mathbf{A}$. The corresponding scalar λ is called an EIGENVALUE of $\mathbf{A}$.

Note: By convention, in numerical work the eigenvectors of a matrix are normalized to have unit length.

eighth-power law, 8th-power law – *in aeroacoustics* see LIGHTHILL'S U^8 POWER LAW.

eikonal – *in geometrical acoustics* a position-dependent function $\tau(\mathbf{x})$ that implicitly defines the shape of WAVEFRONTS, in a medium whose properties are independent of time. Thus propagation of a transient pressure pulse is described by $p(\mathbf{x}, t) = A(\mathbf{x})f[t - \tau(\mathbf{x})]$, where f is an arbitrary function; single-frequency propagation is described by $p = A(\mathbf{x})e^{j\omega(t-\tau)}$. In both cases $A(\mathbf{x})$ is a real function of the position vector $\mathbf{x}$, whose value varies slowly on a wavelength scale.

Physically, $\tau(\mathbf{x}) = T_1$ and $\tau(\mathbf{x}) = T_2$ define two wavefront surfaces in space, separated by a travel time $T_2 - T_1$ that is the same for all ray paths connecting the two surfaces. *Units* s.

Einstein rule, Einstein summation convention – alternative names for the SUMMATION CONVENTION (applied to expressions in cartesian subscript or matrix subscript notation), whereby a repeated subscript implies summation (**A EINSTEIN 1916**).

elastic constants – coefficients that appear in a stress-strain relation for an elastic solid material. Examples are YOUNG'S MODULUS and POISSON'S RATIO for an isotropic material.

elastic deformation – *of a solid material* a deformation such that the material returns to its original shape when the load is removed. The term is sometimes also understood to imply REVERSIBLE (i.e. lossless) deformation.

elastic foil absorber – see MEMBRANE ABSORBER, note (2).

elastic solid – a material that returns to its original configuration after all stresses are removed. The term is sometimes also understood to imply that the material is lossless.

elastic waves – mechanical waves that propagate in elastic solid or a compressible fluid. See also shear waves, compressional waves.

electrical analogies – historically, acousticians have made use of electrical circuit theory as an aid to designing, and analysing, linear acoustic systems that can be adequately represented by LUMPED ELEMENT models. Each acoustic lumped element is assigned an ***electrical analogue***; the electrical analogue elements are then joined in an ***equivalent circuit***, whose behaviour corresponds to that of the acoustical system. Electrical analogies are particularly useful in modelling electromechanical systems, such as microphones and loudspeakers, where different types of element are combined.

Note: In *mobility analogies* force or pressure are represented by current, while in *impedance analogies* they are represented by potential. The table below summarizes the two alternative approaches. Listed in pairs alongside the electrical quantities potential (V) and current (I) are their acoustical and mechanical counterparts for each type of analogy, with symbols F, p, u, U denoting force, pressure, velocity, and volume velocity. As an example, the mechanical quantities that equate to (V, I) are (F, u) in the impedance analogy, and (u, F) in the mobility analogy.

Electrical quantity	Acoustical or mechanical equivalents (impedance)	(mobility)
V	p, F	U, u
I	U, u	p, F

electric response audiometry – measurement of auditory function by means of recorded electrical potentials, evoked by applying acoustic or bone-conduction stimuli to the ear.

electrodynamic shaker – a vibration generator driven by electromagnetic (Lorentz) forces.

electromechanical transducer – a TRANSDUCER that converts a mechanical input signal (e.g. an acceleration or pressure) into an electrical output signal, or vice versa. Examples are an accelerometer, a hydrophone, and a loudspeaker.

emission time coordinates – *for a point acoustic source moving relative to a uniform medium* the observer coordinates measured relative to the position of the source at the time of emission. They are the components of the relative position vector **s**, defined by

$$\mathbf{s} = \mathbf{s}(\mathbf{x}, t) = \mathbf{x} - \mathbf{y}_0(t_e);$$

here **x** is the observer position (assumed to be fixed relative to the medium), t is the time at which the sound is observed, and $\mathbf{y}_0(t_e)$ is the source position at the time the sound was emitted.

end correction – (1) *of an opening at either end of a short uniform pipe or duct* the length of an imaginary extension to either end of the duct, such that the product $\rho S l_{\text{eff}}$, where l_{eff} is the total length including end corrections, gives the correct value for the ACOUSTIC INERTANCE of the duct at low frequencies. Here ρ is the fluid density, and S is the cross-sectional area of the duct. The quantity l_{eff} is called the ***effective length*** of the duct. *Units.* m.

Note (1): The end correction is a lumped-element concept; compare ACOUSTIC RADIATION IMPEDANCE, note (1).

Note (2): The end correction of an opening, and the related acoustic inertance, are determined by the kinetic energy of the local oscillatory flow.

Note (3): The physical reason for the end correction can be seen by considering the limiting case of an aperture in a thin panel. The geometric duct length is small in this case (equal to the panel thickness), and most of the kinetic energy associated with oscillatory flow through the aperture is due to the external fluid motion. The end corrections account for this energy; each is of the same order as the aperture diameter.

end correction – (2) *at the open end of a long uniform pipe or duct* the end reflection process at low frequencies, for plane waves that propagate along a uniform rigid-walled duct and arrive at an open end, is equivalent (at a specified frequency) to extending the duct by an amount δ called the end correction, and imposing a *negative* real reflection coefficient at the new end position. Thus if a standing-wave pressure field is measured in the duct at a given frequency, δ is the axial distance beyond the end of the duct at which the next pressure minimum would appear, if the interior standing-wave pattern were extrapolated. *Units* m.

Note (1): The end correction depends on frequency, but at low frequencies (i.e. wavelengths λ long compared with the perimeter L of the duct cross-section) it tends to a constant value. For a circular pipe of internal radius a and wall thickness h, the end correction as defined above is given by

$$\delta/a = 0.6133 - 0.1168(ka)^2 + O(ka)^4 \quad (h/a \to 0),$$

$$\delta/a = 0.8217 - 0.367(ka)^2 + O(ka)^4 \quad (h/a \to \infty).$$

The second equation corresponds to an open end with a large rigid baffle or flange. It differs from the "classical" estimate ($\delta/a \approx 8/3\pi = 0.849$) because the latter is based on the assumption of a uniform particle velocity over the mouth of the tube, which is not valid here.

Note (2): The presence of an outflow can significantly alter the end correction at low Helmholtz numbers (i.e. $ka \ll 1$), depending on whether the STROUHAL NUMBER $S = \omega a/U = ka/M$ is large or small. Here U is the mean flow velocity down the pipe towards the open end ($U > 0$), and M is the flow Mach number U/c. For example, for low Mach number flow exiting from an unflanged thin-walled pipe,

$$\delta/a\,(M \to 0,\ ka \to 0) \approx 0.6133\,(S \gg 1);\ 0.2554\,(S \ll 1).$$

Note (3): In the low-frequency limit, the equivalent PRESSURE REFLECTION COEFFICIENT at the new end position approaches -1. The reflection process is then equivalent to applying a PRESSURE-RELEASE boundary condition at that position.

endolymph – the fluid that fills the central channel (*scala media*) of the COCHLEA, between the two partitioning membranes that run along its length (basilar membrane and Reissner's membrane). Compare PERI-LYMPH.

energy – (1) *in acoustics* abbreviation for ACOUSTIC ENERGY or VIBRATIONAL ENERGY. *Units* J.

energy – (2) *in physics, especially continuum mechanics* a thermodynamic property of a system, whose existence follows from the FIRST LAW OF THERMODYNAMICS. In fluid dynamics, the energy of a fluid is divided into kinetic energy and internal energy. *Units* J.

energy – (3) *of a transient signal* the integral of the squared signal magnitude with respect to time:

$$E = \int_{-\infty}^{\infty} |x(t)|^2 dt.$$

See also ENERGY AUTOSPECTRAL DENSITY.

energy absorption coefficient – *of a boundary in an acoustic field, at a specified frequency* an alternative term for the SOUND POWER ABSORPTION COEFFICIENT. *Units* none.

energy absorption exponent – *of an acoustic medium at a specified frequency* an alternative term for the ENERGY ATTENUATION COEFFICIENT. *Units* m^{-1}.

energy acoustics – a generic term for methods of acoustical calculation that describe sound fields in terms of their energy or energy density, and in which phase information is lost. Thus energy acoustics takes no acount of INTERFERENCE phenomena. See STATISTICAL ROOM ACOUSTICS, STATISTICAL ENERGY ANALYSIS.

energy attenuation coefficient – *for plane progressive sound waves at a given frequency* the coefficient m in the spatial attenuation factor e^{-mx}, which describes the reduction of intensity (or mean square pressure) with distance, x, in the propagation direction due to energy dissipation in the medium. It is related to α, the attenuation coefficient for acoustic pressure, by $m = 2\alpha$. *Units* m^{-1}.

energy autospectral density – *of a transient continuous signal* the quantity

$$E_{xx}(f) = |X(f)|^2,$$

where $X(f)$ is the FOURIER TRANSFORM of the transient signal $x(t)$ at frequency f. The integral of $E_{xx}(f)$ over all frequencies (positive and negative) gives the ENERGY (3) of the signal, and is assumed to be finite.

If $x(t)$ is a RANDOM SIGNAL, the EXPECTED VALUE $E\{|X(f)|^2\}$ defines the energy autospectral density of the signal.

Note: Some authors use the term *energy autospectral density function* for the quantity defined above. Compare DENSITY FUNCTION.

energy bandwidth – (1) *of a resonant response curve in which the squared gain factor of a linear system is plotted against frequency* the area under the response curve, normalized by the height of the resonance peak. Mathematically, if the squared gain factor of the system is $\Gamma(f)$, the energy bandwidth of the system response is

$$B_e = \frac{1}{\Gamma_{max}} \int_0^\infty \Gamma(f) df.$$

The energy bandwidth of a BANDPASS FILTER is referred to as the *equivalent rectangular bandwidth* or *equivalent noise bandwidth* of the filter. *Units* Hz.

energy bandwidth – (2) *of a narrowband signal* a measure of the frequency bandwidth based on the equation above, with $\Gamma(f)$ interpreted as the spectral density of the signal. *Units* Hz.

energy cross-spectral density – *of two transient signals* the quantity

$$S_{xy}(f) = |X^*(f)Y(f)|^2,$$

where $X(f)$, $Y(f)$ are the FOURIER TRANSFORMS of the transient signals $x(t)$ and $y(t)$, evaluated at frequency f.

Note (1): If $x(t)$ and $y(t)$ are RANDOM SIGNALS, the EXPECTED VALUE $E\{|X^*(f)Y(f)|^2\}$ defines the energy cross-spectral density of the two signals.

Note (2): Some authors use the term *energy cross-spectral density function* for the quantity defined above. Compare DENSITY FUNCTION.

energy decay coefficient – *for free oscillation of a damped linear system in a single mode* the coefficient λ in $e^{-\lambda t}$, the factor that describes the decay of the system energy with time, t. It is related to the AMPLITUDE DECAY COEFFICIENT, δ, by $\lambda = 2\delta$. *Units* s^{-1}.

energy decay time – *for free oscillation of a damped linear system in a single mode* the time taken for the energy of free oscillations to decay by a factor $1/e$. It is the TIME CONSTANT τ in the energy decay factor $e^{-t/\tau}$, where t denotes time. See also DECAY RATE. *Units* s.

Note: The energy decay time is related to the ENERGY DECAY COEFFICIENT, λ, by $\tau = 1/\lambda$.

energy density – *in acoustics* abbreviation for ACOUSTIC ENERGY DENSITY. *Units* J m^{-3}.

energy-density spectrum – *of a transient continuous signal* an alternative term for the ENERGY AUTOSPECTRAL DENSITY of the signal. Other equivalent terms are *energy spectral density* and *energy spectrum*.

energy flux density – (1) *in a three-dimensional wave field* a general term for the power crossing a specified surface per unit area; an equivalent term is *power density*. Specific terms used in particular disciplines are *acoustic intensity* (acoustics) and *radiant flux density* (optics). *Units* W m^{-2}.

energy flux density – (2) *in underwater acoustics* an equivalent term for the TIME-INTEGRATED INTENSITY at a far-field measurement position. When used as a descriptor of the directional radiation from a transient source, the energy flux density is generally normalized to a unit reference distance. *Units* J m^{-2}, or J m^{-2} @ 1 m.

Note: The *band-limited energy flux density* refers to the contribution to the energy flux density from a specified band of frequencies.

energy flux density level – *in underwater acoustics* a decibel measure of the TIME-INTEGRATED INTENSITY at a far-field measurement position. Abbreviated as *EFDL*. *Units* dB re $(1/\rho c)\mu$Pa^2 s.

energy level – *for transient sound signals* the level of the time-integrated pressure squared, or equivalently the level of the ENERGY (3) contained in the transient pressure signal. In equation form, the energy level L_E of a transient acoustic pressure waveform is

$$L_E = 10\log_{10}\frac{E}{E_{\text{ref}}}, \quad \text{with } E = \int p^2(t)\,dt.$$

Units dB re E_{ref}; $E_{\text{ref}} = (20\ \mu\text{Pa})^2$ s (air and other gases), 1 μPa^2 s (water and other media).

Note: The term *energy level* is used in acoustics, particularly in sonar applications, as a convenient alternative to the equivalent *time-integrated sound pressure squared level*. Compare SOUND EXPOSURE LEVEL, which differs by being subjectively weighted for frequency content.

energy reflection coefficient – *at a boundary in an acoustic field* an equivalent term for SOUND POWER REFLECTION COEFFICIENT. *Units* none.

energy source level (ESL) – *in underwater acoustics, for a transient sound source* the level of signal ENERGY (3) measured in a given radiation direction (i.e. the pressure-squared integral with respect to time), scaled to a reference distance of $r_{\text{ref}} = 1$ m. In equation form, the energy source level is related to the transient pressure $p(\vec{r}, t)$ at distance $\vec{r}$ by

$$\text{ESL} = 10 \log_{10} \left[\frac{\vec{r}^2 \int p^2(\vec{r}, t) dt}{r_{\text{ref}}^2 p_{\text{ref}}^2 t_{\text{ref}}} \right].$$

For additional comments, see the entry for SOURCE LEVEL. *Units* dB re 1 μPa^2 s @ 1 m, equivalent to expressing the level of $\vec{r}^2 \int p^2(\vec{r}, t) dt$ in dB re 1 μPa^2 m^2 s.

Note (1): When it is required to express the energy source level in spectral form, the appropriate units are dB re 1 μPa^2 s Hz^{-1} @ 1 m, equivalent to dB re 1 μPa^2 m^2 s Hz^{-1}. No standard nomenclature appears to have been established for this quantity; a suggestion is ***energy source spectrum level***.

Note (2): To convert between ESL and the radiated acoustic energy per unit solid angle in SI units (denoted by E_Ω and measured in J sr^{-1}, or equivalently J m^{-2} @ 1 m), the relevant equation is

$$\text{ESL} = 10 \log_{10} \frac{\rho c E_\Omega}{p_{\text{ref}}^2 r_{\text{ref}}^2 t_{\text{ref}}};$$

here ρ and c are the density and sound speed at the source location.

energy spectral density – (1) *of a transient signal* an equivalent term for ENERGY AUTOSPECTRAL DENSITY.

energy spectral density – (2) *in underwater acoustics, for a transient signal* the ENERGY SPECTRAL DENSITY (1) of the radiated far-field pressure signal at a specified point, denoted by $E_{pp}(f)$ where f is the frequency (positive or negative). Measurements are normally presented in the form of SINGLE-SIDED energy spectral densities, $\bar{E}_{pp}(f)$; the latter is defined for positive frequencies only, and equals $2 E_{pp}(f)$. *Units* Pa^2 s Hz^{-1}.

energy spectral density level (ESD) – *in underwater acoustics, for a transient signal* the level of the far-field pressure ENERGY SPECTRAL DENSITY, for a given frequency and measurement position. In equation form, the energy spectral density level is given by

$$\text{ESD}(f) = 10 \log_{10} \left[\frac{\bar{E}_{pp}(f)}{\bar{E}_{\text{ref}}} \right], \quad (f = \text{frequency}).$$

Here $\bar{E}_{pp}(f)$ is the single-sided energy spectral density, and $\bar{E}_{\text{ref}}$ is the reference value of 1 μPa^2 s Hz^{-1}. *Units* dB re 1 μPa^2 s Hz^{-1}.

energy theorem – alternative term for PARSEVAL'S THEOREM.

energy–time curve – *for the characterization of linear time-invariant systems* the result of processing the filtered IMPULSE RESPONSE FUNCTION of a system so as to yield the INSTANTANEOUS AMPLITUDE (or envelope) response as a function of time, with phase information suppressed. An equivalent term is ***time delay response***. Mathematically, the energy–time curve (***ETC***) is a plot of $|y(t) + jy_H(t)|^2$, where $y(t)$ is the filtered impulse response at time t and $y_H(t)$ is its Hilbert transform. Note that $y(t) + jy_H(t)$ is the ANALYTIC SIGNAL based on the real signal $y(t)$.

Note: The ETC does not involve averaging or integrating $[y(t)]^2$ over time, as in a SCHROEDER PLOT. To the extent that the filter bandwidth allows, it is therefore able to resolve fine detail in the time domain. This capability can be useful in identifying reflections in acoustic systems, for example. See also TIME DELAY SPECTROMETRY, which is a related time and frequency domain technique.

energy transmission coefficient – *at a boundary in an acoustic field* an equivalent term for SOUND POWER TRANSMISSION COEFFICIENT. *Units* none.

energy velocity – (1) *for plane progressive waves in a free field* the ratio of the acoustic intensity vector, **I**, to the acoustic energy density, w. For sound waves in a lossless stationary medium, the energy velocity equals the sound speed. *Units* m s^{-1}.

energy velocity – (2) *for a progressive-wave field in a uniform waveguide* the ratio of the sound power transmitted along the waveguide to the mean acoustic energy, per unit length, contained in the waveguide. Compare GROUP VELOCITY. For periodic waveguides, see FOSTER'S REACTANCE THEOREM (2). *Units* m s^{-1}.

engine order – (1) a frequency that is an integer multiple of the shaft rotational speed in an engine; (2) sound or vibration occurring at such a frequency.

ensemble – (1) the degree to which an orchestra plays well together, starting and finishing notes simultaneously when required. Good ensemble depends on the musicians being able to hear each other clearly, though performing experience and good conducting obviously also contribute. See also BALANCE and BLEND.

ensemble – (2) a set of signals, usually all of the same type and all having the same record length, assembled for statistical analysis.

ensemble average – *in signal processing* an average taken over an ENSEMBLE. In the case of a STOCHASTIC PROCESS, an ensemble average is an average over all possible SAMPLE FUNCTIONS.

ensonify – to irradiate a target, or energize an acoustic medium, with fluid-borne sound. Hence *ensonification*, the process of irradiating a target or medium with sound. Also spelt *insonify*.

enthalpy – a thermodynamic property of a system, useful for describing energy flows across open boundaries (see note below). The enthalpy of mass m of fluid, with volume $\mathcal{V}$ and uniform pressure P, is $H = E + P\mathcal{V}$ where E is the fluid's INTERNAL ENERGY. *Units* J.

Note (1): The *specific enthalpy* (enthalpy per unit mass of fluid) is more often used in fluid dynamics and acoustics. In particular, the rate of energy flow across a fixed control surface S, in an inviscid compressible fluid flow, is $\int (h + \frac{1}{2}u^2)\rho \mathbf{u} \cdot \mathbf{n}\, dS$ evaluated over S. Here $\mathbf{u}$ is the fluid velocity, with magnitude u; h is the specific enthalpy; ρ is the fluid density; and $\mathbf{n}$ is the unit vector normal to S.

Note (2): Small changes in pressure, P, and absolute temperature, T, produce a change in specific enthalpy given by the differential relation

$$dh = \frac{1 - \alpha T}{\rho} dP + C_p dT;$$

here α is the thermal expansivity of the fluid, ρ is the density, and C_p is the constant-pressure specific heat.

entrained mass – *of an accelerating rigid body in an incompressible fluid* an equivalent term for ATTACHED MASS or VIRTUAL MASS. *Units* kg.

entropy – a thermodynamic property of a system, whose existence follows from the second law of thermodynamics. Under equilibrium conditions, the entropy (S) of a system can change only through heat transfer to or from the system:

$$dS = (1/T)\, dQ.$$

Here dQ is an incremental energy transfer to the system in the form of HEAT, and T is the ABSOLUTE TEMPERATURE of the system. *Units* J K^{-1}.

Note: The *specific entropy* (entropy per unit mass of fluid) is more often used in fluid dynamics and acoustics. In particular, the specific entropy s, in unsteady compressible flow of an ideal fluid, is constant following a fluid particle; see EULER EQUATIONS.

envelopment – *in concert-hall acoustics* an abbreviation for LISTENER ENVELOPMENT.

environmental acoustics – the science of outdoor sound, its sources, and its propagation in both natural and urban environments; also the application of such knowledge to the control of noise pollution. Hence **environmental acoustician**, an expert or professional adviser in the field of ~.

eolian tones – alternative spelling of AEOLIAN TONES.

epidemiology – (1) scientific study of the occurrence, distribution, and control of disease (including industrial and environmental health problems), in a given population. This involves investigating, through statistical data, both the INCIDENCE of a disease – such as noise-induced hearing loss among men in a certain age range – and also its PREVALENCE at any given time.

epidemiology – (2) the factors that control the presence or absence of a specific disease (e.g. *epidemiology of noise-induced hearing loss*). Compare AETIOLOGY.

EPNL – abbreviation for EFFECTIVE PERCEIVED NOISE LEVEL. *Units* dB re $(20 \,\mu\text{Pa})^2 \cdot 10$ s.

equal-loudness contours – *for pure tones* contours of equal judged loudness, presented on a graph of sound pressure level against frequency. The sound may be presented to the observer either via earphones, or in a free field (in which case frontally incident sound is used). The sound pressure level on the vertical axis is that at the eardrum in the first case, and that in the incident field (as measured midway between the ears) in the second.

equilibrium sound speed – the sound speed calculated on the assumption that thermodynamic equilibrium is maintained during the acoustic motion. In a relaxing medium, the equilibrium sound speed is the speed at which sound propagates in the low-frequency limit (i.e. $\omega\tau \ll 1$, where τ is the RELAXATION TIME and ω is the angular frequency). *Units* m s^{-1}.

equivalent absorption area – *of a sound-absorbing surface or object placed in a diffuse field* a measure of the equivalent totally-absorbing surface area. It is defined as the ratio of the sound power absorbed by the surface or object to the irradiation density, or normal incident intensity. In mathematical terms, the equivalent absorption area A is given by

$$A = \frac{W_{\text{abs}}}{cw/4},$$

where W_{abs} is the sound power absorbed, c is the sound speed and w is the mean ENERGY DENSITY of the diffuse sound field. For a definition of **equivalent absorption area of a room**, see ROOM ABSORPTION. *Units* m^2.

Note: The equivalent absorption area of a test sample of sound-absorbing material may be estimated by introducing the sample into a reverberant room and measuring the increase in room absorption; see also EYRING

ABSORPTION COEFFICIENT, SABINE ABSORPTION COEFFICIENT. However, this type of measurement does not uniquely characterize the sample, since a truly diffuse incident field is not attainable in practice; the result depends on the form of the sound field existing in the room before the change.

equivalent bandwidth – *of a resonant response curve* abbreviation for ***equivalent noise bandwidth*** or ***equivalent rectangular bandwidth***. For a definition, see ENERGY BANDWIDTH (1). *Units* Hz.

equivalent circuit – *for an acoustic or electroacoustic system* see CIRCUIT ANALOGUE, ELECTRICAL ANALOGIES.

equivalent continuous A-weighted sound pressure level – *for airborne sounds that are non-stationary with respect to time* an EQUIVALENT CONTINUOUS SOUND PRESSURE LEVEL formed by applying A-WEIGHTING to the original signal before squaring and averaging. Also known as ***equivalent continuous sound level***. Symbol $L_{\text{Aeq},T}$. *Units* dB re $(20\ \mu\text{Pa})^2$.

equivalent continuous sound level – *for airborne sounds that are non-stationary with respect to time* an alternative term for EQUIVALENT CONTINUOUS A-WEIGHTED SOUND PRESSURE LEVEL. *Units* dB re $(20\ \mu\text{Pa})^2$.

equivalent continuous sound pressure level – *of a non-stationary sound pressure signal defined between specified time limits* the sound pressure level of a notional unvarying sound that, for the same specified duration T, has the same signal energy as the original signal. The resulting mean square pressure is expressed mathematically by

$$\langle p^2 \rangle_T = \frac{1}{T} \int_{t_0}^{t_0+T} p^2(t)\,dt \qquad \text{(for the time interval from } t_0 \text{ to } t_0 + T\text{),}$$

where $p(t)$ is the instantaneous sound pressure. An alternative term is ***time-average sound pressure level***. Symbol $L_{\text{eq},T}$. *Units* dB re p_{ref}^2.

Note: If $L_\tau(t)$ is the sound pressure level based on a short-duration running average of $p^2(t)$, with the averaging time constant τ much less than T, then the equation

$$L_{\text{eq},T} \approx 10 \log_{10} \left\{ \frac{1}{T} \int_{t_0}^{t_0+T} 10^{L_\tau(t)/10}\,dt \right\}$$

gives a close practical approximation to the equivalent continuous sound pressure level.

equivalent noise bandwidth, equivalent rectangular bandwidth – *of a resonant response curve* alternative terms for ENERGY BANDWIDTH (1). *Units* Hz.

equivalent viscous damping coefficient – *of a linear single-degree-of-freedom system whose driving-point resistance varies with frequency* the value of the driving-point mechanical resistance, $R_{\text{mech}}(f_{\text{res}})$, at the resonance frequency of the system. The resonance frequency f_{res} is defined by the reactive part of the driving-point impedance becoming zero: $X_{\text{mech}}(f_{\text{res}}) = 0$. *Units* N s m^{-1}.

Note: For modelling purposes, it can be convenient to represent a real system by an idealized system with VISCOUS DAMPING. If the idealized system is chosen to have a viscous damping coefficient $R_{\text{mech}}(f_{\text{res}})$, then under a given sinusoidal force input the two systems will dissipate the same energy per cycle at resonance.

equivoluminal waves – *in a continuous medium* an equivalent term for SHEAR WAVES. Also known as ROTATIONAL WAVES.

ergodic – a term used to describe a stationary time-dependent STOCHASTIC PROCESS for which time averaging is equivalent to ensemble averaging.

ergodicity – *of a stochastic process* implies that all statistical properties of a time-dependent STOCHASTIC PROCESS $X(t)$ can be discovered from a single sample, or realization, $x(t)$, provided the sample duration is sufficiently long. Ergodicity is possible only if the stochastic process is stationary with respect to time (see STATIONARITY). On the other hand, a stationary process need not be ergodic.

error – if a quantity x is approximated by $\hat{x}$, the error associated with the approximation is $\delta = \hat{x} - x$.

ESD – *in underwater acoustics* abbreviation for the ENERGY SPECTRAL DENSITY LEVEL of a transient pressure signal. *Units* dB re 1 μPa^2 s Hz^{-1}.

ESL – *in underwater acoustics, for a transient sound source* abbreviation for ENERGY SOURCE LEVEL. *Units* dB re 1 μPa^2 s @ 1 m, or dB re 1 μPa^2 m^2 s.

essential boundary conditions – *for a second-order differential equation* conditions that are imposed on the value of the solution at points on the boundary. Also known as DIRICHLET BOUNDARY CONDITIONS. Other names sometimes used are **kinematic**, **displacement**, or **geometric** boundary conditions.

Note: For a higher-order differential equation of order $2m$ (where m is an integer), essential boundary conditions are those that involve the solution and its derivatives up to order $(m - 1)$. Compare NATURAL BOUNDARY CONDITIONS.

ESWL – abbreviation for *extracorporeal shock-wave lithotripsy*. See LITHOTRIPSY.

Euler equations – the IDEAL-FLUID equations

$$Ds/Dt = 0, \quad \rho D\mathbf{u}/Dt = -\nabla P$$

that (along with the CONTINUITY EQUATION) describe the motion of a lossless compressible fluid (**L EULER 1741**). Here s is the specific ENTROPY, $\mathbf{u}$ is the fluid velocity, and D/Dt denotes the MATERIAL DERIVATIVE.

Eulerian coordinates – coordinates defined by a fixed spatial frame of reference. Compare LAGRANGIAN COORDINATES.

Euler–Lagrange equation – see CALCULUS OF VARIATIONS.

Euler's constant – the limit, as n tends to infinity, of

$$1 + \frac{1}{2} + \frac{1}{3} + \cdots + \frac{1}{n} - \ln n.$$

Its value is $0.577\,215\,66\ldots$. An alternative name is the ***Euler–Mascheroni constant***.

Euler's formula – the equation

$$e^{jz} = \cos z + j \sin z,$$

where z is any number, real or complex. Also known as ***Euler's identity***.

Eustachian tube – see MIDDLE EAR.

evanescent – spatially decaying; used, for example, to describe AXIALLY PROGRESSIVE WAVES whose propagation wavenumber is imaginary.

evanescent mode – *in underwater acoustics* a mode of propagation, in a stratified model of the ocean with non-absorbing boundaries, whose radial eigenvalue or horizontal propagation wavenumber is purely imaginary. Compare CUTOFF, VIRTUAL MODE.

evanescent wave – a wave (usually at a single frequency) that decays exponentially in a specified coordinate direction, without change of phase. An alternative term for an evanescent wave is ***inhomogeneous wave***.

even function – a function whose value is unaltered by changing the sign of the argument. Thus if $f(z)$ is an even function, $f(-z) = f(z)$; note that the argument z may be real or complex. An example of an even function is $\cos z$.

exceedance level – a term occasionally used to denote an L_N PERCENTILE LEVEL (meaning a sound pressure level that is exceeded for N percent of the time at a given location). *Units* dB re p_{ref}^2.

excess attenuation – *in outdoor or underwater sound propagation* the part of the total attenuation, over a given transmission path, that is not accounted for by energy spreading (i.e. the transmission loss minus the DIVERGENCE LOSS). The definition may be further narrowed by excluding other stated contributions to the overall attenuation. *Units* dB.

excess attenuation coefficient – *at a given frequency in a composite medium* the amount by which the actual attenuation coefficient of plane acoustic waves in, for example, an emulsion or suspension exceeds the attenuation coefficient due to INTRINSIC ABSORPTION. The mechanisms of excess attenuation include scattering of sound by particles or droplets into thermal and viscous modes; see THERMAL ABSORPTION. *Units* Np m^{-1}.

excitation – an external input (e.g. an oscillatory force or displacement) applied to a system. An alternative term is INPUT; see also RESPONSE.

excursion – *at a point on a vibrating system* the maximum displacement of the point from its equilibrium position. *Units* m.

Note: The excursion either side of equilibrium may not be the same (e.g. for non-sinusoidal time dependence); it then becomes necessary to specify the direction of the excursion in order to avoid ambiguity.

expansivity – *for a fluid* abbreviation for VOLUME THERMAL EXPANSIVITY. *Units* K^{-1}.

expectation operator – an operator, denoted by the symbol $E\{\ldots\}$, that acts on a RANDOM VARIABLE to give the MEAN. It is defined mathematically by

$$E\{X\} = \int_{-\infty}^{\infty} xf(x)dx$$

where X is the random variable and $f(x)$ its PROBABILITY DENSITY FUNCTION. The quantity $E\{X\}$ is called the ***expected value*** of the random variable X.

Note: For an ergodic RANDOM SIGNAL that is stationary with respect to time, the expected value is equivalent to the long-duration time average: see AVERAGE.

expected value – *of a random variable* the result of applying the EXPECTATION OPERATOR to the variable in question.

exponential decay – a decay curve with respect to time that has the form $e^{-t/\tau}$; alternatively a decay curve with respect to a position coordinate, x, that has the form e^{-mx}. Here t is time and τ, m are constants.

exponential horn

exponential horn – a HORN whose cross-sectional area is proportional to e^{mx}, where x is the axial distance measured from the input end and m is the FLARE CONSTANT.

extended reaction – a property of a boundary, signifying that it responds in a non-local manner to an imposed unsteady surface pressure distribution. A boundary of this type is called a *surface of extended reaction*.

Note: The response of a boundary to surface pressures is non-local whenever surface waves can propagate along the boundary. Compare LOCAL REACTION.

extension – *of a spring or one-dimensional extensible element* the increase in length, relative to a reference state (usually unstressed). According to HOOKE'S LAW, the extension is linearly related to the applied load. *Units* m.

external auditory meatus – alternative medical term for the EAR CANAL.

external ear – the first part of the auditory transmission chain; it comprises the PINNA and EAR CANAL and terminates at the eardrum.

external ear canal – an alternative term for EAR CANAL.

external force – *on a system* a force applied to the system by its surroundings, or equivalently, a rate of momentum transfer to the system from outside. *Units* N.

exteroreceptor – a sensory receptor, or biological transducer, located on the external surface of the body that permits detection and processing of stimuli from the environment (sound, touch, smell, taste, heat, light).

extinction cross-section – the sum of the ABSORPTION CROSS-SECTION and the SCATTERING CROSS-SECTION. It represents the power lost from an incident beam by a scattering object or inhomogeneity, per unit incident intensity. *Units* m^2.

extremum – a maximum or minimum value.

Eyring absorption coefficient – *of a sample of sound-absorbing material* the STATISTICAL ABSORPTION COEFFICIENT deduced from reverberation-time measurements via the EYRING EQUATION (see REVERBERATION TIME EQUATIONS). Specifically, the Eyring absorption coefficient of a test sample of surface area S_{test} is

$$\alpha_s = (S/S_{\text{test}}) \, (e^{-A_0/S} - e^{-A_1/S}) + (1 - e^{-A_0/S}),$$

where A_0 is the ROOM ABSORPTION of the test room without the sample, A_1 is the room absorption measured with the sample installed, and S is the total area of the room boundaries. In the limit $A_1/S \ll 1$, the Eyring and Sabine absorption coefficients are the same. *Units* none.

Note: In the equation above, the sample is assumed to cover up a fraction (S_{test}/S) of the absorption originally present in the test room.

Eyring average absorption coefficient – *for a room* the absorption coefficient averaged over the room surfaces, $\bar{\alpha}$, deduced from the Eyring equation $A = -S\ln(1 - \bar{\alpha})$. Here A is the total ROOM ABSORPTION and S is the room surface area. Thus $\bar{\alpha} = 1 - e^{-A/S}$. *Units* none.

Eyring equation for reverberation time – see REVERBERATION TIME EQUATIONS.

F

face velocity – *of the flow through a layer of porous material* the volume flowrate per unit surface area at the face of the layer. See also NORMAL SPECIFIC ACOUSTIC IMPEDANCE. *Units* m s^{-1}.

far field – *for free-field radiation into a lossless fluid* the region sufficiently far from a specified source that:

(1) the sound pressure varies as $(1/r)f(t - r/c)$ with distance, r, measured from the centre of the source, and

(2) the instantaneous radial particle velocity approaches $1/\rho c$ times the sound pressure.

Here f denotes any function, c is the speed of sound, and ρc is the CHARACTERISTIC IMPEDANCE of the fluid. See also SPHERICAL PROGRESSIVE WAVE, SOMMERFELD RADIATION CONDITION.

far-field criteria – rules for determining where the far field of a source begins. Further information may be found under GEOMETRIC NEAR FIELD, HYDRODYNAMIC NEAR FIELD, FRESNEL NEAR FIELD, FRAUNHOFER REGION.

fast field program – *for point sources radiating into a horizontally stratified medium* an efficient algorithm for calculating the acoustic far field of a source, in a medium whose sound speed and density vary in one coordinate direction only (e.g. depth, in ocean acoustics). It uses a fast Fourier transform (FFT) to compute the required Hankel transform asymptotically, in order to obtain the pressure field from the DEPTH-DEPENDENT GREEN'S FUNCTION.

Note: If the depth-dependent Green's function is denoted by $g(k_r; z, z_0)$, where k_r is horizontal wavenumber and (z, z_0) are the receiver and source depths, the pressure field at range r due to a unit-strength point source is given by

$$\tilde{p}(r; z, z_0) = \int_0^\infty g(k_r; z, z_0) J_0(k_r r) k_r \, dk_r.$$

By writing the Bessel function as $J_0 = \frac{1}{2}[H_0^{(1)} + H_0^{(2)}]$, neglecting $H_0^{(1)}$ (for time factor $e^{j\omega t}$) on the basis that it corresponds to incoming waves, and using the asymptotic expansion of $H_0^{(2)}$ for large arguments, the acoustic pressure at ranges of more than a few wavelengths may be approximated as follows:

$$\tilde{p}(r; z, z_0) \approx e^{j\pi/4} \sqrt{\frac{1}{2\pi r}} \int_0^\infty g(k_r; z, z_0) \sqrt{k_r}\, e^{-jk_r r} \, dk_r.$$

With due attention to problems of sampling, aliasing, and the poles of g on the real k_r axis, a discrete version of this integral can be efficiently calculated by means of an FFT.

fast Fourier transform – an implementation of the DISCRETE FOURIER TRANSFORM that offers computational advantages over a direct evaluation, particularly when the *order* (number N of points per cycle) is large. The algorithm is based on successive factorization; in its most efficient form, N is chosen to be 2^v (v = integer), and the original order-N DFT is reduced to evaluating 2^{v-1} DFTs of order 2. The computation time varies as $N \log_2 N$, compared with N^2 for direct evaluation of the transform.

fast weighting – see SOUND LEVEL.

fatigue – (1) *in materials* the failure of crystalline solids under repeated stress reversals.

fatigue – (2) *in psychoacoustics* see AUDITORY FATIGUE.

feedback – (1) *in a sound reinforcement system* instability that occurs when the output from a microphone is amplified and reproduced by a nearby loudspeaker, which then transmits the signal back to the microphone acoustically with an overall gain greater than 1. The system "rings" at the frequency where the gain is highest and the feedback is in phase. Equivalent terms are *acoustic feedback*, *howlback*, and *howlround*.

feedback – (2) the use of information or signals derived from the output of a system to modify the system input. In *feedback control*, feedback is used to drive the system output closer to some desired value.

feedforward – the use of advance information about the input to a system to modify the system output. If part of the input consists of unwanted disturbances that can be independently detected some time in advance, their contribution to the output can then be reduced, or even cancelled; this is the basis of *feedforward control*.

Fermat's principle – *in geometrical acoustics* the statement that the RAY PATHS joining two given points are those for which the phase difference is stationary with respect to infinitesimal path variations. The principle is sometimes referred to as *Fermat's principle of least time* (P DE FERMAT 1657). There is a close connection with the method of STATIONARY PHASE.

Ffowcs Williams–Hawkings equation – (1) *for sound radiation from an unsteady fluid flow with boundaries* a generalization to bounded domains of the Lighthill ACOUSTIC ANALOGY. A restricted domain V is first defined, such that the equations of fluid motion are valid everywhere in V. The Lighthill inhomogeneous wave equation for the density in this region may be written as

$$\left(\frac{1}{c_0^2}\frac{\partial^2}{\partial t^2} - \nabla^2\right)\psi = \frac{\partial^2 T_{ij}}{\partial x_i \partial x_j} \quad \text{(within region } V \text{ bounded by } S\text{)};$$

here ρ, c denote density and sound speed, $\psi = c_0^2(\rho - \rho_0)$, T_{ij} is the LIGHTHILL STRESS TENSOR, and subscript 0 denotes undisturbed-fluid properties. The equation above can break down in the excluded region $\bar{V}$, for example when $\bar{V}$ contains solid bodies. However, the Ffowcs Williams–Hawkings (**FW–H**) equation for the generalized variable $H\psi$ is valid throughout all space; it states that

$$\left(\frac{1}{c_0^2}\frac{\partial^2}{\partial t^2} - \nabla^2\right)(H\psi) = \frac{\partial^2(HT_{ij})}{\partial x_i \partial x_j}$$
$$- \frac{\partial}{\partial x_i}([\rho u_i(u_n - v) + \ell_i]\delta)$$
$$+ \frac{\partial}{\partial t}([\rho(u_n - v) + \rho_0 v]\delta).$$

In the FW–H equation the last two terms on the right are surface dipole and source distributions on the boundary S separating $\bar{V}$ from V. Symbol H is the Heaviside unit step function $H(n)$, where n is a local coordinate normal to the boundary surface S; n is zero on S, and increases outwards from $\bar{V}$. Thus $H = 1$ within V, but $H = 0$ in $\bar{V}$. Symbol δ is the Dirac delta function $\delta(n)$, equal to the derivative $H'(n)$.

The boundary surface is allowed to move: it advances into V at speed v. Other symbols are ℓ_i for the force per unit area in the i direction exerted by S on the fluid, due to the difference in stress on the two sides; and u_n for the fluid velocity in the n direction, adjacent to the boundary (at $n = 0+$).

Note (1): The Ffowcs Williams–Hawkings equation allows the contents of any subregion $\bar{V}$ to be represented by undisturbed fluid, together with a distribution of sources and dipoles on the interface S that separates $\bar{V}$ from V. Inside the subregion $\bar{V}$, all variables take their undisturbed values and the acoustic field is by definition zero. The force per unit area ℓ_i can then be replaced by

$$\ell_i = (P_{ij} - P_0 \delta_{ij})\hat{n}_j,$$

where P_{ij} is the compressive stress tensor, P_0 is the undisturbed pressure, and $\hat{n}_j$ are components of the unit vector normal to S (pointing into the acoustic region V).

Note (2): It follows from the note above that any homogeneous boundary condition may be imposed on surfaces inside $\bar{V}$, without affecting the solution. However, it is often convenient to regard $\bar{V}$ as boundary-free, and to use the free-field Green's function to solve for ψ in the acoustic region V.

Ffowcs Williams–Hawkings equation – (2) *for sound radiation from a moving rigid body* a variant of the general FW–H equation (described in the previous entry), in which the boundary surface S has a fixed shape and volume: thus S can represent the surface of a rigid body. The final surface term in the general equation is rewritten as

$$\frac{\partial}{\partial t}(\rho_0 v \delta) = -\frac{\partial}{\partial x_i}(\rho_0 \bar{a}_i [1 - H]) + \frac{\partial^2}{\partial x_i \partial x_j}(\rho_0 \bar{u}_i \bar{u}_j [1 - H]).$$

Here $\bar{u}_i$ are the components of a rigid-body velocity field within $\bar{V}$ that corresponds to the motion of S, and $\bar{a}_i$ are the corresponding accelerations. The product $\rho_0 \bar{a}_i$ can be interpreted as the applied force distribution that would be required to accelerate the contents of $\bar{V}$, if the region were filled with fluid of density ρ_0 moving with the rigid-body velocity $\bar{u}_i$.

FFT – abbreviation for FAST FOURIER TRANSFORM.

field – (1) abbreviation for SOUND FIELD.

field – (2) a generic term for the entire spatial distribution, at any instant over a specified region, of some quantity (such as acoustic pressure). A basic assumption is that the quantity concerned is a single-valued function of position.

The quantity being described – in this example the acoustic pressure – is called the *field variable*. For example, the velocity of a fluid flow is described by the vector velocity field $\mathbf{u}(\mathbf{x}, t)$; the time argument, t, is included as a reminder that most fields in acoustics are unsteady.

Note: The field variable may be either a scalar or a vector quantity. The names *scalar field* and *vector field* are used to denote what type of variable is being described.

field incidence – *in room acoustics* an empirical modification of RANDOM INCIDENCE, in which the distribution of incident wave angles is cut off at a certain angle to the normal (often taken to be 78°, i.e. $\cos^{-1} 0.2$); near-grazing wave angles are therefore absent. The resulting distribution represents field observations in room acoustics more closely than the DIFFUSE incident field model. Specifically, it brings the theoretical equation for predicting the transmission loss of a partition into closer agreement with measurements made under reverberant conditions. See FIELD INCIDENCE TRANSMISSION COEFFICIENT.

field incidence absorption coefficient – *of a sound-absorbing surface* the directionally-averaged absorption coefficient given by

$$\alpha_{\text{field}} = \frac{2}{1 - \cos^2 \theta_{\max}} \int_0^{\theta_{\max}} \alpha(\theta) \sin\theta \cos\theta \, d\theta.$$

Here $\alpha(\theta)$ is the absorption coefficient for plane sound waves incident at angle θ, and θ_{max} is the cutoff angle for FIELD INCIDENCE (i.e. approximately 78°). *Units* none.

field incidence mass law – *for a partition* a TRANSMISSION LOSS prediction equation based on the FIELD INCIDENCE model for the incident sound field, combined with a limp wall model for the partition. The transmission loss predicted by the field incidence mass law is

$$R_{field} = R_0 - 5 \text{ dB},$$

where R_0 is the transmission loss predicted by the MASS LAW for an incidence angle of 0°. *Units* dB.

field incidence transmission coefficient – *of a partition* the directionally-averaged sound power transmission coefficient obtained under FIELD INCIDENCE conditions. The same integral expression applies as for the FIELD INCIDENCE ABSORPTION COEFFICIENT, but with the absorption coefficient $\alpha(\theta)$ replaced by the transmission coefficient $\tau(\theta)$:

$$\tau_{field} = \frac{2}{1 - \cos^2 \theta_{max}} \int_0^{\theta_{max}} \tau(\theta) \sin \theta \cos \theta \, d\theta.$$

Units none.

field variable – any quantity, vector or scalar, that is expressed as a FIELD (2).

filter – a linear device for producing a frequency-selective INSERTION LOSS, in electrical or acoustical transmission lines; hence any linear device or process that modifies the frequency spectrum of a signal. See also ACOUSTIC FILTER.

filter bandwidth – the frequency bandwidth within which a BANDPASS FILTER has near-zero INSERTION LOSS. The filter bandwidth may be quoted as an ENERGY BANDWIDTH (1), or a HALF-POWER BANDWIDTH (1), or a ***nominal bandwidth***. This last measure is defined as the difference between the upper and lower nominal CUTOFF FREQUENCIES (i.e. the ***band-edge frequencies***).

Units The numerical bandwidth may be expressed in any of three ways: (a) in Hz; (b) as a percentage of the passband CENTRE FREQUENCY; (c) as a logarithmic interval, in octaves or decades, between the upper and lower band-edge frequencies.

finite-amplitude wave – a wave whose amplitude is large enough to cause departures from linear propagation theory. Acoustic waves can exhibit finite-amplitude behaviour both locally and cumulatively: LOCAL NONLINEARITY occurs when the acoustic Mach number $u_{max}/c = \varepsilon$ is not small, while CUMULATIVE NONLINEARITY occurs when the Gol'dberg number

$\beta \varepsilon k_0/\alpha$ is not small. Here k_0/α is the ratio of the acoustic wavenumber to the linear attenuation coefficient (at a typical frequency), and β is the COEFFICIENT OF NONLINEARITY for the fluid.

finite difference method – an approach to solving differential equations whereby variables are located at discrete points (in space and time) on a computational mesh. Finite difference methods are distinguished from FINITE VOLUME METHODS by being based on the differential, rather than integral, form of the equations; partial derivatives of the flow variables in space and time are approximated by algebraic functions of neighbouring mesh values.

finite element method – a numerical algorithm for obtaining approximate solutions to variational problems. The domain of interest is divided into subdomains referred to as *elements*. The spatial dependence of the field variables within each element is approximated by an algebraic function of position, with the unknown nodal values as parameters, and substituted into the variational principle. This leads to a system of algebraic equations that can be solved to obtain the unknown nodal values. Functions used to approximate the geometry of each element are known as *shape functions*, and if the shape functions are identical to the functions used to interpolate the field variables, the elements are referred to as *isoparametric elements*.

Note: The finite element method can be viewed as an extended application of the RAYLEIGH–RITZ METHOD. Thus it introduces interpolation functions that satisfy the essential boundary conditions exactly, and then finds solutions for the unknown nodal values such that the governing differential equations and the natural boundary conditions are satisfied approximately. Compare BOUNDARY ELEMENT METHOD.

finite impulse response – having an IMPULSE RESPONSE FUNCTION that is zero beyond some finite time value. In particular, the term is used to describe a DIGITAL FILTER whose output is determined by a finite number of earlier values of the input. The filter's IMPULSE RESPONSE FUNCTION therefore terminates after a finite number of terms.

finite volume method – *in computational fluid dynamics* the governing equations of fluid dynamics (i.e. the compressible Navier–Stokes equations) can be discretized into algebraic equations in different ways. Discretization of the partial differential form of the equations leads to the FINITE DIFFERENCE METHOD (FDM), while discretization of the integral form of the equations leads to the finite volume method (FVM). The governing equations can also be discretized using the FINITE ELEMENT METHOD (FEM).

The finite volume method (also known as the *control volume method*) results from considering fluxes of mass, momentum, energy, and entropy into and out of a finite-sized fixed region of space – called a ***control cell*** or control volume – and applying the appropriate conservation principle. In practice some FVMs and FDMs are very similar.

FIR – abbreviation for FINITE IMPULSE RESPONSE.

first harmonic, second harmonic, etc. – see HARMONIC (1).

first law of thermodynamics – states that when a closed system is altered adiabatically, the total work associated with the change of state is the same for all possible processes between the two given equilibrium states; i.e. the work done is the same for all adiabatic processes that connect two equilibrium states of a closed system.

Note: A consequence of the first law is that changes in the total energy, E_{tot}, of a system can be defined and measured:

$\Delta E_{tot} = W$ (W = work done on adiabatic system).

The existence of INTERNAL ENERGY as a state variable, and the quantification of HEAT through the generalized relation

$\Delta E_{tot} = Q + W$ (for a non-adiabatic system)

are corollaries. Here Q represents an energy transfer to the system in the form of heat.

first sound – see SUPERFLUIDITY.

flanking transmission – *between two enclosures separated by a partition* the transmission of acoustic energy via paths (either structural or acoustical) that bypass the partition.

flare constant – the coefficient m in the equation $A \propto e^{mx}$, that describes the variation of cross-sectional area A with axial distance x in an EXPONENTIAL HORN. *Units* m^{-1}.

flexural wave – an equivalent term for BENDING WAVE.

flow–acoustic interaction – a generic term for all the phenomena that occur when sound waves interact with a mean flow field; it includes scattering and refraction of sound by the mean flow, and destabilization of the flow by sound.

flow noise – (1) an alternative term for AERODYNAMIC SOUND.

flow noise – (2) a non-propagating (or hydrodynamic) near-field pressure associated with turbulent flow, particularly over a solid boundary; an example is **microphone flow noise** produced by wind flow over the microphone housing. Flow noise in this near-field sense is a HYDRODYNAMIC NEAR FIELD; the corresponding far-field radiation is called AERODYNAMIC SOUND.

flow resistance – *for steady flow through a porous element* the ratio of the pressure drop across the element to the volume velocity flowing through it, under conditions of steady low-speed flow. Also known as *static flow resistance*. Compare SPECIFIC FLOW RESISTANCE, DYNAMIC FLOW RESISTANCE. *Units* Pa s m^{-3}.

Note: It is assumed that the flow resistance is independent of the volume velocity at low speeds. Compare FORCHHEIMER'S LAW.

flow resistivity – *for steady flow through bulk porous material* an intrinsic property of the porous material, defined as the unit-thickness SPECIFIC FLOW RESISTANCE; it is given by

$$\sigma = r/h,$$

where r is the specific flow resistance of a uniform layer of thickness h. Equivalently, the flow resistivity is the factor σ in the local relation

$$\sigma \bar{u}_x = -\frac{\partial P}{\partial x},$$

where $\partial P/\partial x$ is the mean-flow pressure gradient in the x direction, and $\bar{u}_x$ is the volume velocity per unit total cross-sectional area in that direction. In principle, σ can be different for different directions in the material. Compare DARCY'S LAW, DYNAMIC FLOW RESISTIVITY. *Units* Pa s m^{-2}.

Note: It is assumed that the flow resistivity is independent of $\bar{u}$ at low speeds. Compare FORCHHEIMER'S LAW.

fluid – a continuous material that cannot support a shear stress when at rest. In a normal fluid, individual molecules move independently (rather than being bonded into permanent clusters), and every element is therefore capable of unlimited deformation without fracture. Both gases and liquids belong to the general category of fluids.

fluid dynamics – the science of fluids in motion.

fluid loading – (1) *in acoustics* the unsteady surface pressure produced on a solid boundary, when the boundary vibrates in a fluid. For small-amplitude surface vibration the relation between fluid pressure and surface normal velocity is linear, and the local response may be expressed in the frequency domain as a *specific fluid loading impedance* (pressure divided by surface velocity). However, this quantity depends critically on the surface velocity distribution; compare WAVE IMPEDANCE.

The total fluid loading force on a vibrating rigid piston may likewise be expressed as a ***fluid loading impedance*** (force divided by piston velocity); compare PISTON RADIATION IMPEDANCE. For further information, see RADIATION IMPEDANCE, MODAL RADIATION IMPEDANCE.

fluid loading – (2) *in aero- or hydrodynamics* the pressure and shear stress on a solid boundary, e.g. an aircraft wing surface or ship's propeller, due to relative motion through a surrounding fluid. Fluid loading may be either steady or unsteady.

flutter – see AEROELASTICITY.

flutter echo – a succession of ECHOES that arrive at equal or nearly equal spacing in time, when an acoustic space is excited by an impulsive sound source. A flutter echo with a time delay of order 10–20 ms between individual reflections has a characteristic audible "twang".

Note: Flutter echoes are found only in rooms with a high degree of symmetry, such that certain ray paths repeat after a small number of reflections. An example is the sequence of echoes that occurs when a source is placed between two parallel reflecting planes, whose dimensions are large compared with their distance apart.

focusing gain – *of a focused acoustic transducer* an equivalent term for AMPLITUDE FOCAL GAIN. *Units* none.

foil absorber – an equivalent term for MEMBRANE ABSORBER.

folding frequency – *of a discretely sampled signal* one-half of the sampling frequency; also known as the ***Nyquist frequency*** or ***Shannon frequency***. See ALIASING.

forced response – the response of a passive system to a time-stationary input. For mechanical systems, the equivalent term ***forced vibration*** is commonly used.

Note: In the case of transient excitation, the response of a linear system may be viewed in either of two ways:

(1) In the *frequency domain*, where each frequency component of the excitation produces a forced response at the input frequency;

(2) In the *time domain*, as a superposition of decaying free oscillations, triggered by a succession of impulses that make up the transient input. See CONVOLUTION.

forced vibration – *of a mechanical system* an equivalent term for FORCED RESPONSE.

force transmissibility – *of a vibration isolator excited by a force input* the ratio of output to input force amplitude, under single-frequency excitation. Compare MOTION RESPONSE. *Units* none.

Forchheimer's law – *for slow viscous flow through a porous solid* an empirical equation that describes deviations from DARCY'S LAW, in steady flow at finite Reynolds numbers. According to Forchheimer's law, the SPECIFIC FLOW RESISTANCE of a porous layer changes linearly with the seepage velocity $\bar{u}$. In equation form,

$$r = r_0 + C|\bar{u}|, \quad (r_0 = \text{limiting value at low Reynolds number})$$

where C is a constant (**P FORCHHEIMER 1901**).

Note: There is strong evidence from experiments and numerical simulations that $r - r_0$ has an initial quadratic dependence on $\bar{u}$, beginning at Reynolds numbers of order unity (based on pore size). At Reynolds numbers of order 10, the quadratic dependence gives way to Forchheimer's law.

formant – one of a set of broad resonances in the human voice mechanism, or in any sound-generating device, that impart a distinctive colour to the sound produced. The term is usually applied to the vowel sounds of speech, where the formants correspond to acoustic resonances of the vocal tract.

forward transform – see INTEGRAL TRANSFORM.

forward wave – *in a one-dimensional standing wave system* refers to a travelling-wave component whose propagation direction is away from a source, or towards a passive boundary; equivalent to an incident wave. Compare BACKWARD WAVE.

Foster's reactance theorem – ◆ (1) *applied to lossless acoustic fields in a two-port system with rigid boundaries and uniform acoustic pressures across each port* states that if the system is connected at port 2 to a reactive acoustic impedance jX_2, then the input acoustic reactance, X_1, at port 1 varies with frequency according to

$$U_{\text{rms},1}^2 \frac{dX_1}{d\omega} = 2E + U_{\text{rms},2}^2 \frac{dX_2}{d\omega}, \quad (e^{j\omega t} \text{ time factor}).$$

Here U_1 and U_2 are the volume velocities across ports 1 and 2 (counted as positive in the direction from 1 to 2), and E is the time-averaged acoustic energy in the system. The same equation, with the second term on the right deleted, applies to the reactance X_1 of a one-port reactive termination.

Note: The original version of Foster's theorem was derived for electrical networks (**R M FOSTER 1924**). The theorem given here, and the related theorem below, are consequences of the lossless acoustic equations for a uniform fluid.

Foster's reactance theorem – ◆ (2) *applied to lossless acoustic fields in a periodic waveguide with rigid boundaries* states that when the waveguide contains a single-frequency propagating BLOCH WAVE, the PHASE of the pressure at port 2, relative to port 1, varies with frequency according to

$$\frac{\partial \varphi_{12}}{\partial \omega} = \frac{-E}{W_{12}}, \quad (e^{j\omega t} \text{ time factor}).$$

Here φ_{12} is the relative phase, W_{12} is the time-averaged power flowing through the system from 1 to 2, and E is the time-averaged acoustic energy in the system.

Fourier amplitude spectrum – see AMPLITUDE SPECTRUM.

Fourier analysis – the representation of a periodic function as a series of HARMONIC COMPONENTS (**J B J FOURIER 1807**).

If $f(t)$ is a periodic, continuous, real function with period T, it may be represented either as a series of real cosine and sine terms,

$$f(t) = a_0 + \sum_{n=1}^{\infty} (a_n \cos n\omega t + b_n \sin n\omega t)$$

where ω is the FUNDAMENTAL angular frequency $2\pi/T$; or equivalently as a series of complex exponential terms,

$$f(t) = \sum_{n=-\infty}^{\infty} c_n e^{jn\omega t}.$$

The dc terms are the same in the two cases, i.e. $a_0 = c_0$. For positive n, the coefficients are related by

$$a_n = c_n + c_{-n}; \qquad b_n = j(c_n - c_{-n}); \qquad c_n = \tfrac{1}{2}(a_n - jb_n).$$

Note (1): The **Fourier coefficients** a_n, b_n are determined from $f(t)$ as follows:

$$a_0 = \frac{1}{T} \int_0^T f(t)\, dt \quad \text{(mean value over one cycle);}$$

$$a_n = \frac{2}{T} \int_0^T f(t) \cos n\omega t\, dt, \quad b_n = \frac{2}{T} \int_0^T f(t) \sin n\omega t\, dt \quad (n \geq 1).$$

Note (2): A complex signal $f(t)$ can also be represented as a Fourier series of complex exponential terms. The following equation gives c_n for the general case:

$$c_n = \frac{1}{T} \int_0^T f(t)\, e^{-jn\omega t}\, dt.$$

Note (3): For real signals, the **complex Fourier amplitudes** c_n have the symmetry property $c_{-n} = c_n^*$, where * denotes the complex conjugate. If they are written in polar form as $|c_n|e^{j\varphi_n}$, it follows that $|c_{-n}| = |c_n|$ and $\varphi_{-n} = -\varphi_n$.

Note (4): The magnitudes $|c_n|$ and phases φ_n, plotted against frequency, are called the AMPLITUDE SPECTRUM and PHASE SPECTRUM of the periodic signal $f(t)$.

Fourier–Bessel integral – refers to the integral representation

$$f(x) = \int_0^\infty F(\kappa) J_\mu(\kappa x) \kappa \, d\kappa, \quad (0 \le x < \infty);$$

$$F(\kappa) = \int_0^\infty f(x) J_\mu(\kappa x) x \, dx.$$

These relations are also known as HANKEL TRANSFORMS.

Note: The representation of $f(x)$ as an INTEGRAL TRANSFORM with kernel $J_\mu(\kappa x)$ may be viewed as a generalization of the FOURIER INTEGRAL: the Hankel transforms of $f(x)$, of order $-1/2$ and $1/2$, yield the Fourier cosine and sine transforms of $x^{1/2}f(x)$.

Fourier–Bessel modal analysis – *in polar coordinates* if $f(r, \theta)$ is a real function of the continuous real variables (r, θ), with $0 < r < \infty$ and $-\pi < \theta \le \pi$, then the representation

$$f(r, \theta) = \sum_{m=-\infty}^{\infty} e^{-jm\theta} \int_0^\infty F_m(\kappa) J_m(\kappa r) \kappa \, d\kappa$$

is called a Fourier–Bessel modal analysis of the function. The inverse representation is

$$F_m(\kappa) = \int_0^\infty f_m(r) J_m(\kappa r) r \, dr,$$

where $f_m(r)$ are the Fourier coefficients of $f(r, \theta)$:

$$f_m(r) = \frac{1}{2\pi} \int_0^{2\pi} f(r, \theta) e^{-jm\theta} \, d\theta.$$

The first stage of the modal decomposition process is a Fourier analysis at fixed r, and the second stage is an order-m Hankel transform.

Fourier component – *of a periodic signal or waveform* a waveform component that corresponds to one term of the FOURIER SERIES into which the periodic waveform can be analysed. An equivalent term is **harmonic component**.

Fourier integral – the integral representation

$$f(x) = \frac{1}{\pi} \int_0^\infty F_c(k) \cos kx \, dk + \frac{1}{\pi} \int_0^\infty F_s(k) \sin kx \, dk,$$

with

$$F_c(k) = \int_{-\infty}^{\infty} f(y) \cos ky \, dy \quad \textbf{\textit{(Fourier cosine transform)}},$$

$$F_s(k) = \int_{-\infty}^{\infty} f(y) \sin ky \, dy \quad \textbf{\textit{(Fourier sine transform)}}.$$

Here x and k are real, with $-\infty < x < \infty$ and $0 \leq k < \infty$.

Note: The Fourier integral representation may also be written in exponential form instead of using sine and cosine functions; see the entry under FOURIER TRANSFORM.

Fourier phase spectrum – see PHASE SPECTRUM.

Fourier series – an expansion of a periodic continuous function into a series of sine and cosine terms, or alternatively a series of complex exponential terms. For details, see FOURIER ANALYSIS.

Fourier's integral theorem – states that for real ω, t, and τ,

$$\frac{1}{\pi} \int_0^\infty d\omega \int_{-\infty}^\infty f(t) \cos \omega(t-\tau) dt = \begin{cases} f(\tau) & \text{for } f(t) \text{ continuous at } t = \tau; \\ \frac{1}{2}[f(\tau_+) + f(\tau_-)] & \text{otherwise}. \end{cases}$$

Here $f(t)$ is any function such that $\int_{-\infty}^{\infty} |f(t)| \, dt$ is finite, and $f(t)$, $f'(t)$ are assumed piecewise continuous over the whole range of t. The theorem is the basis of the FOURIER TRANSFORM pair.

Fourier's law of heat conduction – the statement that the heat flux due to conduction is proportional to the gradient of the temperature (**J B J FOURIER 1807**):

$$\mathbf{q} = -\kappa \nabla T.$$

Here the vector $\mathbf{q}$ is defined as the heat flow rate in the material per unit cross-sectional area, and T is the local temperature. For an isotropic material, the constant κ is a scalar called the **thermal conductivity** of the material.

Fourier spectrum – *of a transient signal* see COMPLEX FOURIER SPECTRUM.

Fourier transform – the INTEGRAL TRANSFORM

$$X(f) = \int_{-\infty}^{\infty} x(t) \exp(-j2\pi ft) dt$$

that converts a continuous time-domain signal $x(t)$ into its frequency-domain equivalent $X(f)$. The inverse transform is

$$x(t) = \int_{-\infty}^{\infty} X(f) \exp(j2\pi ft) df,$$

and the transform pair is represented schematically by

$x(t) \leftrightarrow X(f)$ (Fourier transform pair).

For the Fourier transform of a finite sequence, see DISCRETE-TIME FOURIER TRANSFORM.

Note (1): Real signals that are *odd* functions of time have Fourier transforms that are *imaginary odd* functions of frequency. Conversely, real signals that are *even* functions of time have Fourier transforms that are *real even* functions of frequency.

Note (2): The Fourier transform of the time-reversed signal $x(-t)$ is $X^*(f)$. The Fourier transform of the time-shifted signal $x(t+\tau)$ is $e^{j\omega\tau}X(f)$, where $\omega = 2\pi f$.

Note (3): It is often convenient to define the Fourier transform using ω, rather than f, as the variable conjugate to t. Unfortunately the notation in this case is not standardized. The most straightforward approach is to define $X(\omega) = X(f)$, which gives

$$X(\omega) = \int_{-\infty}^{\infty} x(t) e^{-j\omega t} dt,$$

$$x(t) = \frac{1}{2\pi} \int_{-\infty}^{\infty} X(\omega) e^{j\omega t} d\omega.$$

Alternatives to defining $X(\omega) = X(f)$ are $X(\omega) = (1/2\pi)X(f)$, and $X(\omega) = (1/2\pi)^{1/2}X(f)$. Whichever version is used, the appropriate changes must be made to the inverse transform definition.

Note (4): The **spatial Fourier transform** in one dimension,

$$U(k) = \int_{-\infty}^{\infty} u(x) e^{jkx} dx,$$

occurs frequently in acoustics. The WAVENUMBER k is analogous to the angular frequency ω, and the spatial transform converts a field $u(x)$ into its wavenumber-domain equivalent $U(k)$. See also TWO-DIMENSIONAL SPATIAL FOURIER TRANSFORM.

Note (5): The e^{jkx} sign convention for the forward spatial transform is introduced to simplify the description of waves: positive k signifies propagation in the $+x$ direction, for positive frequencies.

Fourier transform (2D) – see TWO-DIMENSIONAL SPATIAL FOURIER TRANSFORM.

fraction of critical damping – *for a single-degree-of-freedom system with viscous damping* an equivalent term for DAMPING RATIO. *Units* none.

frame – the solid framework, or **skeleton**, of a cellular or fibrous material.

> Note: For fibre blankets or open-cell foam, all the spaces or pores are interconnected. Compare RETICULATED FOAM.

Fraunhofer approximation – *for far-field radiation from a distributed source* a far-field approximation that yields an asymptotic value of the free-field Green's function, either in the time or the frequency domain (J VON FRAUNHOFER 1823). Suppose a field point at *A*, whose vector position is **x**, receives sound from a source element at *B*, whose vector position is **y**. The vector origin *O* is placed within the source region. We introduce the notation

$$R = OA = |\mathbf{x}|, \qquad r = BA = |\mathbf{x} - \mathbf{y}|;$$

then the Fraunhofer approximation consists of

- Replacing the spreading factor $1/r$ by $1/R$;
- Replacing the travel time r/c by $(R - \mathbf{e}\cdot\mathbf{y})/c$, where **e** is the unit vector $\mathbf{x}/R$ in the radiation direction and c is the sound speed.

Note (1): A specific example of the Fraunhofer approximation in the frequency domain is the equation

$$\frac{e^{-jkr}}{r} \approx \frac{1}{R} e^{-jk(R - y\sin\theta)}, \qquad (k = \text{acoustic wavenumber})$$

which occurs in calculations of sound radiation from line arrays. Here spherical polar coordinates are used, with their origin in the line array and with the polar axis perpendicular to the array; *A* is at $(R, \theta, 0)$ and *B* is at $(y, \pi/2, 0)$.

Note (2): In order to assess the error involved in the Fraunhofer approximation above, we write

$$\frac{e^{-jkr}}{r} = \frac{1}{R} e^{-jk(R - y\sin\theta)} F(y/R, ky, \theta);$$

asymptotic expansion of the correction factor F in powers of y/R yields

$$F = 1 + \frac{y}{R}\sin\theta - j\frac{ky^2}{2R}\cos^2\theta + O(y/R)^2.$$

The second term on the right is associated with the GEOMETRIC NEAR FIELD. The third term, proportional to jky^2/R, is associated with the FRESNEL

NEAR FIELD; it can be incorporated as an additional phase correction, called the **Fresnel correction**, to yield the improved approximation

$$\frac{e^{-jkr}}{r} \approx \frac{1}{R} e^{-jk(R - y\sin\theta + \Delta)}, \qquad \Delta = (y\cos\theta)^2/2R.$$

Fraunhofer region – *for single-frequency radiation into a free field from an extended spatially-coherent source* the region further from the source than the RAYLEIGH DISTANCE $l_{\text{trans}}^2/\lambda$. Here l_{trans} is the maximum dimension of the source, measured transverse to the direction of sound radiation, and λ is the acoustic wavelength (assumed much less than l_{trans}).

Note: In this situation the Rayleigh distance is much greater than either l_{trans} or λ, and the Fraunhofer region coincides with the FAR FIELD.

free boundary – *of a solid* a stress-free boundary.

free field – *in acoustics* an environment free of scattering or reflecting boundaries, so that outgoing waves never return towards the source region.

free-field microphone – a microphone designed to have near-constant sensitivity, as a function of both frequency and incident wave direction, when placed in a free field and exposed to plane progressive sound waves.

free-field radiation – refers to outgoing-wave radiation from a source into a FREE FIELD.

free-field room – see ANECHOIC CHAMBER.

free oscillation – *of a system* the oscillation that continues after any external excitation has been removed. Consists of NATURAL MODES of the system, each vibrating at their own NATURAL FREQUENCY. For mechanical systems, the equivalent term *free vibration* is more commonly used.

free progressive wave – a progressive wave propagating in a region free from sources or scattering boundaries.

free-stream – *in fluid mechanics* refers to flow conditions outside any wakes or boundary layers: ~ *velocity*, ~ *turbulence*.

free surface – *of a liquid* a liquid surface in contact with air (or other gas). A free water surface is commonly modelled as stress-free, i.e. as a PRESSURE-RELEASE BOUNDARY, for waterborne sound.

free vibration – *of a mechanical system* an equivalent term for FREE OSCILLATION.

free wave – abbreviation for FREE PROGRESSIVE WAVE.

frequency – (1) the number of cycles per unit interval of time, for a sinusoidally time-varying quantity; the repetition rate of any cyclic event. See also FREQUENCY COMPONENT. *Units* Hz.

frequency – (2) *in statistics, with reference to quantized or grouped data* the number of occurrences at a given quantized value, or within a particular BIN or CLASS INTERVAL. *Units* none.

frequency analysis – the process of partitioning the energy of a transient signal, or the power of a stationary signal, into FREQUENCY BANDS. The process may be carried out digitally, or by means of an analogue device. An equivalent term is ***spectral analysis***. See also BANDPASS FILTER.

frequency band – a range of frequencies whose upper and lower limits are specified, and which includes all the frequencies between these limits. See also OCTAVE BAND, ONE-THIRD OCTAVE BAND. *Units* Hz.

frequency bandwidth – see BANDWIDTH. *Units* Hz.

frequency component – *of a time-stationary signal* any signal component that varies sinusoidally with time; e.g. *power-supply* ∼ *or mains* ∼ (at 60 or 50 Hz).

frequency discrimination – *of pure tones* the ability of a listener to detect changes in the frequency of a pure-tone sound.

frequency distribution – *of a finite data sample whose individual values are either quantized, or grouped into defined categories or bins* the paired values, (n_i, x_i) or (n_i, i), associated either with the separate quantized data values x_i, or with the separate bins i. Here i is an index that runs from 1 to L; n_i is the associated FREQUENCY (2); and the total number of observations in the data set, given by $N = \Sigma n_i$, is typically much greater than L.

Note (1): A ***histogram*** of the data is produced by plotting n_i versus i, or else n_i versus x_i, as appropriate.

Note (2): If the frequencies n_i are normalized by N, the result is a ***normalized frequency distribution*** analogous to the PROBABILITY DENSITY FUNCTION of a random variable.

frequency-domain response function – an alternative term for the FREQUENCY RESPONSE FUNCTION of a linear system.

frequency interval – *between two given frequencies* an equivalent term for LOGARITHMIC FREQUENCY INTERVAL, i.e. the logarithm of the frequency ratio. *Units* oct, dec.

frequency modulation – see MODULATION.

frequency resolution – *of a discrete Fourier transform* the spacing between individual frequency components in the transformed signal. Its value is the reciprocal of the repetition period of the signal, including added zeros.

frequency response function – (1) *of a linear time-invariant system* the complex ratio of the phasors representing the system output and input, under single-frequency excitation. If the input $x(t)$ and output $y(t)$ are the real parts of $\hat{x}e^{j\omega t}$ and $\hat{y}e^{j\omega t}$, respectively, then the frequency response function at angular frequency ω is

$$H(\omega) = \hat{y}/\hat{x}.$$

Mathematically, the frequency response function $H(\omega)$ of a linear system is related to the IMPULSE RESPONSE FUNCTION $h(t)$ through a Fourier transform: thus

$$H(\omega) = \int_{-\infty}^{\infty} h(t)\, e^{-j\omega t}\, dt = \int_{0}^{\infty} h(t)\, e^{-j\omega t}\, dt \quad [h(t) = 0 \text{ for } t < 0],$$

where the last step follows from causality. According to the equation above, $H(\omega)$ is the convolution of $h(t)$ with the unit-amplitude single-frequency input $e^{j\omega t}$, evaluated at time zero. The function $H(\omega)$ may also be derived from the system response to transient or random excitation, as described in definition (2) below.

Note: Compare TRANSFER FUNCTION. Although some authors use the terms *transfer function* and *frequency response function* interchangeably, in control engineering and linear systems theory the transfer function is defined in the complex s-plane and is not limited to real frequencies. In the present volume, this distinction is maintained to avoid confusion.

frequency response function – (2) *between any two signals that are linearly related* the frequency response function between two *transient* signals $x(t)$ and $y(t)$ is given by the ratio of their respective Fourier transforms $X(f)$ and $Y(f)$:

$$H_{xy}(f) = \frac{Y(f)}{X(f)}.$$

For two *random time-stationary* signals, the corresponding definition is

$$H_{xy}(f) = \frac{S_{xy}(f)}{S_{xx}(f)}.$$

Here $S_{xx}(f)$ is the autospectral density of $x(t)$, and $S_{xy}(f)$ is the cross-spectral density of $x(t)$ and $y(t)$.

frequency selectivity – *in psychoacoustics* the ability of the auditory system to distinguish sounds that occupy neighbouring frequency bands, when the sounds are heard simultaneously.

frequency spectrum – *of a signal* a generalized term for the distribution of either the signal power, or the signal energy, with respect to frequency. See AUTOSPECTRAL DENSITY, ENERGY AUTOSPECTRAL DENSITY.

frequency-weighted – refers to a signal whose spectral components have been weighted according to some specified function of frequency. For example, an A-weighted mean square sound pressure has had its power spectrum frequency-weighted by the A-weighting filter shape, before being integrated over frequency.

frequency-weighted sound pressure level – *for airborne sounds* a sound pressure level based on applying a standard frequency weighting to an acoustic pressure signal. Compare WEIGHTED SOUND PRESSURE LEVEL, which usually implies both time and frequency weighting. *Units* dB re $(20\,\mu\text{Pa})^2$.

frequency weighting – *applied to sound pressure measurements* modification of the frequency spectrum of an acoustic signal by means of an analogue or digital filter having one of the standardized response characteristics known as A, B, C, etc. The A-WEIGHTING filter is the most commonly used. See SOUND LEVEL METER.

frequency weighting function – a frequency-dependent weighting factor, applied in the frequency domain to some property of a signal.

frequency windowing – application of a bandpass filtering process to a signal.

Fresnel correction – *to the far-field Green's function given by the Fraunhofer approximation* an approximate phase correction, quadratic in the transverse distance of the source element from a line joining the origin and observer, that becomes significant at distances from the source less than the RAYLEIGH DISTANCE.

Note: For an explicit expression, see note (2) under FRAUNHOFER APPROXIMATION.

Fresnel diffraction – an equivalent term for near-field diffraction.

Fresnel distance – alternative term for the RAYLEIGH DISTANCE. *Units* m.

Fresnel lens – an acoustic or optical flat lens, in which the continuous curved surface of a conventional lens is replaced by a set of shaped concentric grooves on a flat sheet of refracting material (**A J FRESNEL 1822**).

Fresnel near field – *of an extended spatially-coherent source, radiating at a given frequency, under free-field conditions* a region that is far enough from the source to be outside the GEOMETRIC NEAR FIELD, but not far enough for the FRESNEL CORRECTION to be neglected. Typically, this region extends out to some multiple of the RAYLEIGH DISTANCE l_{trans}^2/λ. Here l_{trans} is the maximum dimension of the source, measured transverse to the direction of sound radiation, and λ is the acoustic wavelength. Also known as the *Fresnel region* or the *Fresnel diffraction region*.

Fresnel number, Fresnel radius – see FRESNEL ZONES.

Fresnel zones – *encircling a given point in a spherical wave field* adjacent annular zones, centred on the given point *P*, that divide up the plane transverse to the axis of propagation into regions of alternating phase. The *n*th Fresnel zone is bounded by radii (r_n, r_{n-1}), where

$$r_n^2 = (z + n\lambda/2)^2 - z^2;$$

in this expression z is the distance of *P* from the apparent focus or source of the wavefronts (i.e. their centre of curvature). The **Fresnel number** is any path difference measured in half-wavelengths; in the situation above, it is the number n.

The first Fresnel zone ($n = 1$) has an *inner* radius of zero; its *outer* radius r_1 approaches the asymptotic value $(\lambda z)^{1/2}$ in the far field (i.e. $z \gg \lambda$), and $(\lambda z)^{1/2}$ is therefore called the **Fresnel radius**.

friction damping – an energy dissipation mechanism in which oscillatory motion is resisted by a force of constant magnitude, whose direction is always opposed to the motion. Sometimes called **dry friction damping**, to emphasize the distinction between it and viscous damping. Equivalent term: **Coulomb damping**. Compare STRUCTURAL DAMPING, VISCOUS DAMPING.

Froude number – *for steady flows with a free surface* a dimensionless velocity ratio analogous to MACH NUMBER (1) or (2), but with the sound speed replaced either by the speed of small-amplitude surface gravity waves, or by $\sqrt{gL}$ where L is a length scale characteristic of an object in the flow:

$$\mathrm{Fr} = \frac{u}{\sqrt{gL}} \quad (u = \text{relative flow velocity}),$$

(**W FROUDE 1872**). Some authors define $\mathrm{Fr} = u^2/gL$.

frozen pattern – a space-time pattern that can be made to appear frozen (i.e. unchanging with time), by viewing it in a frame of reference moving at a particular constant velocity. The latter is called the CONVECTION VELOCITY of the frozen pattern.

frozen sound speed – *in a relaxing medium* the sound speed calculated on the basis that internal degrees of freedom of molecules do not participate in the acoustic motion (and are therefore "frozen"). Sound propagates at the frozen sound speed in the high-frequency limit $\omega\tau \gg 1$ (where τ is the relaxation time). *Units* m s^{-1}.

function – one of the basic concepts in mathematics. A function is a relation that maps members of a set X into members of a set Y in a unique manner, such that for every member of X (denoted by x) there is one and only one member of Y (denoted by y). In symbolic form, $f: X \to Y$ defines the function f such that $y = f(x)$.

functional – an OPERATOR that maps a function into a real number. Functionals are typically expressed in terms of derivatives and integrals of functions. As an example, the value of the functional

$$J[f] = \int_{x_1}^{x_2} f(x)\,dx,$$

where x_1 and x_2 are given constants, depends on how the real function $f(x)$ varies over the interval from x_1 to x_2. The notation $J[\ldots]$ indicates that the argument of J is not a number but a function.

functional absorber – a sound-absorbing device that is installed in the interior of a room, rather than forming part of the surface treatment.

fundamental – *used as a noun* an abbreviation for FUNDAMENTAL FREQUENCY (1); i.e. the first harmonic of a periodic signal. See HARMONIC COMPONENT.

fundamental frequency – (1) *of a periodic signal* the first HARMONIC, i.e. the repetition frequency $1/T$ where T is the PERIOD. *Units* Hz.

fundamental frequency – (2) *of an oscillatory system* an informal term for the lowest NATURAL FREQUENCY of the system. See FUNDAMENTAL MODE. *Units* Hz.

fundamental mode – *of an oscillatory system* the mode of free oscillation that has the lowest NATURAL FREQUENCY.

fundamental resonance – *of a linear system* the lowest frequency at which a RESONANCE occurs, for designated input and output variables. *Units* Hz.

fuzzy structure – any vibroacoustic system that consists of a MASTER STRUCTURE plus one or more FUZZY SUBSTRUCTURES.

fuzzy substructure – a set of secondary structures, connected to a deterministic MASTER STRUCTURE, whose features are too poorly characterized (or too complicated) to permit conventional analysis, but which can be treated in a statistical manner.

G

g – symbol for the acceleration due to gravity. *Units* m s^{-2}.

gain – *of a linear system whose inputs and outputs are of the same type* refers either to the GAIN FACTOR of the system, or (more usually) to a decibel measure of the squared gain factor. According to the second definition, if the gain factor is denoted by K, then the system gain is

$G = 20 \log_{10} K.$

Units dB.

gain factor – *of a linear system whose inputs and outputs are of the same type* the magnitude of the FREQUENCY RESPONSE FUNCTION of the system. Applied to any two linearly-related signals that are measured in the same units, the gain factor between signals $x(t)$ and $y(t)$ is the magnitude of the ratio of their Fourier transforms, i.e. $|Y(f)/X(f)|$. An equivalent term is ***gain constant***. Compare SENSITIVITY; see also PHASE FACTOR. *Units* none.

gas – a fluid that does not form a free surface, but expands to fill the space available. The volume of the component molecules is a small fraction of the total volume. See also CRITICAL STATE.

gas constant – the ratio $P\mathcal{V}/T$ in the low-density limit, for a given quantity of gas. Here P and T are the absolute pressure and temperature, and $\mathcal{V}$ is the volume of gas.

Note: Versions of the gas constant in common use are the gas constant per mole, and the gas constant per unit mass of gas. The former is the ***molar gas constant*** or ***universal gas constant***, R_0. The latter is the ***specific gas constant***, R.

Note: $R_0 = 8314.51$ J kmol^{-1} K^{-1}; $R(\text{air}) = 287.05$ J kg^{-1} K^{-1}.

gas mixture relations – *for a mixture of ideal gases* relations that connect the gas constant, specific heats, and molar mass to the mixture composition; they are consequences of the GIBBS–DALTON LAW.

Note (1): The following gas mixture relations allow the acoustic properties of a mixture of ideal gases to be calculated from the component mass fractions y_i. The subscript mix denotes the mixture properties: e.g. γ_{mix} is the specific-heat ratio for the gas mixture.

(gas constant) $R_{\text{mix}} = \sum y_i R_i$ (R_i = gas constant, component i)

(specific-heat ratio) $\gamma_{\text{mix}} - 1 = \dfrac{R_{\text{mix}}}{\sum y_i R_i \left(\frac{1}{\gamma-1}\right)_i}$ ($\gamma = C_p/C_v$)

(molar mass) $\bar{m}_{\text{mix}} = R_0/R_{\text{mix}}$ (R_0 = molar gas constant).

Note (2): The following alternative relations allow calculation in terms of the component mole fractions x_i:

(molar mass) $\bar{m}_{\text{mix}} = \sum x_i \bar{m}_i$ ($\bar{m}_i$ = molar mass, component i)

(gas constant) $R_{\text{mix}} = (1/\bar{m}_{\text{mix}}) R_0$ (R_0 = molar gas constant)

(specific-heat ratio) $\gamma_{\text{mix}} = \dfrac{\sum x_i \left(\frac{\gamma}{\gamma-1}\right)_i}{\sum x_i \left(\frac{1}{\gamma-1}\right)_i}$.

gating – application of a time window to data during capture.

Gaussian distribution – a generic term that covers a family of PROBABILITY DENSITY FUNCTIONS (pdf's), ranging from UNIVARIATE to MULTIVARIATE, all of which are called Gaussian (C F GAUSS *1777–1855*). The Gaussian pdf family is important because many naturally occurring random phenomena are statistically distributed in this way, as a consequence of the central limit theorem. Three specific examples of Gaussian pdf's are given below.

(1) For a single Gaussian random variable with zero mean, denoted by X with typical realization x, the expression

$$f_X(x) = \dfrac{1}{(2\pi)^{1/2} \sigma_X} \exp\left[-\dfrac{1}{2}\left(\dfrac{x}{\sigma_X}\right)^2\right]$$

gives the univariate pdf of the random variable in terms of its STANDARD DEVIATION σ_X. The name ***normal distribution*** is also used to describe this type of pdf.

(2) For a pair of jointly-Gaussian random variables with zero mean, denoted by (X, Y) with typical realization (x, y),

$$f_{XY}(x, y) = \dfrac{1}{2\pi \sigma_X \sigma_Y \sqrt{1-\rho^2}} \exp\left[-\dfrac{1}{2}(\xi^2 + \eta^2 - 2\rho\xi\eta)\right]$$

gives the bivariate joint pdf of the two variables in terms of their CORRELATION COEFFICIENT ρ and standard deviations σ_X, σ_Y. Here the normalized coordinates ξ, η are defined by

$$\xi = \dfrac{x}{\sigma_X \sqrt{1-\rho^2}}, \quad \eta = \dfrac{y}{\sigma_Y \sqrt{1-\rho^2}}.$$

(3) For a set of N jointly Gaussian random variables $(X_1, X_2, \ldots, X_N)$ with

zero mean, represented by the column vector **X** with typical realization **x**, the general expression

$$f_\mathbf{X}(\mathbf{x}) = (2\pi)^{-N/2} |\det \mathbf{C}|^{-1/2} \exp\left[-\frac{1}{2}\mathbf{x}^T \mathbf{C}^{-1}\mathbf{x}\right]$$

gives the multivariate joint pdf of the N variables in terms of their normalized-covariance matrix **C**. The elements of **C** are

$$C_{ij} = (\sigma_i \sigma_j)^{-1} E\{X_i X_j\},$$

where σ_i is the standard deviation of X_i. Note that the diagonal elements of **C** are all unity. In the pdf expression above, the column vector **x** stands for the set of values $\{x_i\}$, and $\mathbf{x}^T$ is its TRANSPOSE.

Note (1): If in example (2) the correlation coefficient ρ is zero, or if in (3) all the off-diagonal elements C_{ij} are zero, then the Gaussian joint pdf reduces to the product of the individual univariate pdf's. This implies that the random variables are STATISTICALLY INDEPENDENT and have zero covariance at all orders. However, the link between zero correlation and statistical independence applies only to random variables whose joint distribution is Gaussian.

Note (2): The pdf expressions in (1)–(3) above may easily be generalized to variables with non-zero mean. Thus in example (1), x/σ_X would be replaced with $(x - \mu_X)/\sigma_X$, where $\mu_X = E\{X\}$.

Gaussian noise – a stationary RANDOM SIGNAL with the property that when pairs of values, denoted by $x(t)$, $x(t + \tau)$, are used to form a JOINT PROBABILITY DENSITY FUNCTION, the result is a bivariate GAUSSIAN DISTRIBUTION. It also follows that the UNIVARIATE probability density function of $x(t)$ is Gaussian. Here τ is an arbitrary time separation, and the statistics are averaged with respect to t.

Note (1): Gaussian noise that is ***Gaussian in the strict sense*** will also have Gaussian joint statistics at all higher orders. This means that a sample of N values measured at different times will have an N-fold multivariate pdf, averaged over time with the separations held fixed, that is of Gaussian form.

Note (2): The frequency components of a Gaussian noise signal have INITIAL PHASES that are randomly distributed over a 2π interval.

Gaussian random variable – any RANDOM VARIABLE that follows a GAUSSIAN DISTRIBUTION. If two or more variables have a Gaussian JOINT PROBABILITY DENSITY FUNCTION, the set of variables is called ***jointly Gaussian***.

Gaussian stochastic process – a STOCHASTIC PROCESS whose ENSEMBLE-AVERAGED statistics are Gaussian. If the process is denoted by $X(t)$, this means that its N-fold joint pdf, formed from the random variables

$\{X(t_1), X(t_2), \ldots, X(t_N)\}$ measured at N different times, will follow a Gaussian distribution for all orders N.

Gaussian white noise – GAUSSIAN NOISE that has a flat power spectrum. See also WHITE NOISE.

Gauss's divergence theorem – states that the outward flux of a vector field **u** over a closed surface is equal to the volume integral of the divergence, div **u**, over the region enclosed by the surface (C F GAUSS 1813). Also known as the *divergence theorem* and *Ostrogradski's formula* (M V OSTROGRADSKI 1831). In symbols,

$$\int_{\partial R} \mathbf{u} \cdot \mathbf{n}\, dS = \int_R \operatorname{div} \mathbf{u}\, dV.$$

Here R is the region to which Gauss's theorem is being applied; ∂R is the surface that bounds this region, and **n** is the unit normal to ∂R pointing out of R. The vector field **u**, and its spatial derivatives, are assumed to be continuous within R.

Note: Various corollaries of Gauss's theorem exist, in which the gradient or curl operator appears in the volume integral. If P is a scalar field, for example the pressure in a fluid,

$$\int_{\partial R} P \mathbf{n}\, dS = \int_R \nabla P\, dV;$$

this version yields the standard expression $-\nabla P$ for the pressure-gradient force per unit volume on a fluid. Also, for any vector field **u**,

$$\int_{\partial R} \mathbf{n} \times \mathbf{u}\, dS = \int_R \nabla \times \mathbf{u}\, dV.$$

Both these alternative forms follow from the original version.

generalized Burgers equation – *for finite-amplitude cylindrical or spherical progressive waves* see BURGERS EQUATION.

generalized coordinates – *for an N-degree-of-freedom system* any set of N independent displacement coordinates q_i ($i = 1$ to N) that can be used to define the displacement of the system, relative to some initial or reference configuration.

generalized forces – *for an N-degree-of-freedom system* a set of N forces Q_i ($i = 1$ to N), that correspond to a given set of GENERALIZED CO-ORDINATES q_i ($i = 1$ to N). The forces are defined such that the work done on the system in any small displacement is $\delta W = \Sigma Q_i \delta q_i$.

generator – an alternative term for SOURCE (1); e.g. *voltage generator, force generator*.

geoacoustic model – *in underwater acoustics* a set of parameters describing the acoustic properties of the sea bed. The model provides the compressional wave (and often shear wave) propagation speed and attenuation coefficient as functions of depth, along with the density.

geoacoustics – the study of mechanical wave propagation in the earth's crust, mantle, core, and oceans.

geometrical acoustics – an approximate description of high-frequency sound propagation in slowly-varying media, with wavefronts treated as locally plane and progressive. Slowly-varying means that the relative changes in density and sound speed over a distance of one wavelength are small. The locally-plane approximation is valid only when the wavelength is much less than the wavefront radius of curvature ($\lambda \ll R$). Also known as RAY THEORY.

geometrical theory of diffraction – the extension of GEOMETRICAL ACOUSTICS or geometrical optics to include *edge-diffracted rays* (i.e. rays produced at edges, corners and vertices of boundary surfaces), and creeping rays (i.e. rays incident tangentially on a smooth convex surface that "bend" to follow the surface). The field close to each diffracting edge or surface is calculated using a local diffraction model, in which the incident field is represented by plane wavefronts; the resulting diffracted field is then propagated away from the boundary using ray theory. The theory was originally invented for optics, and finds many applications in acoustics.

geometric centre frequency – *of a filter* an alternative term for MIDBAND FREQUENCY. *Units* Hz.

geometric near field – *of a finite-size source, radiating under free-field conditions* describes the region close to the source within which the distances of different source elements from the field point cannot be treated as equal. Specifically, the ($1/r$) weighting factor on radiated pressure cannot be adequately approximated by ($1/R$); here r is the distance of the field point from a typical source element, and R is the distance of the field point from the centre of the source region.

geophone – an accelerometer used to measure ground vibration, usually in the frequency range 5 Hz to 100 Hz.

Note: Seismic measurements in water or mud are made using HYDROPHONES, sometimes called *pressure geophones* in this context.

Gibbs–Dalton law – *for a mixture of ideal gases* states that each component of the mixture acts as if the others were absent and that component occupied the total volume of the container, at the same temperature. It follows that:

(1) Any extensive property of the gas mixture other than volume – e.g. the mixture internal energy – is determined by adding the contributions for each component, evaluated at the mixture volume and temperature;

(2) The mixture pressure is the sum of the component pressures evaluated in the same way.

The last statement is called ***Dalton's law of additive pressures*** (J DALTON 1801; J W GIBBS 1878).

Gibbs phenomenon – an overshoot that occurs whenever a waveform containing a discontinuity of slope is represented by a FOURIER SERIES. For example, Fourier series representation of a square wave introduces narrow spikes at the corners of the waveform: the spikes are extensions of the vertical sides of the waveform. They project upwards and downwards by about 9 percent of the peak-to-peak amplitude.

Note (1): The *error region* becomes vanishingly narrow as the number of terms in the Fourier series tends to infinity, but the *error magnitude* remains finite at each point on the waveform where the slope is discontinuous.

Note (2): The Gibbs phenomenon is not limited to Fourier series representations; it also occurs when spatial EIGENFUNCTIONS, such as mode shape functions in acoustic waveguides, are used as BASIS FUNCTIONS to represent a spatial distribution.

Gol'dberg number – *in nonlinear acoustics* a dimensionless number that indicates the importance of nonlinear waveform steepening relative to viscothermal diffusion, in periodic plane-progressive sound waves of finite amplitude (**Z A GOL'DBERG 1956**). It is defined as

$$\Gamma = \frac{1}{\alpha l_d},$$

where l_d is the SHOCK-FORMATION DISTANCE calculated as if the medium were lossless. The factor α is the ATTENUATION COEFFICIENT for small-amplitude sound waves, evaluated at the fundamental frequency f_0 of the periodic waveform. See also FINITE-AMPLITUDE WAVE. *Units* none.

Note: If loss mechanisms other than viscothermal diffusion contribute significantly to the attenuation at frequencies of order f_0, a generalized Gol'dberg number can still be defined based on the actual $\alpha(f_0)$ value, provided α varies approximately as f^2 up to a frequency of at least $5f_0$ (so as to encompass the first few harmonics).

grazing angle – *for plane waves arriving at a plane boundary* the angle that the incident wavenormal direction makes with the boundary surface.

More generally, in geometrical acoustics applied to a stratified medium, the term refers to the angle that the local wavenormal makes with the plane of stratification. *Units* rad (but commonly expressed in degrees).

grazing incidence – the limiting situation in which plane waves arrive at a plane boundary with an ANGLE OF INCIDENCE approaching 90° (i.e. the GRAZING ANGLE approaches zero).

Green function, Green's function – a point-input response function (G GREEN 1828). In the examples below, G is the Green's function for a point monopole source, t is time, r is distance from the source, and $\mathbf{x}$ is position relative to the source in 3-dimensional space.

(1) $\quad G = \dfrac{1}{4\pi r}\delta(t - r/c)$

is the causal free-field solution of the scalar wave equation driven by an impulsive point source,

$$\Box^2 \varphi = -\delta(\mathbf{x})\delta(t)$$

(see D'ALEMBERTIAN OPERATOR).

(2) $\quad G = \dfrac{1}{4\pi r} e^{-jkr}$

is the outgoing-wave free-field solution of the forced Helmholtz equation

$$(\nabla^2 + k^2)\varphi = -\delta(\mathbf{x}).$$

These examples describe acoustic radiation from a point source in a uniform unbounded medium. In case (2) the source is single-frequency, with $e^{j\omega t}$ time dependence.

Green's theorem – a corollary of GAUSS'S DIVERGENCE THEOREM that is useful for developing GREEN'S FUNCTION solutions of the HELMHOLTZ EQUATION, and also for demonstrating the orthogonality of Helmholtz equation eigenfunctions under appropriate boundary conditions (G GREEN 1828). Suppose p, q are two scalar fields defined within a region R bounded by a surface ∂R, and let $\mathbf{n}$ denote the unit outward normal on ∂R. The usual form of Green's theorem in this notation is

$$\int_R (p\nabla^2 q - q\nabla^2 p)dV = \int_{\partial R}(p\nabla q - q\nabla p)\cdot \mathbf{n}dS,$$

which relates a volume integral over region R to a surface integral over the boundary.

Note: The region R need not be 3-dimensional; if R is defined in 2 dimensions, for example, then the integral over ∂R is a line integral.

ground attenuation – *in outdoor sound propagation between points near the ground* the additional attenuation attributable to propagation at near-grazing angles over soft ground or snow. Also referred to as ***ground absorption***. *Units* dB.

ground effect – *in outdoor sound propagation* the influence of the ground on horizontal propagation; over distances of 25 m and above, this can be significant at audio frequencies, particularly when reflection at near-grazing angles is involved.

group delay – *for a linear time-invariant system* the frequency-dependent parameter, with dimensions of time, defined by

$$T_g(f) = -\frac{1}{2\pi}\frac{d\varphi(f)}{df};$$

here $\varphi(f)$ is the PHASE-SHIFT FUNCTION of the system at frequency f. The group delay can be either positive or negative; the term ***anomalous group delay*** is sometimes used to describe a negative value of $T_g(f)$. *Units* s.

Note: If a MINIMUM-PHASE SYSTEM is excited by a narrowband transient pulse with centre frequency f_0, the output will have a similar waveform shape to the input, but will be delayed by $T_g(f_0)$. However, apart from this special case the group delay does not have a simple physical interpretation.

grouped data – see BINS.

group velocity – *for single-frequency plane waves propagating in a uniform medium* in a general anisotropic medium, the free-wave DISPERSION RELATION may be expressed as

$$\omega = \Phi(k_1, k_2, k_3),$$

where ω is the complex angular frequency and (k_1, k_2, k_3) are wavenumber components in a rectangular coordinate system.

For real wavenumbers in a lossless medium, $\Phi(k_1, k_2, k_3)$ is real and the group velocity corresponds to the ENERGY VELOCITY. The group velocity in that case, for plane waves propagating in direction **n** with wavenumber k, is the velocity vector $\mathbf{v}(\mathbf{n}, k)$ whose components are

$$v_i = \left.\frac{\partial \Phi}{\partial k_i}\right|_{\mathbf{k}=\mathbf{n}k} \quad (i = 1, 2, 3).$$

Units m s^{-1}.

Note (1): For 1D propagation, say in the x_1 direction, the transverse Laplacian $\nabla_\perp^2$ in the (x_2, x_3) plane reduces to a constant multiplier. The dispersion relation may then be inverted to give a (possibly complex) propagation wavenumber as a function of frequency:

$$k_1 = F(\omega), \quad (\omega \text{ real}).$$

Two situations for which the function $F(\omega)$ takes complex values are lossy media and waveguides with sound-absorbent walls. The one-dimensional group velocity in such cases is defined by

$$v_1(\omega) = \mathrm{Re}\left(\frac{\partial F}{\partial \omega}\right).$$

Note (2): The group velocity for one-dimensional wave propagation – as in a waveguide – can be interpreted as the speed of a localized disturbance, formed by adding together a set of waves covering a narrow range of frequencies.

Grüneisen parameter – *for a fluid* a dimensionless number used in thermoacoustics, proportional to the thermal expansivity (**E A GRÜNEISEN 1926**). Its definition is

$$\eta = \frac{\alpha c^2}{C_p}, \text{ where } \alpha = \frac{1}{V}\left(\frac{\partial V}{\partial T}\right)_P.$$

The factor α is the volume thermal expansivity of the fluid (defined above in terms of specific volume V, temperature T and pressure P); c is the sound speed; and C_p is the specific heat at constant pressure. Also known as the **Grüneisen coefficient**. *Units* none.

Note: An equivalent definition of the Grüneisen parameter is

$$\eta = \frac{\partial(\ln T)}{\partial(\ln \rho)}\bigg|_s,$$

where ρ is the density and subscript s indicates that the derivative is evaluated isentropically. If η is a constant, the isentropic temperature–volume relation is $T \propto \rho^\eta$; compare ISENTROPIC EXPONENT.

gust response – see AEROELASTICITY.

H

Haas effect – an equivalent term for the PRECEDENCE EFFECT (**H HAAS 1951**).

habituation – a reduction in an individual's response to a specified stimulus that arises from repeated exposure to the stimulus. The term is usually applied to behavioural response, in contrast to ADAPTATION which refers to neurophysiological response.

hair cell – a specialized type of nerve cell, found in the COCHLEA and in the organs of balance. There are two types of hair cells in the cochlea, arranged in rows along the length of the cochlear spiral. The *inner hair cells* form a single row and are primarily responsible for converting mechanical motion into nerve impulses in the afferent, or incoming, auditory pathway. In contrast, the three (or occasionally four) rows of *outer hair cells* produce mechanical forces that amplify the vibration of the BASILAR MEMBRANE.

half-life – *of a quantity that decays exponentially with respect to time* the time taken for the value to fall by a factor 1/2. *Units* s.

half-power bandwidth – (1) *of a resonant response curve in which the squared gain factor of a linear system is plotted against frequency* the frequency separation Δf between the 3 dB down points on either side of the resonance peak, i.e. where the curve has fallen to half its peak value.

half-power bandwidth – (2) *of a signal with a peaked power spectrum* the frequency separation Δf between the points on either side of the spectral peak where the spectrum level is 3 dB down. Compare ENERGY BANDWIDTH. *Units* Hz.

half-power beamwidth – *of a source or transducer with a peaked directional response* the angular difference in direction between the 3 dB down points either side of the peak (in a specified plane through the beam axis, if the beam is not axisymmetric). *Units* degrees.

half-space – a semi-infinite region of 3-dimensional space, bounded by an infinite plane. For example, $y > 0$ defines a half-space in terms of the rectangular coordinates (x, y, z).

Hamiltonian – *in mechanics* a function $H(q_i, \dot{q}_i)$ of the GENERALIZED COORDINATES q_i of a conservative system, and of their time derivatives $\dot{q}_i$, that corresponds to the sum of the kinetic and potential energy of the system (**W HAMILTON 1834**). *Units* J.

Hamilton's principle – a VARIATIONAL PRINCIPLE stating that of all the possible motions of a CONSERVATIVE SYSTEM during a fixed interval of time from t_1 to t_2, the actual motion is such that the FUNCTIONAL $\int_{t_1}^{t_2} L\, dt$, where L is the LAGRANGIAN of the system, is a minimum with respect to all possible neighbouring paths joining prescribed initial and final configurations.

Note: The Lagrangian $L = T - V$ allows for the influence of external CONSERVATIVE FORCES, since these can be included in the potential energy function V. Hamilton's principle can be extended to include non-conservative forces as well, by writing

$$\int_{t_1}^{t_2} (\delta L + \delta W_{nc})\, dt = 0,$$

in which δW_{nc} represents the virtual work done on the system by non-conservative forces in a virtual displacement.

Hankel functions – the cylinder functions of the third kind, denoted by $H_\mu^{(1)}(z)$, $H_\mu^{(2)}(z)$ where z is the argument of the function and μ is the order (H HANKEL 1869). See CYLINDER FUNCTIONS.

Hankel transform – the order-μ Hankel transform of the function $f(x)$ is the function $F(\kappa)$ defined by the integral transform relation below:

$$F(\kappa) = \int_0^\infty x f(x) J_\mu(\kappa x)\, dx, \qquad \text{for } \mu > -\frac{1}{2}.$$

Here J_μ is the BESSEL FUNCTION of order μ. At all points where the original function is continuous, $f(x)$ may be recovered from $F(\kappa)$ via the *inverse Hankel transform*, defined by

$$\int_0^\infty \kappa F(\kappa) J_\mu(\kappa x)\, d\kappa = f(x), \qquad \text{for } \mu > -\frac{1}{2}.$$

Note (1): The following PARSEVAL RELATION applies, where $F(\kappa)$, $G(\kappa)$ denote the Hankel transforms of the functions $f(x)$, $g(x)$:

$$\int_0^\infty x f^*(x) g(x)\, dx = \int_0^\infty \kappa F^*(\kappa) G(\kappa)\, d\kappa.$$

Note (2): Hankel transforms are used in acoustics to propagate axisymmetric sound fields in the axial direction, in a similar manner to that described in the note under TWO-DIMENSIONAL SPATIAL FOURIER TRANSFORM.

Hanning window – *in signal processing* a function used for WINDOWING; it is defined in the time domain by

$$w(t) = \begin{cases} \dfrac{1}{2}\left(1 + \cos\dfrac{2\pi t}{T}\right), & -\dfrac{1}{2}T \leq t \leq \dfrac{1}{2}T; \\ 0, & \text{otherwise.} \end{cases}$$

Here t denotes time, and T is the window length.

hard boundary – a boundary that imposes zero normal velocity on an adjacent acoustic medium. Also known as a *rigid boundary*. Compare NEUMANN BOUNDARY CONDITION.

harmonic – (1) *used as a noun* an abbreviation for HARMONIC COMPONENT, meaning an individual frequency component of a PERIODIC signal. The fundamental frequency f_0 is called the *first harmonic*, and in general the frequency component at nf_0 is called the nth harmonic.

harmonic – (2) *used as an adjective* sinusoidally-varying. Hence *time-harmonic* means single-frequency. Compare the different use of the term in HARMONIC FUNCTION.

harmonic component – an individual FOURIER COMPONENT of a periodic waveform or signal; one term of a FOURIER SERIES. The Fourier component that has the same period as the waveform is called the *fundamental* or *first harmonic*, and in general the Fourier component that repeats n times in one waveform period is called the nth harmonic.

harmonic distortion – (1) *produced by a nonlinear system* the occurrence of output signal components at $2f_0$, $3f_0$, etc. in response to a single-frequency input signal at frequency f_0. In general, if the output contains such components, the signal is said to have suffered harmonic distortion. The output signal components at $2f_0$, $3f_0$, etc. are called *harmonic distortion terms*.

harmonic distortion – (2) *expressed numerically* an abbreviation for TOTAL HARMONIC DISTORTION. *Units* %.

harmonic function – *in mathematics* a generic term for any function that satisfies the differential equation $\ddot{q}(t) + \omega^2 q(t) = 0$ (simple harmonic motion in time), or its spatial equivalent $\nabla^2 \varphi(\mathbf{x}) + \kappa^2 \varphi(\mathbf{x}) = 0$ (spatial harmonic functions in 1, 2, or 3 dimensions). Often abbreviated to *harmonic*, as in *spherical surface harmonic*.

harshness – *in automotive engineering* a generic term for any intrusive noise produced inside a vehicle, in the approximate frequency range 250 Hz to 2.5 kHz, that has some audibly distinct characteristic (e.g. a cluster of 3 or 4 consecutive engine orders at a higher level than the others). See also BOOM (1).

HATS – abbreviation for HEAD-AND-TORSO SIMULATOR.

He – symbol for HELMHOLTZ NUMBER; also written He. *Units* none.

head-and-torso simulator – a device simulating the geometry of the median adult human head and torso, and to some extent its acoustic diffraction properties. The head-and-torso simulator, abbreviated as ***HATS***, extends downwards from the top of the head to the waist. The head is equipped with two pinna simulators and at least one EAR SIMULATOR; the microphone in the latter is placed at the position corresponding to the eardrum.

The generic term ***manikin*** is also used for head-and-torso simulators.

head-related transfer function – *for a particular head and a specified incident sound field* the complex ratio of the pressure at one ear to the incident-field pressure measured midway between the ears, expressed as a function of frequency. The relevant pressure in the numerator is the acoustic pressure on the outer surface of the tympanic membrane. *Units* none.

head wave – *at a plane interface between two media* a transient reflected wavefront produced when an outgoing spherical or cylindrical wave is incident on the interface from the slow side, with a local wavenormal direction close to one of the CRITICAL ANGLES FOR TRANSMISSION. The head wave amplitude falls off with horizontal distance (r) like $1/r^2$ for the spherical wave case, and its point of contact with the interface propagates at the wave speed in the faster medium.

The transmitted field associated with the head wave decays exponentially into the faster medium. Compare LATERAL WAVE.

For reflection of fluid-borne sound from a solid there are two types of head wave, that propagate along the interface at speeds equal to the shear and compressional wave speeds in the solid medium.

hearing aid – a prosthetic device to aid auditory perception of sound. The output signal is normally presented acoustically; alternative methods are cranial bone vibration, and direct electrical stimulation of the auditory nerve (cochlear implant). The term is also applied to devices that exploit other sensory modalities, e.g. vibrotactile aids.

hearing impairment – *at a specified frequency* the amount in decibels by which an individual's hearing threshold level exceeds a standard value. The related term THRESHOLD SHIFT implies before-and-after comparison, whereas hearing impairment commonly assumes a notional starting point such as AUDIOMETRIC ZERO. *Units* dB.

Note: Hearing impairment has a quantitative meaning, as distinct from the qualitative term HEARING LOSS. The latter is used loosely to mean a symptom of hearing disorder, and is often modified by descriptors: e.g. *noise-induced* ∼, *conductive* ∼, *sensorineural* ∼, etc.

hearing level – *in pure-tone audiometry, for a specified method of auditory stimulus presentation at a given frequency* the signal level produced by the stimulus in a specified ear simulator or similar device, minus the appropriate (standardized) REFERENCE EQUIVALENT THRESHOLD LEVEL. In pure-tone air-conduction audiometry, for example, the stimulus is provided by an earphone; the hearing level is then given by subtracting the reference equivalent threshold sound pressure level from the sound pressure level produced by the earphone in an ear simulator. For a correctly calibrated audiometer, the hearing level equals the dial setting. Compare HEARING THRESHOLD LEVEL. *Units* dB.

hearing loss – *in audiology* a generic term for hearing difficulty that is caused by disease of, or damage to, the auditory system. See CONDUCTIVE HEARING LOSS, SENSORINEURAL HEARING LOSS, NOISE-INDUCED HEARING LOSS. For a quantitative measure of loss, see HEARING IMPAIRMENT.

hearing protector – a generic term for a noise-excluding device worn on the head or in the ear canal; includes some types of helmet, as well as EARMUFFS and EARPLUGS.

hearing threshold level – *in pure-tone audiometry, for a specified method of auditory stimulus presentation at a given frequency* the threshold of hearing at that frequency, expressed as a HEARING LEVEL in decibels. *Units* dB.

Note: Hearing threshold level depends on the ear under test, whereas hearing level depends only on the audiometric test equipment.

heat – a transfer of ENERGY between a system and its surroundings, by means of a temperature difference. *Units* J.

heat flux vector – the rate of heat flow per unit cross-sectional area. See also FOURIER'S LAW OF HEAT CONDUCTION. *Units* W m^{-2}.

Heaviside unit step function – the generalized function defined by

$H(x) = 0$ for $x < 0$;

$H(x) = 1$ for $x > 0$.

Its value at $x = 0$ is not defined. Alternative names are ***unit step function***, ***Heaviside step function***, and ***Heaviside function*** (**O HEAVISIDE 1892**).

Note: The alternative notation $u(x)$ is more common in signal processing, control, and systems contexts.

helicity – see POLARIZATION.

Helmholtz equation – any differential equation of the form $\nabla^2 q + k^2 q = 0$, in which k is a constant (real or complex) and the dependent variable q is a function of position in one or more space dimensions (**H L F VON HELMHOLTZ 1860**). The terms *vector* $\sim$, *scalar* $\sim$ are used to distinguish the cases where q is a vector or a scalar, respectively.

Helmholtz instability – *in inviscid fluid dynamics* an alternative name for KELVIN–HELMHOLTZ INSTABILITY.

Helmholtz integral formula – a solution to the HELMHOLTZ EQUATION over a given domain V, written as a surface integral over the domain boundary S:

$$C(\mathbf{x})\varphi(\mathbf{x}) = \int_S \left(G \frac{\partial \varphi}{\partial n} - \varphi \frac{\partial G}{\partial n} \right) dS(\mathbf{y}).$$

Here φ is the field variable, and G is the free-field Green's function $\exp(-jkR)/R$ in three dimensions; R denotes the distance $|\mathbf{x} - \mathbf{y}|$ between points $\mathbf{x}$ and $\mathbf{y}$. This relation holds for all points $\mathbf{x}$ on a piecewise smooth closed surface S bounding the volume V; n is the outward normal coordinate. The coefficient C is equal to the solid angle subtended by the closed surface at the point $\mathbf{x}$; it equals 2π when $\mathbf{x}$ lies on a smooth part of the surface S, compared with 4π when $\mathbf{x}$ lies within V. In the latter case the equation is known as the ***Helmholtz theorem*** or ***Helmholtz's representation***, and is a special case of Green's third identity (**H L F VON HELMHOLTZ 1860**).

Helmholtz number – *in acoustics* a dimensionless measure of size in relation to acoustic wavelength (**H L F VON HELMHOLTZ 1821–1894**). Any linear dimension, L, can be expressed nondimensionally as a Helmholtz number by defining

He $= kL$ (k = acoustic wavenumber)

     $= 2\pi L/\lambda$ (λ = wavelength).

Symbol He. *Units* none.

Helmholtz resonance absorber – a sound absorber consisting of a Helmholtz resonator, or an array of resonators, mounted (for example) in the wall of a room.

Helmholtz resonator – a rigid cavity, filled with air or other compressible fluid, and with a small opening (called the neck). Its fundamental natural frequency is approximated by $\omega = c\sqrt{S/l_{\text{eff}} V}$, where ω is the angular frequency, c is the sound speed of the enclosed fluid, S and l_{eff} are the area and effective length of the neck, and V is the cavity volume.

hemi-anechoic room – a test room with a hard reflecting floor, whose other surfaces are anechoic. Compare SEMI-ANECHOIC ROOM.

hereditary material – a material whose CONSTITUTIVE RELATION is independent of time; thus the material neither ages, nor suffers mechanical degradation (e.g. work hardening) as a result of its loading history. An equivalent term is *non-ageing material*.

Hermitian matrix – a square matrix whose transpose is the complex conjugate of the original matrix. Thus if matrix **A** is Hermitian,

$\mathbf{A}^T = \mathbf{A}^*$.

Also known as a *self-adjoint matrix*.

Hermitian transpose – *of a matrix* the complex conjugate of the transposed matrix. Also called the *adjoint matrix*. Written as $\mathbf{A}^H$, where **A** denotes the original matrix (**C HERMITE** ~ 1860).

hertz – the unit of FREQUENCY in the SI system; 1 hertz ≡ 1 cycle per second (**H HERTZ** *1857–1894*). *Unit symbol* Hz.

high intensity focused ultrasound – *in clinical practice* the use of focused pulses of ultrasound to burn away unwanted tissue in non-invasive surgery (abbreviated *HIFU*).

high-pass filter – a FILTER designed to remove unwanted low-frequency components from a signal. If the filter removes all frequencies below a lower limit f_c, then f_c is called the *lower cutoff frequency* of the filter.

Hilbert transform – an INTEGRAL TRANSFORM defined by

$$f_H(x) = \frac{1}{\pi} \wp \int_{-\infty}^{\infty} \frac{f(y)}{x - y} dy.$$

Here $f(x)$ is any function of x, and $f_H(x)$ is its Hilbert transform (**D HILBERT** *1924*). The integral is interpreted as a Cauchy principal value (symbol $\wp$). The inverse relation is

$$f(x) = -\frac{1}{\pi} \wp \int_{-\infty}^{\infty} \frac{f_H(y)}{x - y} dy,$$

and the transform pair is represented schematically by

$f(x) \leftrightarrow f_H(x)$ (Hilbert transform pair).

Examples are

$\cos x \leftrightarrow \sin x$; $\sin x \leftrightarrow -\cos x$.

Some authors use a minus sign in the forward transform definition, rather than in the inverse transform; the convention above is that generally adopted in signal processing and systems texts.

Note (1): Applying the Hilbert transform twice in succession restores the original function with its sign reversed: if $f(x) \leftrightarrow f_H(x)$, then $f_H(x) \leftrightarrow -f(x)$ (symmetry relation).

Note (2): The Hilbert transform may also be viewed as a convolution of $f(x)$ with $1/\pi x$:

$$f_H(y) = \left[\left(\frac{1}{\pi x}\right) * f(x)\right](y).$$

Note (3): Applying a Hilbert transform to a function of time, i.e. $f(t) \to f_H(t)$, is equivalent in the frequency domain to introducing a phase lag or delay of one-quarter period. The phase of each frequency component (positive or negative) is shifted by $-\pi/2$ (positive frequencies) or $+\pi/2$ (negative frequencies), using the $e^{j\omega t}$ convention. The respective FOURIER TRANSFORMS $F(\omega)$, $F_H(\omega)$ are related by

$$F_H(\omega) = \begin{cases} e^{-j\pi/2}F(\omega) = -jF(\omega) & (\omega > 0) \\ e^{j\pi/2}F(\omega) = +jF(\omega) & (\omega < 0), \end{cases}$$

but the energy spectrum of the signal is unaffected since $|F_H(\omega)| = |F(\omega)|$.

histogram – see FREQUENCY DISTRIBUTION.

HLA – in *underwater acoustics* abbreviation for horizontal line array.

holomorphic function – *of a complex variable* a function, $f(z)$, is called holomorphic in a region D of the complex plane if it can be expanded locally as a Taylor series in $(z - z_0)$, where z_0 is any point within D. An ANALYTIC FUNCTION of z has this property.

homentropic – *in fluid mechanics* of uniform specific entropy.

homogeneous – uniform, usually with respect to position. A *homogeneous medium* is a medium whose properties are spatially uniform.

homogeneous boundary conditions – boundary conditions that are homogeneous in the solution variable u. Examples are the DIRICHLET BOUNDARY CONDITION $u = 0$, and the NEUMANN BOUNDARY CONDITION $\partial u/\partial n = 0$, where $\partial/\partial n$ denotes the normal gradient operator. A more general example, which includes both the above as special cases, is the IMPEDANCE BOUNDARY CONDITION

$$\alpha u + \beta \frac{\partial u}{\partial n} = 0 \quad (\alpha, \beta \text{ constants}).$$

Note: All passive boundary conditions encountered in linear acoustics fall in this category.

homogeneous equation – an equation of the form $F(u) = 0$, in which every term on the left contains the unknown variable u or its derivatives; here F is a function or an operator. See also HOMOGENEOUS SOLUTION.

Note: Homogeneous equations describe the *unforced* motion of a system or medium.

homogeneous plane wave – *in a lossless medium* a freely-propagating plane-wave field in 3 dimensions that consists of plane WAVEFRONTS all propagating in one direction. Such waves can be described, at a single frequency, by a WAVENUMBER VECTOR with purely real components in all three coordinate directions. Compare INHOMOGENEOUS PLANE WAVE.

homogeneous solution – starting from any NON-HOMOGENEOUS DIFFERENTIAL EQUATION, one may obtain a corresponding ***homogeneous equation*** by deleting those terms that do not contain the solution variable. Solving the homogeneous equation yields a so-called *homogeneous solution* of the original differential equation, also known as the ***complementary function***. If the equation is of order n, its homogeneous solution in general form contains n arbitrary coefficients.

Note: Provided the homogeneous equation is linear in the solution variable, the general solution of the *non-homogeneous* equation is obtained by adding a PARTICULAR INTEGRAL to the homogeneous solution. In any particular problem, the values of the coefficients in the homogeneous solution are then chosen to satisfy the required boundary or initial conditions.

Hooke's law – a statement of LINEARITY for static deformation of elastic bodies or materials (**R HOOKE 1676**). A continuous material that obeys Hooke's law has local strain components that depend linearly on the local stress components, and vice versa.

horizontal distance, horizontal range – *in underwater or atmospheric acoustics* the horizontal component of the distance between source and receiver; in CYLINDRICAL COORDINATES (r, ϕ, z) with the z axis vertical and passing through the source, it corresponds to the cylindrical radius r. Also abbreviated as *range*. *Units* m.

horizontally stratified – an acoustic medium is said to be horizontally stratified if its properties are constant over horizontal planes, but vary in the vertical direction (either continuously, or in discrete layers). See also STRATIFIED MEDIUM.

horn – a waveguide whose properties (particularly the cross-sectional area) are arranged to vary monotonically from one end to the other, in order to produce a resistance in the lowest-order waveguide mode, Re(Z), that is many times larger at the driver end than at the opposite end. The definition of the horn impedance Z depends on the horn type: for ACOUSTIC HORNS the lowest-order mode is approximately a plane wave with $Z = p/U$ (average pressure divided by volume velocity), while for SOLID HORNS $Z = F/u$ (axial force divided by average axial velocity).

Since in either case the power fed into the horn is Re(Z) times the mean square of the input quantity (either U or u), the horn allows more power to be extracted from the same input excitation.

horn loudspeaker – loudspeaker in which the driver or diaphragm is coupled to the external medium by an ACOUSTIC HORN.

horn mouth – *for an acoustic horn* the output end of the horn, where the cross-sectional area is largest. See ACOUSTIC HORN.

horn throat – *for an acoustic horn* the input end of the horn, where the cross-sectional area is smallest. See ACOUSTIC HORN.

horn transmission factor – the ratio, τ, of the net power transmitted by an acoustic horn to the corresponding power in a straight tube having the same cross-sectional area as the horn throat, and excited in the same way as the throat of the actual horn. For example, if the throat excitation at a given frequency is equivalent to a piston with prescribed velocity amplitude, then τ is the specific radiation resistance (real part of the specific PISTON RADIATION IMPEDANCE) at the throat, normalized by the characteristic impedance ρc. An older term is ***horn transmission coefficient***. *Units* none.

Howe's equation for vortex–entropy sound – an acoustic analogy equation for the stagnation enthalpy, B, that attributes aerodynamic sound generation to vorticity and entropy perturbations in an otherwise irrotational and homentropic ideal-fluid flow. It may be written as

$$\frac{D}{Dt}\left(\frac{1}{c^2}\frac{DB}{Dt}\right) - \hat{\nabla}\cdot\nabla B =$$

$$\hat{\nabla}\cdot(\omega \times \mathbf{u} - T\nabla s - \tilde{\mathbf{f}}) + \frac{\partial}{\partial t}\left(\frac{\alpha T}{C_p}\frac{Ds}{Dt}\right) + \frac{D}{Dt}\left\{\frac{1}{c^2}\left(T\frac{Ds}{Dt}+\mathbf{u}\cdot\tilde{\mathbf{f}}\right)\right\}.$$

Here $\mathbf{u}$ denotes the fluid velocity and $\omega = \nabla \times \mathbf{u}$ is the vorticity; c, T, s, α, C_p are the sound speed, absolute temperature, specific entropy, thermal expansivity, and constant-pressure specific heat of the fluid; $\tilde{\mathbf{f}}$ is the viscous force per unit mass on a fluid element; and the modified gradient operator is defined by

$$\hat{\nabla} \equiv \left(\nabla + \frac{1}{\rho c^2}\nabla P\right).$$

The equation is general and exact, apart from the assumption that the isentropic compressibility $1/\rho c^2$ depends only on the pressure P (as in a perfect gas).

Note: In an unbounded flow, with vorticity, entropy gradients, and viscous forces confined to a limited region, the stagnation enthalpy B outside the vortical region is related to the velocity potential φ by

$$B - B_\infty = -\dot{\varphi} \qquad \text{(dot denotes time derivative)}.$$

The Howe equation then reduces to the equation for nonlinear sound propagation in irrotational homentropic flow:

$$\frac{D}{Dt}\left(\frac{1}{c^2}\frac{D\dot{\varphi}}{Dt}\right) - \hat{\nabla}\cdot\nabla\dot{\varphi} = 0.$$

Huygens' principle – *for transient sound fields in a lossless acoustic medium at rest* a kinematic construction, derived from KIRCHHOFF'S FORMULA in the limit of short wavelengths, that relates the position of an advancing wavefront at time t_2 to the position at an earlier time t_1. Each point on the earlier wavefront is regarded as a secondary source, which emits a secondary circular wavefront of radius $c(t_2 - t_1)$ where c is the propagation speed of the wavefront. The envelope of the secondary wavefronts is the new wavefront at time t_2 (**C HUYGENS 1678**).

hydrodynamic near field – a region of a sound field in which the particle velocity and the pressure are nearly in quadrature. Typically, in such regions the peak particle velocity is much greater than $p_{max}/\rho c$, where p_{max} is the peak acoustic pressure and ρ, c are the fluid density and sound speed.

The hydrodynamic near field of a COMPACT source or scattering object typically extends a distance of $\lambda/4$ from the source, where λ is a typical acoustic wavelength. Within this region the motion is effectively incompressible.

Note: A large radiating surface can also have a hydrodynamic near field, if its surface motion is dominated by wavenumber components larger than the acoustic wavenumber; the field within a distance $\lambda/4$ from the surface is then mainly EVANESCENT.

hydroelasticity – the equivalent of AEROELASTICITY, but for the case where the fluid is a liquid (typically water), rather than a gas (typically air).

hydrophone – a transducer similar to a MICROPHONE, but designed to measure acoustic pressures in water or other liquids.

hypersonic – *in acoustics* refers to the frequency range where acoustic waves become heavily damped during transmission, meaning that the attenuation per wavelength is no longer small. The term can be applied to sound in any homogeneous medium; in gases at 1 bar pressure, the hypersonic range is typically from 10 GHz upwards.

hysteretic damping – see STRUCTURAL DAMPING.

I

i – one of the square roots of −1; the *i* notation is due to Euler. Alternatively written as *j*, as in this dictionary.

IACC – abbreviation for INTERAURAL CROSS-CORRELATION COEFFICIENT. *Units* none.

ideal bandpass filter – a BANDPASS FILTER with zero INSERTION LOSS between the passband limits, and infinite insertion loss at frequencies outside the passband.

ideal fluid – a fluid that has zero viscosity and thermal conductivity. In general, ideal fluids are free from all dissipative phenomena. The motion of an ideal fluid is governed by the EULER EQUATIONS.

ideal gas – a gas whose equilibrium volume, pressure, and temperature are related by the IDEAL GAS LAW. The specific heat capacities of an ideal gas are functions of temperature in general; compare PERFECT GAS.

Note: There is a risk of confusion between the terms ideal gas and IDEAL FLUID; in fact they refer to entirely different idealizations.

ideal gas law – the equation

$$P\mathcal{V}/T = \text{constant}.$$

Here (P, T) are the absolute pressure and temperature of a sample of gas and $\mathcal{V}$ is its volume. The ideal gas law is followed by all gases in the low-density limit, provided the sample contains a sufficiently large number of molecules. See also GAS CONSTANT.

identity matrix – a DIAGONAL MATRIX whose diagonal elements are equal to 1.

IEC – abbreviation for ***International Electrotechnical Commission***.

IIR – abbreviation for INFINITE IMPULSE RESPONSE.

IL – (1) abbreviation for INSERTION LOSS. *Units* dB.

IL – (2) abbreviation for INTENSITY LEVEL. *Units* dB re 1 pW m^{-2}.

Im – operator symbol for the IMAGINARY PART of a complex number.

image – *of a source in a plane reflecting boundary* an abbreviation for IMAGE SOURCE.

image source – *as a model for reflection of sound by a plane uniform boundary* a replica of an actual source, located at the opposite point of symmetry (i.e. in the mirror image position) with respect to the boundary. The reflection model consists of the actual source and the image source in a free field.

Note: By assigning an appropriately modified phase and amplitude to the image source, any desired plane-wave reflection coefficient on the boundary can be simulated by the image-source combination. However, the simple image representation described above can only simulate boundaries whose reflection coefficient is independent of angle of incidence. This means that an exact image representation is available, in practice, only for rigid or pressure-release boundaries.

imaginary axis – any complex number $z = x + jy$ may be represented graphically by a point in the xy plane, with x plotted horizontally and y vertically. The vertical axis is then called the imaginary axis.

imaginary part – any complex number z may be represented as $z = x + jy$, where x and y are both real. The quantity y is called the imaginary part of z, denoted by Im z or $\mathcal{I}z$. It equals $(z - z^*)/2j$.

immission – see NOISE IMMISSION LEVEL.

immittance – see ACOUSTIC IMMITTANCE.

impedance – *in acoustics* the complex ratio of two phasors, the first of which is a pressure or force and the second of which is a velocity or volume velocity. See ACOUSTIC $\sim$, SPECIFIC ACOUSTIC $\sim$, DIFFERENTIAL $\sim$, MECHANICAL $\sim$, RADIATION $\sim$, SPECIFIC RADIATION $\sim$, TRANSFER $\sim$.

Note: Equivalently, an impedance is a FREQUENCY RESPONSE FUNCTION in which pressure p or force F (in a given direction) is regarded as the output, and velocity u (in a given direction) or volume velocity U (across a specified surface) is regarded as the input.

impedance boundary condition – *for single-frequency sound fields* a linear relation between the acoustic pressure and its normal gradient, that holds at each point on a boundary. It is expressed mathematically by

$$\alpha p + \beta \frac{\partial p}{\partial n} = 0,$$

where p is the pressure, $\partial/\partial n$ denotes the normal gradient, and α, β are constants at each position but may vary over the boundary. Also known as a ***Robin boundary condition*** (G ROBIN 1886). Compare HOMOGENEOUS BOUNDARY CONDITION.

impedance matching – (1) *for acoustic waves incident on a boundary* matching the normal specific acoustic impedance, z_s, of the boundary (at a given frequency) to the real CHARACTERISTIC IMPEDANCE z_{char} of the incident wave system normal to the boundary. This has the effect of maximizing the SOUND POWER TRANSMISSION COEFFICIENT, since the latter is given (for real values of characteristic wavenumber and impedance) by

$$\alpha_t = \frac{4\operatorname{Re}\zeta}{|1+\zeta|^2}, \qquad (\zeta = z_s/z_{\text{char}}).$$

Maximizing α_t requires $\zeta = 1$, i.e. the impedances z_s and z_{char} need to be matched for maximum power transmission.

Note: The description above also applies to plane-wave transmission at duct discontinuities, with appropriate redefinitions of z_s and z_{char}.

impedance matching – (2) *for lumped-element systems* matching the load impedance to the complex conjugate of the source internal impedance, thereby maximizing the power supplied from a source to a passive load.

This can be described mathematically as follows. Suppose a linear lumped-element system consists of a single-frequency source with internal impedance Z_i, connected at A to an external load impedance Z_e. The response of the load at the driving point A is given by

$$v = \frac{F}{Z_i + Z_e}.$$

For example, the expression above describes the driving-point velocity response to a force F connected to both impedances. Given F and Z_i, maximization of the power supplied to the load requires $Z_e = Z_i^*$. This is the general impedance-matching result.

An equivalent result applies if the source consists of a volume velocity U, driving both an internal admittance Y_i and an external load admittance Y_e. The pressure at the driving point is

$$p = \frac{U}{Y_i + Y_e},$$

and the power supplied is a maximum when $Y_e = Y_i^*$.

Definition (1) of impedance matching, given for waves incident on a boundary, is a special case of the present definition. It corresponds to $Z_i = Z_{\text{char}}$, where Z_{char} is the CHARACTERISTIC IMPEDANCE of the incident waves in the normal direction. Maximization of the absorbed power, for a given incident wave amplitude, is seen to require $Z_e = Z_{\text{char}}^*$. Since Z was assumed real, this amounts to making the boundary impedance equal to the wave-system characteristic impedance.

impedance matrix – (1) *for a one-port acoustical system, e.g. a termination in a multimode acoustic waveguide* an abbreviation for TERMINATION IMPEDANCE MATRIX. *Units* $\text{Pa}\,\text{s}\,\text{m}^{-1}$.

impedance matrix – (2) *for a network with two or more ports, or any extended system with inputs and responses at two or more discrete points* an abbreviation for MULTIPORT IMPEDANCE MATRIX.

Note: The term *impedance matrix* in the multiport sense is standard terminology in electrical engineering. The main reason in acoustics for preferring the longer terms *termination* ~ or *multiport* ~ is to avoid confusion between the two different meanings (1) and (2) above.

impedance matrix – (3) *for the driving-point response of a mechanical system* an abbreviation for DRIVING-POINT IMPEDANCE MATRIX. *Units* N s m^{-1}.

impedance ratio – a normalized impedance; it usually refers to a SPECIFIC ACOUSTIC IMPEDANCE at a boundary, or in a standing-wave field, divided by the CHARACTERISTIC IMPEDANCE of the acoustic medium. *Units* none.

impedance-translation theorem – a relationship that connects the impedances at two points in a standing-wave system. For standing waves that have the same CHARACTERISTIC WAVENUMBER k and CHARACTERISTIC IMPEDANCE Z_{char} for either direction of propagation (e.g. plane waves in a pipe, but not in a horn or flow duct), the impedance-translation theorem is conveniently expressed in terms of the impedance ratio $Z(x)/Z_{\text{char}} = \zeta(x)$:

$$\zeta(x-d) = \frac{\zeta(x) + j\tan kd}{1 + j\zeta \tan kd}.$$

Here x is the coordinate in the direction of incident-wave propagation, and $Z(x)$ is the impedance looking in the x direction; the time factor is $e^{j\omega t}$.

impedance tube – a uniform rigid-walled tube filled with fluid, with a transducer at one end that excites plane acoustic waves. Axial standing waves, set up by reflection from the other end (called the termination), are measured using a traversing microphone or two or more fixed microphones, in order to determine the termination impedance. Also known as a STANDING WAVE TUBE.

improperly expanded jet – a locally-supersonic gas jet produced by operating a nozzle at a supercritical pressure ratio different from its design value. Standing shock waves are produced in the nozzle and the downstream jet, that can interact with unsteady flow in the jet pipe – or with turbulence in the mixing region – to generate noise.

impulse – the time integral of a transient force, given by

$$\mathbf{J} = \int_{-\infty}^{\infty} \mathbf{F}(t)\, dt \quad \text{or} \quad \mathbf{J} = \int_{t_1}^{t_2} \mathbf{F}(t)\, dt.$$

Here $\mathbf{F}(t)$ is the force, and the integral is taken either over the entire time-history, or between specified limits (as in the second integral above). The impulse vector represents the total MOMENTUM transferred by the force during that time. *Units* m kg s^{-1} ≡ N s.

impulse response – abbreviation for IMPULSE RESPONSE FUNCTION.

impulse response function – *of a linear time-invariant system* the function $h(t)$ that gives the system output, when the system is driven by a DELTA-FUNCTION input at time $t = 0$. Provided the system is STABLE and CAUSAL, $h(t)$ is zero for all negative values of t. Its Fourier transform with respect to t is the system FREQUENCY RESPONSE FUNCTION.

Note: The system response to an *arbitrary* transient input that begins at time t_0 may be found by SUPERPOSITION: it is the sum of the responses to a sequence of elementary inputs, beginning at t_0 and ending at the current time t. See CONVOLUTION.

incidence – *of a particular disease or disability* the number of new cases arising per year in a specified population, usually normalized to a standard population size of 10^5. Also known as ***incidence rate***.

incidence angle – (1) *in acoustics* the ANGLE OF INCIDENCE. *Units* rad (but commonly expressed in degrees).

incidence angle – (2) *in aerodynamics* the angle between the incident flow direction and the chord of a blade or airfoil. *Units* rad (but commonly expressed in degrees).

incident field – *with respect to a scattering object or reflecting boundary* the sound field that would be present if the scatterer or boundary were removed to allow free-field propagation.

incoming wave – a progressive sound wave that propagates inward towards a focus; the opposite of an outgoing wave. Examples are a SPHERICAL WAVE that propagates in towards the origin, and a CYLINDRICAL WAVE that propagates in towards the axis. An alternative term is ***converging wave***.

incompressible flow – an idealized fluid flow in which the density of every fluid particle remains constant over time. A consequence is that the divergence of the fluid velocity field is zero at every point in the flow: div **u** = 0.

Note (1): A flow can still be incompressible even when different fluid particles have different densities, as when a river (fresh water) flows into the sea (salt water of higher density).

Note (2): Diffusion, either of heat or of chemical components, can blur density differences over time. Strictly, the incompressible flow description is violated when this occurs. Compare INCOMPRESSIBLE FLUID.

incompressible fluid – a fluid whose SOUND SPEED is infinite.

incremental – a term used for small changes in the properties of a medium that are caused by a sound field. For example, the *incremental density* is the difference $(\rho - \rho_0)$, where ρ is the instantaneous density at a point and ρ_0 is the ambient or undisturbed density.

independent random variables – see STATISTICALLY INDEPENDENT.

index – *in acoustics* a generic term for the decibel representation of a ratio or factor, as in *directivity* ∼, *radiation* ∼, *sound reduction* ∼.

Note: This convention is not always followed, particularly where terms are borrowed from fields outside acoustics (e.g. *refractive index, modulation index*).

index of refraction – an equivalent term for REFRACTIVE INDEX. *Units* none.

induced mass – *of an accelerating rigid body in an incompressible fluid* an equivalent term for ATTACHED MASS or VIRTUAL MASS. *Units* kg.

inertance – see ACOUSTIC INERTANCE. Compare ACCELERANCE. *Units* kg m^{-4}.

inertia base – (1) *in vibration isolation* a massive block on which a machine is mounted, to increase the isolated mass and reduce the MOTION TRANSMISSIBILITY.

inertia base – (2) a massive block used as a platform for a shaker.

inertial cavitation – refers to bubble oscillations in which the bubble collapse phase is dominated by inertial forces in the liquid, rather than by bubble pressure or interfacial phenomena. Inertial cavitation typically involves the bubble volume changing by a factor of 10 or more. See also CAVITATION.

inertial frame of reference – a frame of reference in which Newton's first law applies; i.e. the velocity of a particle of fixed mass remains constant over time, unless there are forces acting on the particle.

infinite impulse response – describes a DIGITAL FILTER whose output is determined by an infinite number of earlier values of the input. Its IMPULSE RESPONSE FUNCTION therefore contains an infinite number of non-zero terms.

infrasonic frequency – any frequency in the range (approximately) 1 to 20 Hz, which covers the frequencies characteristic of whole-body vibrational response and also extends up to the lower boundary of the audio frequency range.

infrasonic waves – *in the Earth's atmosphere* long-period INTERNAL WAVES of period 10 s or more, also known as acoustic–gravity waves, that are typically generated by seismic vertical motion of the Earth's crust (compare TSUNAMI), or by unstable large-scale atmospheric motion.

infrasound – sound at frequencies below (roughly) 15–20 Hz, and therefore normally inaudible by humans.

inhomogeneous boundary conditions – boundary conditions that contain at least one term in which the solution variable does not appear.

Note: In acoustics, all boundary conditions in which a disturbance is imposed at the boundary fall in this category.

inhomogeneous differential equation – an alternative term for NON-HOMOGENEOUS DIFFERENTIAL EQUATION.

inhomogeneous medium – a medium whose properties are spatially non-uniform. In acoustics, the term implies that the local ambient density and sound speed are functions of position, and likewise – if the medium is moving – the mean flow velocity.

inhomogeneous plane wave – *in a lossless medium* a single-frequency PROGRESSIVE WAVE for which the WAVENUMBER VECTOR $\mathbf{k} = (k_x, k_y, k_z)$ is purely imaginary in one direction, and purely real in the plane normal to that direction. Surfaces of constant amplitude and of constant phase are therefore orthogonal. Mathematically, the usual propagation factor $\exp j(\omega t - \mathbf{k} \cdot \mathbf{x})$ still applies; but the fact that $\mathbf{k}$ has an imaginary component means the wave amplitude varies with position $\mathbf{x}$.

As an example, it is always possible to orientate the x, y, z axes so that k_z is zero; the Helmholtz equation for free-wave propagation then gives

$$k_x^2 + k_y^2 = k_0^2$$ (where k_0 is the acoustic wavenumber).

It follows that if either of the component wavenumbers (k_x, k_y) is greater in magnitude than k_0, an inhomogeneous plane wave is produced.

Note: Such a wave is generated when subsonic surface waves, travelling on an infinite plane boundary, radiate into a fluid-filled half-space. The wave motion is EVANESCENT, i.e. it decays exponentially with distance away from the boundary.

initial conditions – *applied to the time-dependent solution of a differential equation* conditions imposed at the start of the solution (typically at time $t = 0$). If the differential equation is of second order in time t, initial conditions are usually applied to the solution variable and its first time derivative.

initial phase – (1) *of a time-harmonic signal* an alternative term for PHASE ANGLE. It equals the INSTANTANEOUS PHASE (1) of the signal at time $t = 0$. *Units* rad.

initial value problem – a mathematical problem in which values of the unknown variable u, and/or its derivatives, are specified at some initial time (e.g. $t = 0$), and a differential equation has to be solved to find u for all later times. This is possible only if the problem is CAUSAL; it must also be integrable (i.e. not lead to chaotic solutions with sensitive dependence on initial conditions).

inner ear – this term, and its less common equivalent *internal ear*, are used by some authors to denote the COCHLEA.

Alternatively, the inner ear may refer to the entire fluid-filled innermost part of the ear, situated within the temporal bone of the skull. This larger structure embraces the cochlea and VESTIBULAR SYSTEM; it takes the form of a hollow sac of complex interior geometry called the ***membraneous labyrinth***, suspended inside a bony cavity (the ***bony labyrinth***).

input – (1) *to a physical or analogue system* a physical excitation applied to such a system: e.g. *input force*. Also, a continuous signal on which an ANALOGUE SYSTEM operates.

input – (2) *to a digital system* a sequence of numbers on which a DIGITAL SYSTEM operates.

input impedance – *of a linear two-port system* the IMPEDANCE of the system viewed from the input port.

insert earphone – a small EARPHONE that is inserted in the ear canal.

insertion gain – see INSERTION LOSS. *Units* dB.

insertion loss – *of a component in a linear transmission system* the decrease in signal level, at a point in the transmission path downstream of that component, due to inserting the component between the source and the point in question. See also TRANSMISSIBILITY. Symbol IL. *Units* dB.

Note (1): If the signal level increases, the amount of increase is called the ***insertion gain***.

Note (2): The insertion loss of a component in a transmission system is not an inherent property of that component, but depends additionally on the load (or termination) impedance and the source impedance. In the special case where the termination and the source are both anechoic, the insertion loss and the TRANSMISSION LOSS (1) of the component are the same.

insonate – to irradiate with sound, especially at ultrasonic frequencies; e.g *insonate a target*. Possibly formed by analogy with *insolate* (= to irradiate with solar energy). Compare ENSONIFY.

insonify – alternative spelling of ENSONIFY. Hence *insonification*, the process of irradiating a target or medium with sound.

instantaneous acoustic energy density – the instantaneous sum of the kinetic energy density and the potential (or compressional) energy density, at a point in an acoustic field. Compare ACOUSTIC ENERGY DENSITY, which normally refers to the time-averaged value. *Units* $J\,m^{-3}$.

instantaneous acoustic intensity – the instantaneous energy flow per unit area at a point in an acoustic field. For sound waves in a stationary medium, it is given by

$$\mathbf{I}(t) = p(t)\mathbf{u}(t),$$

where $p(t)$ is the acoustic pressure and $\mathbf{u}(t)$ is the particle velocity vector at time t. An alternative term is ACOUSTIC ENERGY FLUX VECTOR. *Units* $W\,m^{-2}$.

Note (1): Compare ACOUSTIC INTENSITY, which refers to the time-averaged intensity. See also REACTIVE INTENSITY, note (3).

Note (2): When a mean flow is present, the instantaneous acoustic intensity may be generalized as follows:

$$\mathbf{I}(t) = p\mathbf{u}' + (\mathbf{M}\cdot\mathbf{u}')(\mathbf{M}p + \rho c\mathbf{u}') + \mathbf{M}\frac{p^2}{\rho c}.$$

The intensity component perpendicular to the mean flow, for example, is related to the fluctuating velocity component in that direction by

$$I_\perp(t) = (p + \rho c\mathbf{M}\cdot\mathbf{u}')u'_\perp,$$

where subscript $\perp$ denotes the component normal to the mean flow. In these equations $\mathbf{u}'$ is the fluid velocity fluctuation (previously denoted by $\mathbf{u}$), ρ and c are the density and sound speed, and $\mathbf{M}$ is the mean flow Mach number vector.

The $\mathbf{I}(t)$ expression above, combined with an appropriate definition of energy density, leads to a CONSERVATION LAW for acoustic energy in irrotational homentropic flows.

instantaneous amplitude – *of a quasi-monochromatic signal* the time-dependent modulus $|\check{x}(t)|$, where $\check{x}(t)$ denotes the ANALYTIC SIGNAL that corresponds to the actual signal $x(t)$.

instantaneous angular frequency – *of a quasi-monochromatic signal* the time derivative of the INSTANTANEOUS PHASE, $\varphi(t)$:

$$\omega(t) = \frac{d}{dt}[\varphi(t)].$$

Units rad s^{-1}.

instantaneous envelope – *of a quasi-monochromatic signal* an equivalent term for the INSTANTANEOUS AMPLITUDE of the signal.

instantaneous frequency – *of a quasi-monochromatic signal* the INSTANTANEOUS ANGULAR FREQUENCY, $\omega(t)$, divided by 2π. For example, the signal $A\cos[2\pi f_0 t + \theta(t)]$ has an instantaneous phase $2\pi f_0 t + \theta(t)$; its instantaneous frequency is therefore $f_0 + (1/2\pi)d\theta/dt$. See also MODULATION. *Units* Hz.

Note: The term is sometimes used to mean the instantaneous angular frequency in radians per second.

instantaneous phase – (1) *of a single-frequency signal or wave field* the argument of the cosine function that describes the signal or field. For example the signal $\cos(\omega t + \alpha)$, where t is time and (α, ω) are real constants, has instantaneous phase $(\omega t + \alpha)$ at time t. Compare PHASE ANGLE. *Units* rad.

Note: If an acoustic pressure disturbance p propagates as a WAVE in the x direction, with $p = A\cos(\omega t - kx + \alpha)$ where (A, k) are positive constants, the instantaneous phase of the wave at position x and time t is $(\omega t - kx + \alpha)$; a *point of constant phase* moves at speed ω/k in the x direction. Generalizing to three dimensions, $p = A\cos(\omega t - \mathbf{k}\cdot\mathbf{x} + \alpha)$ represents waves that propagate in the direction of the wave vector $\mathbf{k}$.

instantaneous phase – (2) *of a quasi-monochromatic signal* the time-dependent phase of the ANALYTIC SIGNAL $\check{x}(t)$ that corresponds to the actual signal $x(t)$. Since $\check{x}(t)$ is defined as $x(t) + jx_H(t)$, where $x_H(t)$ is the Hilbert transform of the signal $x(t)$, the instantaneous phase $\varphi(t)$ is given by

$$\varphi(t) = \tan^{-1}[x_H(t)/x(t)] = \text{Im}[\ln \check{x}(t)].$$

Units rad.

instantaneous power – *in acoustics* abbreviation for INSTANTANEOUS SOUND POWER. *Units* W.

instantaneous sound power – the instantaneous rate of energy flow across a given surface. The ∼ *of a source* is the instantaneous rate of energy output from the source into the surrounding medium. See also INSTANTANEOUS ACOUSTIC INTENSITY. *Units* W.

instantaneous sound pressure – equivalent term for SOUND PRESSURE. See also EFFECTIVE SOUND PRESSURE. *Units* Pa.

integral transform – an operation of the form

$$\int K(y, x) f(x) \, dx = F(y)$$

that maps a function of x into a different function of y. The integral over x is always between definite limits that are specified according to the type of transform. A schematic representation of the transform relation above is

$$f(x) \to F(y);$$

if there is a unique inverse it is represented by

$$f(x) \leftarrow F(y).$$

This leads to the notion of an ***integral transform pair***, such that $F(y)$ is the ***forward transform*** of $f(x)$ and $f(x)$ is the ***inverse transform*** of $F(y)$. Schematically, the relationship is expressed by

$$f(x) \leftrightarrow F(y).$$

See FOURIER TRANSFORM, HANKEL TRANSFORM, HILBERT TRANSFORM, LAPLACE TRANSFORM.

Note: The direction of the arrows in the equations above is significant; the arrow pointing from left to right denotes the forward transform, and the opposite arrow denotes the inverse transform.

integrated impulse response method – *for measuring the reverberation time and early decay time in a room* a method of arriving at an averaged decay curve for the mean square pressure at a point in a room, based on processing the measured impulse response of the room. The IMPULSE RESPONSE FUNCTION $h(t)$ between two positions in the room is squared and integrated according to the equation below:

$$D(t) = \int_t^{t_{max}} h^2(t') \, dt' = \int_0^{t_{max} - t} h^2(t_{max} - u) \, du.$$

In the second version of the integral, the squared impulse response is integrated backwards from the upper limit t_{max} (which is set by the noise floor of the impulse response). The result is a smoothed decay curve, $D(t)$, for the squared pressure at the receiver position; it is equivalent to the average of infinitely many interrupted-noise decays obtained by the traditional method.

Note (1): An advantage over the traditional INTERRUPTED NOISE METHOD is that no averaging time is involved; the traditional method gives different decay curves for mean square pressure, according to the integration time used.

Note (2): The room impulse response function is normally filtered before squaring, to yield band-limited reverberation times T_B (e.g. for octave or 1/3-octave bands); this makes it unnecessary to specify in detail the rest of the excitation system – all of which in principle contributes to $h(t)$ – because of the narrow frequency range. However, the filter bandwidth B must be kept large in comparison with the room decay rate (i.e. $BT_B \gg 10$), in order to avoid measuring the decay time of the filter.

Note (3): The influence of background noise on the smoothed decay curve $D(t)$ can be minimized by carrying out two separate measurements of $h(t)$, say $h_1(t)$ and $h_2(t)$, and using their product to compute the decay curve:

$$D(t) = \int_t^{t_{max}} h_1(t')\, h_2(t')\, dt'.$$

integrated tone-burst method – the original name given to the INTEGRATED IMPULSE RESPONSE METHOD.

intelligibility – *of speech in a particular listening environment* a qualitative term that describes the ability of the acoustic environment to transmit speech intelligibly, usually expressed relative to perfect listening conditions. Intelligibility can be quantified for a particular speech sample by asking listeners to record their interpretation of what they hear, and processing the data to obtain the PERCENTAGE SYLLABLE ARTICULATION (or the corresponding percentage for ***words*** or ***sentences***).

intensity – *in acoustics* abbreviation for ACOUSTIC INTENSITY. *Units* $\mathrm{W\,m^{-2}}$.

intensity attenuation coefficient – an equivalent term for ENERGY ATTENUATION COEFFICIENT. Also known as ***bulk intensity attenuation coefficient***, especially in ultrasonics. *Units* $\mathrm{m^{-1}}$.

intensity cross-section – a generic term for either SCATTERING CROSS-SECTION or ABSORPTION CROSS-SECTION. *Units* $\mathrm{m^2}$.

intensity focal gain – *of an acoustic transducer that produces a single-frequency focused beam* the square of the AMPLITUDE FOCAL GAIN. Also known as ***intensity gain*** or ***intensity gain factor***. *Units* none.

intensity level – *in a sound field* the quantity L_I defined by

$$L_I = 10\,\log_{10}(I/I_{ref}),$$

where I is the ACOUSTIC INTENSITY in a specified direction and I_{ref} is the reference intensity. See also LEVEL, REFERENCE VALUE. *Units* dB re 1 pW m^{-2}.

interaural cross-correlation coefficient (IACC) – *for a specified listener position in a room* a measure of the similarity between the signals received by the two ears. It is obtained by cross-correlating the impulse response functions of the room measured at the two ear positions, over a sample period that is usually defined as the first 80 ms after the direct sound arrival. The IACC is defined as the peak value of the normalized cross-correlation function between the two signals, over a range of signal delays of ± 1 ms. *Units* none.

Note: IACC is often considered to relate to the various components of SPATIAL IMPRESSION. A low IACC implies large APPARENT SOURCE WIDTH.

interaural time difference (ITD) – the time delay of the acoustic pressure at one ear relative to the pressure at the other ear, for a given frequency and a given source location relative to the head. The relevant acoustic pressures are those measured on the outer surface of the tympanic membrane. *Units* s.

interface wave – *at a boundary between two media* a wave that propagates along the interface, with a disturbance amplitude that decays away from the interface in either direction. Leaky interface waves radiate energy into one or both media, and consequently cannot propagate unattenuated along the interface. True interface waves radiate no energy and propagate without attenuation in lossless media. See RAYLEIGH WAVE, SCHOLTE WAVE, STONELEY WAVE.

interference – *between two or more coherent wave fields that are stationary in time* a consequence of superposition: the combined field has a local mean square value different from the sum of the mean square values of the components. For example, wherever two harmonic fields with the same frequency are in ANTIPHASE, the signal amplitudes *subtract*, and where the component signals are in phase, the signal amplitudes *add*. In the first case, called **destructive interference**, the mean square signal is less than either of the component values; in the second case called **constructive interference**, it is greater.

More generally, interference occurs when any two CORRELATED SIGNALS are added, since the relation $p = p_1 + p_2$ (real instantaneous quantities, not phasors) implies $p^2 = p_1^2 + p_2^2 + 2p_1p_2$. The contribution of the $2p_1p_2$ term to the mean square pressure is responsible for the peaks and troughs in an acoustic STANDING WAVE, for example. If the two signals were uncorrelated, the cross term would have zero time average, and there would be no interference.

interference pattern – see STANDING WAVE (2).

interferometer – *in ultrasonics* see ULTRASONIC INTERFEROMETER.

internal auditory meatus – the passage within the temporal bone that accommodates the cochlear and vestibular nerves, as well as the blood supply to the cochlea and balance organs.

internal ear – an alternative term (not in common use) for INNER EAR.

internal energy – energy that is not kinetic. It includes compressional energy, thermal energy, and energy associated with chemical bonds. See KINETIC ENERGY. *Units* J.

internal resonance – *of a multi-degree-of-freedom system* a situation that exists when a group of natural frequencies is connected by a relation of the form

$$m_1\omega_1 + m_2\omega_2 + \ldots + m_n\omega_n \approx 0 \qquad (m_1, m_2, \text{etc. integers}).$$

Here ω_1, ω_2, etc. are the natural frequencies of the modes that take part in the internal resonance. Strong nonlinear coupling of modes occurs under finite-amplitude excitation when two or more modal frequencies are related in this way.

internal waves – gravity waves at the horizontal interface between any two fluids of different density. More generally, internal waves can propagate in any fluid whose density is horizontally stratified; see BUOYANCY FREQUENCY. The restoring force in all cases is positive or negative buoyancy, encountered as a fluid particle moves downwards or upwards respectively from its equilibrium position.

Note: The propagation of acoustic waves in a horizontally-stratified fluid is influenced by gravity at sufficiently low acoustic frequencies (i.e. frequencies comparable with the buoyancy frequency). Waves in which both gravitational stability and fluid compressibility play a significant part are called ***acoustic–gravity waves***.

interoceptor – a type of sensory receptor located in the viscera that transmits information about bodily movement or acceleration.

interrupted noise method – *for measuring the reverberation time of a room* this traditional method involves exciting the room with a steady-state noise that is abruptly switched off. The sound pressure signal, recorded at a point in the room, is filtered and converted into sound pressure level to produce a decay curve. The reverberation time is based on the rate of decay of level after the sound is switched off. Filtering is usually in octaves or 1/3 octaves.

intimacy – *in concert-hall acoustics* the sense of involvement with, as opposed to detachment from, the musical performance in a concert hall. Intimacy is a property of the sound heard by the listener, and tends to be higher for listening positions closer to the performing platform.

intrinsic absorption – *in a composite medium with inhomogeneities much smaller than an acoustic wavelength* the total absorption of acoustic energy that would occur in each of the component media, if the sound field were not scattered by the inhomogeneities. Specifically, if an emulsion or bubbly liquid contains volume fraction x of the disperse phase (with attenuation coefficient α_2), and volume fraction $1 - x$ of the continuous phase (with attenuation coefficient α_1), the ATTENUATION COEFFICIENT attributed to intrinsic absorption is

$$\alpha_{\text{int}} = x\,\alpha_2 + (1 - x)\alpha_1.$$

in vacuo – literally, in a vacuum. The *in vacuo* modes of a structure are the modes in the absence of fluid loading, and the *in vacuo* natural frequencies are the corresponding modal natural frequencies.

inverse circular functions – the multivalued functions $\cos^{-1} x = u(x)$, $\sin^{-1} x = v(x)$, $\tan^{-1} x = w(x)$; they are defined by $x = \cos u$, $x = \sin v$, $x = \tan w$ respectively. Also known as ***inverse trigonometric functions***. Alternative notations are:

inverse cosine	$\cos^{-1} x = \arccos x$;
inverse sine	$\sin^{-1} x = \arcsin x$;
inverse tangent	$\tan^{-1} x = \arctan x$.

Any one branch of these functions spans a range of π. The value of the function changes by π from one branch to the next (for given x). It is convenient for computational purposes to designate one branch as the ***principal branch*** (e.g. for the $\tan^{-1}$ function, this is $-\pi/2 \le w < \pi/2$); the value of the function along that branch is then called the ***principal value***.

inverse Fourier transform – see FOURIER TRANSFORM.

inverse matrix – the *inverse* of a square matrix $\mathbf{A}$, if it exists, is another square matrix $\mathbf{A}^{-1}$ with the property

$$\mathbf{A}\mathbf{A}^{-1} = \mathbf{A}^{-1}\mathbf{A} = \mathbf{I}.$$

Here $\mathbf{I}$ is the IDENTITY MATRIX of the same size as $\mathbf{A}$.

Note: If $\det \mathbf{A} = 0$, the matrix $\mathbf{A}$ is SINGULAR and no inverse exists.

inverse problems – *in mathematical modelling or data interpretation* problems in which a mathematical model is sought, subject to certain constraints, such that the model output is consistent with a given set of data. An example in underwater acoustics is the problem of determining the parameters of a layered ocean model from a set of sound transmission measurements between a source and an array of receivers; compare MATCHED FIELD PROCESSING.

Inverse problems are the opposite of ***forward*** or ***direct problems***, which involve analysis of a predetermined model in order to predict an outcome. Forward and inverse problems are both encountered in mathematical modelling: a proposed solution to the inverse problem has to be tested by reversing the process, i.e. solving a forward problem, to see if the data are correctly predicted.

inversion – *in the earth's atmosphere* abbreviation for TEMPERATURE INVERSION.

irradiance – *in acoustics* abbreviation for ACOUSTIC IRRADIANCE. *Units* $W\,m^{-2}$.

irradiate – *used as a transitive verb* to expose something (an object or surface) to an acoustic field.

irradiation strength – *in acoustics* an equivalent term for ACOUSTIC IRRADIANCE. *Units* $W\,m^{-2}$.

irreversibility – *of a process* irreversibility in a thermodynamic process literally means that the process is not REVERSIBLE, and implies the presence of DISSIPATION. A positive indicator of irreversibility is the net generation of entropy.

irrotational – refers to a vector field (typically the displacement or velocity in a continuous medium) that can be expressed as the gradient of a scalar potential φ; in mathematical form, $\mathbf{u} = \nabla\varphi$. A consequence is that the curl of the vector field is zero everywhere: $\nabla \times \mathbf{u} = \nabla \times \nabla\varphi = 0$.

irrotational flow – a fluid flow in which the VORTICITY is zero at every point. The velocity field in such a flow can be written as the gradient of a potential:

$\mathbf{u} = \nabla\varphi$ (where φ is called the VELOCITY POTENTIAL).

isentrope – a contour line joining points of equal specific entropy; also, a curve showing the connection between two state variables under conditions of constant specific entropy.

isentropic – *for a system* refers to any change of thermodynamic state in which the ENTROPY of the system remains unchanged. In the particular case of a homogeneous substance, this means the entropy per unit mass (specific entropy) remains unchanged. Fluid compressions and expansions in a sound field are approximately isentropic away from boundaries, implying that small changes in pressure and density following a fluid particle are related by $dP \approx c^2 d\rho$, where c is the sound speed of the fluid.

Note: An ADIABATIC process is isentropic provided it is REVERSIBLE, i.e. non-dissipative.

isentropic equation of state in series form – a Taylor series expansion of pressure in terms of density along an isentrope, in the form

$$P - P_0 = A\,\zeta + B\frac{\zeta^2}{2!} + C\frac{\zeta^3}{3!} + \cdots .$$

Here subscript 0 labels a reference state of the fluid, with pressure P_0 and density ρ_0, and ζ is the relative increase in density defined by

$$\zeta = \frac{\rho - \rho_0}{\rho_0} \qquad \text{(also known as the CONDENSATION).}$$

The coefficients in the series expansion are given by

$$A = \rho_0 \left(\frac{\partial P}{\partial \rho}\right)_{s,0},\ B = \rho_0^2 \left(\frac{\partial^2 P}{\partial \rho^2}\right)_{s,0},\ C = \rho_0^3 \left(\frac{\partial^3 P}{\partial \rho^3}\right)_{s,0},\ \ldots$$

where subscript s indicates that the derivatives are evaluated at constant specific entropy.

Note: For practical purposes, it is useful to divide the original equation through by A, recognizing that $A = \rho_0 c_0^2$ (c = speed of sound in the fluid):

$$\frac{P - P_0}{\rho_0 c_0^2} = \zeta + \frac{B}{A}\frac{\zeta^2}{2!} + \frac{C}{A}\frac{\zeta^3}{3!} + \cdots .$$

The dimensionless ratios B/A, C/A, etc. are called the **nonlinearity parameters of order 2, 3,** etc., for the given reference state (P_0, ρ_0).

isentropic exponent – *for a fluid* the ratio $\kappa = B/P$, where B is the isentropic bulk modulus of the fluid ($B = \rho c^2$, in terms of density ρ and sound speed c) and P is the absolute pressure. *Units* none.

Note (1): The isentropic exponent for air at atmospheric temperatures and pressures remains nearly constant during an ISENTROPIC compression process. The same applies to any gas at densities much less than the critical density (see CRITICAL STATE). Compare TAIT–KIRKWOOD ISENTROPE EQUATION.

Note (2): An equivalent definition of the isentropic exponent is

$$\kappa = \left.\frac{\partial(\ln P)}{\partial(\ln \rho)}\right|_s ;$$

the subscript s indicates that the derivative is evaluated at constant specific entropy. If κ is a constant, the isentropic pressure–density relation is $P \propto \rho^\kappa$. For the equivalent exponent relating temperature and density, rather than pressure and density, see GRÜNEISEN PARAMETER.

isentropic process – *for a system* refers to any change of thermodynamic state in which the total ENTROPY of the system remains unchanged.

Note: An ADIABATIC process is isentropic provided it is REVERSIBLE, i.e. equilibrium is maintained at all times. Reversible processes are non-dissipative.

ISO – abbreviation for ***International Organization for Standardization***.

isochore, isochor – a contour line joining points of equal specific volume; also, a curve showing the connection between two state variables under conditions of constant specific volume.

isolation – the process of reducing the transmission of sound or vibration to a system, by blocking the transmission path with a device that produces a positive INSERTION LOSS. A typical isolation device for airborne sound is a duct silencer; a typical vibration isolator is a resilient element.

isolation efficiency – *of a linear device used for sound or vibration isolation* the quantity $100(1 - T)$, where T is the TRANSMISSIBILITY of the device as installed. *Units* %.

isotherm – a contour line joining points of equal temperature; also, a curve showing the connection between two state variables under conditions of constant temperature.

isothermal process – a process that takes place at constant temperature.

isotropic – the same in all directions; non-directional. Thus a solid material that is isotropic has no preferred directions. See also ANISOTROPIC, TRANSVERSELY ISOTROPIC.

J

j – one of the square roots of -1. Which of the two roots ($-\sqrt{-1}$ or $\sqrt{-1}$) is identified with j depends on the convention chosen for the SQUARE ROOT operator.

jerk – the time derivative of acceleration, i.e. $\dot{\mathbf{v}}$, where $\mathbf{v}$ is the velocity vector.

jet noise – aerodynamic sound produced by unsteady mixing between a jet and the surrounding fluid. See also LIGHTHILL'S U^8 POWER LAW.

jet screech – intense tonal sound radiated from an improperly expanded supersonic jet, as a result of a feedback cycle between the jet nozzle and shock waves in the downstream flow.

joint cdf – abbreviation for JOINT CUMULATIVE DISTRIBUTION FUNCTION.

joint cumulative distribution function – *of the paired random variables X and Y* the function $F(x, y)$ that gives the probability that variable X will be less than or equal to a specified value x, and variable Y will be less than or equal to a specified value y. In symbols,

$$F(x, y) = P(X \leq x, Y \leq y),$$

where $P(A, B)$ is the probability of statements A and B both being true.

The function $F(x, y)$ is related to the JOINT PROBABILITY DENSITY FUNCTION $f(x, y)$ by

$$F(x, y) = \int_{-\infty}^{x} du \int_{-\infty}^{y} dv\, f(u, v).$$

jointly stationary – two stationary random signals, $x(t)$ and $y(t)$, are called jointly stationary if the JOINT PROBABILITY DENSITY FUNCTION of $x(t)$ and the shifted signal $y(t + \tau)$ is independent of time t, for any given value of τ.

joint pdf – abbreviation for JOINT PROBABILITY DENSITY FUNCTION.

joint probability – *of two events* the probability that both events will occur. It is denoted by $P(A, B)$, where A and B are the separate events, or outcomes, whose joint probability is being quantified. Some authors use the equivalent notations $P(A \cap B)$ or $P(AB)$ for $P(A, B)$.

joint probability density function – *of the paired random variables X and Y* the function $f(x, y)$ whose integral gives the JOINT CUMULATIVE DISTRIBUTION FUNCTION:

$$P(X \leq x, Y \leq y) = \int_{-\infty}^{x} du \int_{-\infty}^{y} dv\, f(u, v).$$

Provided X and Y both take a continuous range of values, it follows that

$$P(x < X < x + dx, y < Y < y + dy) = f(x, y)\, dx\, dy;$$

i.e. the probability that X and Y will simultaneously take values in an elementary region dx by dy is $f(x, y)$ multiplied by the area $dx\, dy$. Further, the probability of X, Y lying within any closed region C of the x–y plane is given by the integral $\iint f(x, y)\, dx\, dy$ evaluated over C. The function $f(x, y)$ is related to the JOINT CUMULATIVE DISTRIBUTION FUNCTION $F(x, y)$ by

$$f(x, y) = \frac{\partial^2 F(x, y)}{\partial x\, \partial y}.$$

Note: Joint probability density functions may also be defined for three or more random variables. For example, the joint pdf $f(x, y, z)$ of the three variables X, Y, Z is defined by

$$P(X \leq x,\ Y \leq y,\ Z \leq z) = \int_{-\infty}^{x} du \int_{-\infty}^{y} dv \int_{-\infty}^{z} dw\, f(u, v, w).$$

joint statistics – the STATISTICS (2) of two or more random variables in combination, based on the JOINT PROBABILITY DENSITY FUNCTION of the random variables concerned.

judged perceived noise level – the sound pressure level of an octave band of pink Gaussian noise, centred on 1000 Hz and of duration 2 s, that appears to a normal observer to be equal in noisiness to the sound in question. The comparison signal is required to be a plane progressive wave, incident from directly in front and heard binaurally. Compare LOUDNESS LEVEL. *Units* dB re $(20\, \mu\text{Pa})^2$.

K

Keller's geometrical theory of diffraction – see GEOMETRICAL THEORY OF DIFFRACTION (**J B KELLER 1958**).

kelvin – the SI base unit of temperature. The kelvin scale measures absolute or thermodynamic temperature (**W THOMSON, LORD KELVIN 1851**). *Unit symbol* K.

Kelvin–Helmholtz instability – *in inviscid fluid dynamics* the instability of a VORTEX SHEET to small disturbances. The two fluids on either side of the vortex sheet are assumed to be in steady uniform motion initially; they may have the same or different densities.

Kelvin's theorem – see CIRCULATION.

Kelvin–Voigt model – *of a viscoelastic material with time-independent properties* an alternative term for VOIGT MODEL.

Keulegan–Carpenter number – *for oscillatory fluid flow relative to a solid body* the dimensionless quantity

$$K = \frac{u_{\max} T}{L},$$

where $u_{\max}$ is the maximum relative velocity of the flow past the body, T is the period of the oscillatory flow, and L is a typical length scale of the body (usually measured in the flow direction). *Units* none.

kilomole – a unit for the amount of substance in a system, equal to 10^3 MOLES. *Unit symbol* kmol.

kinaesthesis – stimulation of the sense of motion, by means other than the VESTIBULAR SYSTEM. In the human body, receptors sensitive to force or acceleration are located in the muscles and joints; see PACINIAN CORPUSCLES. Related terms are *kinaesthesia* for the resulting stimulated state, and the adjective *kinaesthetic*.

kinematic viscosity – *of a fluid* the ratio of viscosity to density; denoted by $\nu = \mu/\rho$, where μ is the fluid viscosity and ρ is the fluid density. Compare VISCOSITY. *Units* $m^2 s^{-1}$.

kinesthesis – alternative (*N Am*) spelling of KINAESTHESIS.

kinetic energy – the contribution made to the energy of a system by the macroscopic or bulk motion of its various parts. In fluid mechanics, the kinetic energy of microscopic molecular motion is considered to be part of the INTERNAL ENERGY of the fluid. *Units* J.

Note: The absolute value of the kinetic energy is dependent on the frame of reference chosen.

kinetic energy density – see ACOUSTIC ENERGY DENSITY. *Units* J m^{-3}.

Kirchhoff attenuation coefficient – *for quasi-plane-wave propagation in a rigid-walled tube* see note under LARGE-TUBE ATTENUATION COEFFICIENT. *Units* Np m^{-1}.

Kirchhoff's formula – a general solution of the wave equation $\Box^2 p = -\mathcal{F}$ in a three-dimensional region V, bounded by a closed fixed surface S (G KIRCHHOFF 1883). The solution is expressed in terms of the source density within V, together with the pressure and its derivatives on S:

$$4\pi p(\mathbf{x}, t) = \int_V [\mathcal{F}] \frac{d^3 \mathbf{y}}{r}$$
$$+ \int_S \left[\frac{\partial p}{\partial n} + \frac{\partial r}{\partial n} \left(\frac{1}{c} \frac{\partial}{\partial t} + \frac{1}{r} \right) p \right] \frac{dS(\mathbf{y})}{r}.$$

Here the variable n is a local coordinate normal to S, pointing outward from V; $d^3\mathbf{y}$ is a volume element of V, $dS(\mathbf{y})$ is a surface element of S, $\mathbf{y}$ is the position of either element, and $r = |\mathbf{x} - \mathbf{y}|$. The square brackets in both integrals imply evaluation at the emission time $t - r/c$. Kirchhoff's formula is the time-domain equivalent of the Helmholtz theorem in the frequency domain; see HELMHOLTZ INTEGRAL FORMULA.

Note: Typically the surface S consists of one or more finite closed surfaces plus a larger surrounding surface (which can be placed at infinity to represent free-field radiation). The region V lies outside the smaller finite surfaces, but inside the larger surface.

Kramers–Kronig relations – relations that connect the real and imaginary parts of any CAUSAL frequency-domain response function that is stable, linear, and tends to zero at infinite frequency:

$$H_I(\omega) = -\frac{1}{\pi} \wp \int_{-\infty}^{\infty} \frac{H_R(x)}{\omega - x} dx; \qquad H_R(\omega) = \frac{1}{\pi} \wp \int_{-\infty}^{\infty} \frac{H_I(x)}{\omega - x} dx.$$

Here $H(\omega) = H_R(\omega) + jH_I(\omega)$ is the frequency response function, and the integrals are interpreted as Cauchy principal values (symbol $\wp$). The inverse Fourier transform of $H(\omega)$,

$$h(t) = \frac{1}{2\pi} \int_{-\infty}^{\infty} H(\omega) e^{j\omega t} d\omega, \qquad (t = \text{time}),$$

is the IMPULSE RESPONSE FUNCTION of the corresponding time-invariant stable linear system; its value is zero for all negative t, by causality.

Note (1): The form of the Kramers–Kronig (K–K) relations given above is equivalent to stating that $H_R(\omega)$ is the HILBERT TRANSFORM of $H_I(\omega)$, or equivalently that $-H_I(\omega)$ is the Hilbert transform of $H_R(\omega)$:

$$H_I \leftrightarrow H_R, \qquad H_R \leftrightarrow -H_I \qquad \text{(Hilbert transform pairs)}.$$

Note (2): The K–K relations are equivalent to stating that the response function $H(\omega)$ has no POLES in the complex ω plane on or below the real axis (based on the $e^{j\omega t}$ convention as used in the definition of FOURIER TRANSFORM). Resonant systems with no damping are therefore excluded. The K–K relations also imply that $H(\omega)$ has no ZEROS below the real axis.

Note (3): In physical applications, $h(t)$ is real, which means that $H(-\omega) = H^*(\omega)$. Using this fact allows the K–K relations to be written as

$$H_I(\omega) = -\frac{2}{\pi} \wp \int_0^\infty \frac{\omega H_R(x)}{\omega^2 - x^2} dx, \qquad H_R(\omega) = \frac{2}{\pi} \wp \int_0^\infty \frac{x H_I(x)}{\omega^2 - x^2} dx$$

(H A KRAMERS, R DE L KRONIG 1926, 1942).

Note (4): In order for the integrals to converge at infinite frequency, $H(\omega)$ needs to approach zero sufficiently rapidly:

$$|H(\omega)| \sim |\omega|^{-\varepsilon}, \qquad \varepsilon > 0 \qquad (\text{as } \omega \to \pm\infty).$$

However, if a causal system has a finite (non-zero) value of $H(\infty)$, the K–K relations may be applied to the difference $H(\omega) - H(\infty)$.

Kronecker delta, Kronecker symbol – the quantity δ_{ik}, defined by the following rules for positive integer values of i and k:

$$\delta_{ik} = \begin{cases} 1, & \text{for } i = k; \\ 0, & \text{for } i \neq k, \end{cases} \qquad \text{(L KRONECKER 1866)}.$$

In situations where both i and k range from 1 to N, the values of δ_{ik} may be thought of as the elements of a square matrix, the $N \times N$ IDENTITY MATRIX.

An alternative notation is $\delta_{ik} = \delta[i - k]$. The function $\delta[m]$ defined in this way is called the UNIT PULSE FUNCTION; it is the discrete counterpart of the DIRAC DELTA FUNCTION.

Kundt's tube – *for visualizing acoustic standing waves* a rigid-walled standing-wave tube, whose inner surface is coated with a thin film of fine powder. When an axial standing wave is set up in the tube, acoustic streaming patterns develop; these drive the powder towards points of minimum particle velocity amplitude (velocity nodes, or pressure antinodes). After some time the powder forms regular striations, spaced $\lambda/2$ apart along the tube (where λ is the axial sound wavelength in the tube at the driving frequency). This was one of the earliest laboratory techniques for measuring the speed of sound in gases (**A A KUNDT 1866**).

Kutta condition – *in unsteady flow* a condition applied at a sharp trailing edge to simulate the effects of viscosity, in flow models where viscosity is not explicitly included. The Kutta condition requires that all velocities remain finite in the vicinity of the edge. It was originally developed for use in steady flow (**M W KUTTA 1911**), and has since been extensively adopted for aeroacoustic model studies based on the Euler equations.

KZK equation – *in nonlinear acoustics* a one-way propagation equation for narrow-angle beams of sound, similar to the nonlinear propagation equation (see NPE) but not limited to lossless fluids. In a uniform thermoviscous fluid, with nonlinear effects represented by the COEFFICIENT OF NONLINEARITY β and with diffusive effects represented by the DIFFUSIVITY OF SOUND δ, the KZK equation for sound pressure p is

$$\frac{\partial^2 p}{\partial z \partial \tau} = \frac{c_0}{2}\nabla_\perp^2 p + \frac{\delta}{2c_0^3}\frac{\partial^3 p}{\partial \tau^3} + \frac{\beta}{2\rho_0 c_0^3}\frac{\partial^2 p^2}{\partial \tau^2}.$$

Here z is the coordinate along the beam axis; τ is the retarded time $t - z/c_0$; ρ and c denote density and sound speed, and subscript 0 denotes undisturbed values. The Laplacian $\nabla_\perp^2$ operates in the plane normal to the beam axis, and accounts for beam diffraction in that plane (**E A ZABOLOTSKAYA, R V KHOKHLOV 1969; V P KUZNETSOV 1971**).

L

L~ – symbol for LEVEL, with the subscript used to indicate the physical quantity concerned; e.g. L_p for SOUND PRESSURE LEVEL. Variants of this notation are used in acoustics for SOUND LEVEL and EQUIVALENT CONTINUOUS SOUND LEVEL. Also see the entry under L_N for L_{10}, L_{90}, etc. *Units* dB (with an appropriate REFERENCE VALUE).

laboratory transmission loss – *for a partition separating two reverberant sound fields* an equivalent term for SOUND INSULATION; i.e. the diffuse-field transmission loss of the partition calculated from a level difference measurement. *Units* dB.

Lagrange acceleration formula – *in continuum mechanics* the equation

$$\frac{D\mathbf{u}}{Dt} = \frac{\partial \mathbf{u}}{\partial t} + \boldsymbol{\omega} \times \mathbf{u} + \nabla(\tfrac{1}{2}u^2)$$

(J L LAGRANGE 1781). It relates the acceleration of a material element to the local velocity field $\mathbf{u}(\mathbf{x}, t)$ and its curl, the vorticity $\boldsymbol{\omega}$; here $\mathbf{x}$ is position and t is time. For an alternative formula, see ACCELERATION OF A FLUID ELEMENT.

Lagrange multiplier – a parameter used in a technique for solving problems in which the variables are not independent but are functionally related. For example, let $f(x, y, z)$ be a real-valued function whose maximum or minimum value is to be determined, and suppose that the variables are connected by the equation $\varphi(x, y, z) = 0$. The latter is referred to as a *constraint*. Define an augmented function, also called a ***Lagrange function***, by $F(x, y, z) = f(x, y, z) + \lambda \varphi(x, y, z)$, where λ is a parameter. It can be shown that extrema of the function $f(x, y, z)$, subject to the constraint, coincide with extrema of the augmented function. Such points can be determined by solving the necessary conditions $\partial F/\partial x = 0$, $\partial F/\partial y = 0$, and $\partial F/\partial z = 0$, together with the equation of constraint $\varphi(x, y, z) = 0$, for the values of x, y, z, and λ. The parameter λ is called a Lagrange multiplier. This technique, called the ***method of Lagrange multipliers***, is often used to introduce constraints into VARIATIONAL PRINCIPLES.

Lagrangian – *of a system* is defined by $L = T - V$, where T and V are the KINETIC ENERGY and POTENTIAL ENERGY of the system (J L LAGRANGE 1756).

Lagrangian coordinates – *in a continuous material* a coordinate system embedded in the material, so that each material particle retains a unique set of coordinates as the material deforms. These are commonly chosen to correspond to the spatial coordinates of the particle at some initial time. Compare EULERIAN COORDINATES.

Lagrangian derivative – *in continuum mechanics* an alternative term for the MATERIAL DERIVATIVE.

Lagrangian variable – a variable expressed in LAGRANGIAN COORDINATES.

Lambert's cosine law – the relation $W_{\Omega s}(\theta) \propto \cos\theta$ that describes the directional distribution of the acoustic power scattered from a diffusely reflecting surface, when the sound power absorption coefficient of the surface is zero. Here $W_{\Omega s}(\theta)$ denotes the power per unit solid angle that is scattered or reflected from unit surface area, and the scattering direction θ is measured from the surface normal. Each surface element acts as an independent source of scattered sound power, with a $\cos\theta$ polar angle distribution relative to the normal. See DIFFUSE REFLECTION.

In the case of a partially-absorbing surface, with sound power absorption coefficient $\alpha(\theta)$, the $\cos\theta$ directional distribution for diffuse reflection is modified to $[1 - \alpha(\theta)]\cos\theta$. An equivalent statement is that the ACOUSTIC BRIGHTNESS of a diffusely-reflecting surface varies with direction as $1 - \alpha(\theta)$.

All these results are consequences of RECIPROCITY, given the diffuse reflection hypothesis. However, truly diffuse reflection is seldom approached in practice, despite its frequent use as an idealization in modelling studies.

Lamb waves – *in solid mechanics* waveguide modes in a plate with stress-free faces (H LAMB 1917), consisting of combinations of P WAVES and SV WAVES. The two lowest-order Lamb modes correspond to bending waves (the lowest antisymmetric mode) and in-plane longitudinal waves (the lowest symmetric mode). An alternative term is ***plate waves***.

Note: In the limit as the plate thickness tends to zero (thickness $h \ll b/\omega$, where b is the shear wave speed and ω is the angular frequency), only the two lowest-order modes propagate with real wavenumbers. In-plane longitudinal waves in a thin plate propagate at the ***plate longitudinal-wave speed*** $c_P = [E/\rho(1 - v^2)]^{1/2}$; here E is Young's modulus for the plate material, ρ is the density and v is Poisson's ratio. Bending waves in a thin plate propagate at the ***plate bending-wave speed*** $(\omega\kappa c_P)^{1/2}$, where $\kappa = h/\sqrt{12}$.

laminar flow – fluid flow that does not contain TURBULENCE.

Langevin radiation pressure – see ACOUSTIC RADIATION PRESSURE. *Units* Pa.

Langmuir circulation – *in lakes and oceans* wind-driven circulation currents that can carry small bubbles (diameter 20 µm or less, produced by wave action at the surface) down to depths of order 10 m, and thus influence the propagation of sound in the near-surface layer. Langmuir circulation cells are characterized by closed patterns of circulation in planes normal to the wind axis, with lateral dimensions of order 10 m to 100 m (**I LANGMUIR 1938**).

Laplace's equation – *for a scalar or vector field* the LINEAR PARTIAL DIFFERENTIAL EQUATION $\nabla^2 u = 0$, where u is the FIELD VARIABLE and ∇^2 is the LAPLACIAN operator (**P S LAPLACE *1749–1827***).

Laplace transform – an integral transform applied to a causal function $f(t)$, where t is a real variable ranging from zero to infinity. The forward transform is

$$F(s) = \int_0^\infty f(t)\, e^{-st}\, dt \qquad (s \text{ complex}).$$

Note that if s, called the **Laplace variable**, equals $j\omega$ with ω real, the result is the FOURIER TRANSFORM of $f(t)$. The inverse Laplace transform is defined by

$$f(t) = \frac{1}{2\pi j} \int_{c-j\infty}^{c+j\infty} F(s)\, e^{st}\, ds$$

where c is a real constant, chosen large enough to make the integral converge.

Laplace variable – the variable s in the LAPLACE TRANSFORM definition. *Units* s^{-1} (where the transform is from the time domain).

Laplacian – the operator ∇^2. It is defined for rectangular coordinates (x, y, z) in three space dimensions as $\nabla^2 \equiv \partial^2/\partial x^2 + \partial^2/\partial y^2 + \partial^2/\partial z^2$. An alternative symbol for the operator is $\triangle$.

Note: The modified symbol $\nabla_\perp^2$ is used for the two-dimensional Laplacian operator, in planes perpendicular to a specified axis (subscript $\perp$ denotes perpendicular). Thus if the axis is chosen to be in the x direction, $\nabla_\perp^2 \equiv \partial^2/\partial y^2 + \partial^2/\partial z^2$.

large-tube attenuation coefficient – *for quasi-plane-wave propagation in a rigid-walled tube* an expression for the viscothermal ATTENUATION COEFFICIENT for axial sound propagation in a tube, valid when the internal diameter d is large compared with the VISCOUS PENETRATION DEPTH δ. At angular frequency ω, it is given by

$$\alpha \approx \frac{\omega}{c}\left[\frac{1}{\sqrt{S}}\left(1+\frac{\gamma-1}{\sqrt{\Pr}}\right)+\frac{2}{S}\left(1+\frac{\gamma-1}{\Pr}-\frac{\gamma(\gamma-1)}{2\Pr}\right)\right] \quad (d \gg \delta).$$

Here Pr is the Prandtl number of the fluid, c is the sound speed, γ is the specific-heat ratio, and S is the STOKES NUMBER $(d/\delta)^2$. Units Np m^{-1}.

Note: The leading term in the expression for α is proportional to $\omega^{1/2}$ and is known as the **Kirchhoff attenuation coefficient**, α_K (G KIRCHHOFF 1868):

$$\alpha_K d = \frac{\omega \delta}{c}\left(1+\frac{\gamma-1}{\sqrt{\Pr}}\right).$$

laser velocimeter – a VELOCIMETER (2) that measures velocities non-invasively, by scattering a laser beam from the moving object and comparing the scattered frequency with the frequency of the incident beam. For measuring fluid flow velocities it is necessary to use scattering particles that move with the fluid. Abbreviated **LV**; also known as **laser Doppler velocimeter** (**LDV**).

lateral efficiency – *for an auditorium* an alternative term for EARLY LATERAL ENERGY FRACTION. *Units* none.

lateralization – *in psychoacoustics* the perception of a sound image as if it were located at a particular region within the head. Lateralization (rather than exterior localization) generally occurs when sounds are presented via headphones. Compare LOCALIZATION.

lateral quadrupole – a point quadrupole made up of two parallel DIPOLES of equal and opposite strength, displaced relative to one another in the lateral direction (i.e. transverse to their axes). Equivalently, a lateral quadrupole may be visualized as four equal-strength monopoles of alternating sign, placed at the corners of a rectangle.

lateral-quadrupole source distribution – an ACOUSTIC SOURCE DISTRIBUTION in the form of a double space derivative with respect to cartesian coordinates, where the derivatives are in two orthogonal directions; e.g.

$$\mathcal{F} = \frac{\partial^2 F_{xy}}{\partial x\, \partial y}.$$

Compare AXIAL-QUADRUPOLE SOURCE DISTRIBUTION.

lateral wave – *produced by reflection of a spherical wave from a plane interface separating two media* a contribution to the reflected field that arises when the wave speed is slower inz the medium containing the source (the "upper" medium) than in the lower medium. In terms of rays, the lateral wave corresponds to a ray that travels from the source to the interface at the CRITICAL ANGLE FOR TRANSMISSION, is refracted along the interface, and emerges back into the upper medium at the critical angle. In terms of waves, the lateral-wave wavefront travels along the interface at a phase speed equal to the wave speed in the lower medium. The wavefront is conical, and extends far enough from the interface to merge with the spherical reflected wave from the IMAGE SOURCE.

Note (1): Because the lateral wave travels faster along the interface than the wave speed in the upper medium, it arrives earlier than the specularly reflected wave.

Note (2): The lateral wave amplitude falls off with horizontal range (r) as $1/r^2$, at sufficiently large distances that $r \gg (z + z_0)\tan\theta_{crit}$. Here z_0, z are the heights of the source and observer above the interface, and θ_{crit} is the critical angle for transmission. Because of the more rapid $1/r^2$ decay factor (compared with the $1/r$ spherical-wave dependence), lateral waves often make negligible contribution to the total field far from the source. Compare HEAD WAVE.

leakage – *in spectral analysis of time-stationary signals* the spreading of frequency components that results from WINDOWING the data in the time domain. The signal energy is spread in the frequency domain by the convolution process, so that energy that originates from one part of the frequency range contaminates spectral estimates in other parts of the range. Leakage in the frequency domain can be minimized by choosing a time-domain window whose Fourier transform falls off rapidly to either side of the resolution bandwidth.

Note: An alternative term is *smear*. A similar phenomenon occurs in the time domain, with the roles of time and frequency reversed.

leaky Rayleigh wave – a spatially-damped interface wave that propagates along the boundary between an elastic solid and a compressible fluid, when the RAYLEIGH WAVE speed in the solid is faster than the sound speed in the fluid. The damping is due to acoustic radiation of energy into the fluid. Compare SCHOLTE WAVE.

least squares – a general criterion for fitting an m-parameter model to a set of n data points, with $m < n$; the least-squares criterion states that the model parameters should be chosen to minimize the sum of the squared errors. The ***method of least squares*** is a procedure for finding the coefficients of a polynomial fit to the data, based on the least-squares criterion.

Legendre functions – see note under ASSOCIATED LEGENDRE FUNCTIONS.

Legendre polynomials – a sequence of POLYNOMIALS that are orthogonal over the interval $[-1, 1]$ (**A-M LEGENDRE 1785**). The polynomial of degree n is denoted by $P_n(x)$. The sequence begins as follows:

$P_0(x) = 1$

$P_1(x) = x$

$P_2(x) = \frac{1}{2}(3x^2 - 1)$.

It may be continued indefinitely by means of the recurrence relation

$(n + 1) P_{n+1}(x) = (2n + 1) x P_n(x) - n P_{n-1}(x)$ $(n \geq 1)$.

In acoustics the Legendre polynomials appear in SEPARABLE SOLUTIONS of the scalar wave equation. Specifically, they provide the angular directivity factor when the solution is expressed in spherical polar coordinates, and is required to have axial symmetry about the polar axis.

L_{eq} – symbol for EQUIVALENT CONTINUOUS SOUND PRESSURE LEVEL. *Units* dB re p_{ref}^2.

LEV – in *concert-hall acoustics* an acronym for LISTENER ENVELOPMENT.

level – *of any power-like quantity, typically the mean square of an oscillatory signal* the level of the quantity E, relative to a specified REFERENCE VALUE E_{ref}, is defined in acoustics as ten times the logarithm, to base 10, of the ratio E/E_{ref}. In equation form,

$L_E = 10 \log_{10}(E/E_{ref})$

gives the level of E in **decibels relative to** E_{ref} (abbreviated as dB re E_{ref}, or sometimes as dB//E_{ref}). *Units* dB re E_{ref}.

Note (1): Typically, the quantity E being expressed as a level is the mean square value of an oscillatory signal (or more generally, a second-order statistic). Power-like quantities in acoustics include mean square pressure, mean square voltage, and sound power.

Note (2): Terms like *sound pressure level* or *acceleration level*, although common in acoustics, are technically abbreviations, since they refer to levels calculated from the *mean square value* of the indicated linear quantity. For example, if p denotes sound pressure, with mean square value $\langle p^2 \rangle$ and root mean square p_{rms}, the equation

$L_p = 10 \log_{10} \langle p^2 \rangle / p_{ref}^2 = 20 \log_{10} (p_{rms}/p_{ref})$

defines the SOUND PRESSURE LEVEL. The more logical terminology **sound pressure-squared level** is not normally used.

Note (3): Likewise, by convention sound pressure levels are often quoted as dB re p_{ref} (rather than dB re p_{ref}^2, as is the practice in the present dictionary).

The same applies to levels of other linear acoustic variables (e.g. acceleration, particle velocity, displacement): compare REFERENCE VALUE.

Note (4): Levels may also be quoted for peak values of linear quantities like pressure, acceleration, and force. An example is the ***peak sound pressure level***, defined as

$$L_{pk} = 20 \log_{10}(p_{peak}/p_{ref})$$

where the positive quantity p_{peak} is the PEAK SOUND PRESSURE defined as a magnitude (the instantaneous sound pressure may be either positive or negative). Note that this corresponds to the second L_p definition in note (2), with p_{rms} replaced by p_{peak}.

Note (5): The number $10 \log_{10} x$ may also be written as $\log_d x$, where $d = 10^{1/10}$. The level of E in DECIBELS is therefore definable as $L_E = \log_d(E/E_{ref})$, in the same way that the level in BELS is defined as $\log_{10}(E/E_{ref})$. According to this approach, by using different bases for the log function (e.g. $\log_{10}$ or $\log_2$ in place of $\log_d$), one can define "levels" in units other than decibels. Generally, however, unless another base is explicitly stated, the term LEVEL in acoustics may be taken as implying conventional decibel units.

level difference – (1) a logarithmic measure of the ratio between two power-like quantities of the same type, expressed in decibels. The level difference between E_1 and E_2 is

$$L_{E1} - L_{E2} = 10 \log_{10} E_1/E_2.$$

Units dB.

level difference – (2) *in room acoustics* the difference in level of spatially-averaged mean square reverberant pressure (or reverberant energy density) between two rooms, when one of the rooms is excited by a source of constant acoustic power output. In symbols, the level difference D is given by

$$D = L_s - L_r,$$

where L_s is the level in the source room and L_r is the level in the receiver room. An alternative term is ***sound isolation between rooms***. *Units* dB.

LF – *for an auditorium* abbreviation for EARLY LATERAL ENERGY FRACTION. *Units* none.

l'Hôpital's rule, l'Hospital's rule – a rule for finding the ratio of two functions $f(x)$ and $g(x)$, at points where both of the functions are zero. Suppose both functions pass through zero at $x = x_0$; then l'Hôpital's rule states that their ratio at that point is given by

$$\lim_{x \to x_0} \frac{f(x)}{g(x)} = \lim_{x \to x_0} \frac{f'(x)}{g'(x)},$$

provided the derivatives exist. The rule may be applied repeatedly until a definite result is obtained (G DE L'HOSPITAL 1696).

lift – see AERODYNAMIC FORCE. *Units* N.

lift coefficient – the normalized lift on an object, given by

$$C_L = L/Aq$$

where A is an appropriate area (e.g. the area of a wing), q is the free-stream DYNAMIC PRESSURE, and L is the lift force. *Units* none.

Lighthill's acoustic analogy – see ACOUSTIC ANALOGY (M J LIGHTHILL 1952).

Lighthill stress tensor – the difference between the local momentum flux tensor in a fluid flow, and the acoustic approximation for an ideal uniform medium at rest (the reference medium). The acoustic momentum flux is a normal stress, whose value in any direction is $c_0^2(\rho - \rho_0)$; ρ and ρ_0 are the density and undisturbed density, and c_0 is the undisturbed sound speed. It follows that

$$T_{ij} = \rho u_i u_j + p_{ij} - c_0^2(\rho - \rho_0)\delta_{ij},$$

where p_{ij} is the difference between the compressive stress tensor in the actual fluid and that in the reference fluid:

$$p_{ij} = P_{ij} - P_0 \delta_{ij} \qquad (P_0 = \text{undisturbed pressure}).$$

Units Pa.

Lighthill's U^8 power law – the scaling law that relates the free-field sound power output, W, from a subsonic turbulent fluid flow to a typical flow velocity U (M J LIGHTHILL 1952). In mathematical form, it states that

$$W \propto \rho_0 c_0^{-5} L^2 U^8,$$

where L is a typical dimension and U is a typical velocity in the flow, ρ, c are density and sound speed, and subscript 0 denotes ambient fluid properties. In the case of a turbulent jet, L and U are taken as the nozzle diameter and jet exit velocity.

Note: The U^8 power law is an asymptotic prediction for low Mach numbers ($U \ll c_0$), and is limited to fluid flows of uniform density and sound speed. A U^6 power law applies when sound is generated by turbulent mixing of different-density fluids, such as hot and cold air, provided they have the same isentropic compressibility $1/\rho c^2$. For turbulent flows in which different regions of fluid have different compressibilities (e.g. flows containing bubbles) the radiated power is proportional to U^4.

Lilley's equation – a third-order wave equation proposed as a model for sound radiation from high-speed shear flows (G M LILLEY 1971, 1974). It is obtained by combining the equations of mass and momentum conservation for an inviscid fluid with the equation of state for a PERFECT GAS, and assuming zero thermal conductivity so that $Ds/Dt = 0$ (s = specific entropy). In terms of the nondimensional pressure variable $\Pi = \gamma^{-1}\ln(P/P_\infty)$, where P is the fluid pressure, P_∞ is a reference pressure, and γ is the specific-heat ratio (assumed constant), the equation is

$$\frac{D}{Dt}\left[\frac{D^2\Pi}{Dt^2} - \frac{\partial}{\partial x_i}\left(c^2\frac{\partial\Pi}{\partial x_j}\right)\right] + 2\frac{\partial u_k}{\partial x_j}\frac{\partial}{\partial x_k}\left(c^2\frac{\partial\Pi}{\partial x_j}\right) = -2\frac{\partial u_k}{\partial x_j}\frac{\partial u_i}{\partial x_k}\frac{\partial u_j}{\partial x_i}.$$

Here x_i are cartesian coordinates ($i = 1, 2, 3$), u_i are velocity components, c is the local speed of sound, and D/Dt is the material derivative operator.

Note (1): For practical applications, the left-hand side of the equation is usually linearized about a parallel mean shear flow. The right-hand side is then of second order in the perturbation velocity, and may be regarded as a source term. In this version of Lilley's equation, the perfect-gas assumption is not required.

Note (2): The process described in note (1), with all nonlinear terms discarded, leads to an equation homogeneous in the perturbation pressure p; the equation is used for modelling sound propagation in the presence of a shear flow. For example in an axisymmetric parallel shear flow, with mean velocity components $\{U(r), 0, 0\}$ in the (x, r, ϕ) directions and with mean density $\rho(r)$, the following wave equation for $p(x, r, \phi, t)$ is obtained:

$$\bar{D}\left\{\frac{1}{c^2}\bar{D}^2p - \nabla^2p + \left(\frac{1}{\rho}\frac{d\rho}{dr}\right)\frac{\partial p}{\partial r}\right\} + 2\frac{dU}{dr}\frac{\partial^2 p}{\partial x\,\partial r} = 0.$$

Here (x, r, ϕ) are axial, radial, and azimuthal coordinates in a cylindrical coordinate system, and $\bar{D}$ stands for the linearized material derivative operator $(\partial/\partial t + U\partial/\partial x)$.

Note (3): Solution of the equation above is facilitated by Fourier transformation with respect to x and t, and Fourier analysis into azimuthal modes. The result is an ordinary differential equation in r, sometimes known as the **Pridmore-Brown equation** (D C PRIDMORE-BROWN 1958).

limit cycle – *for a nonlinear system that exhibits instability* a nonlinear periodic oscillation of the system that neither grows nor decays in amplitude. A limit cycle is represented in the PHASE PLANE by a periodic orbit.

limp wall – a partition idealized as having inertia but no stiffness. For plane sound waves incident obliquely on a limp wall, the SOUND POWER TRANSMISSION COEFFICIENT (2) is

$$\tau(\theta) = \frac{1}{1 + \left(\dfrac{\omega m \cos\theta}{2\rho c}\right)^2} \qquad \text{(at angular frequency } \omega\text{),}$$

where θ is the angle of incidence, ρc is the characteristic impedance of the fluid on either side, and m is the partition mass per unit area. See MASS LAW.

linear acoustics – the branch of acoustics limited to small-amplitude signals or oscillations, so that the relation between any two oscillatory quantities (displacement, stress, force etc.) is independent of amplitude. Linear sound waves propagate independently of one another, without interaction.

linear equations – see NON-HOMOGENEOUS SET OF LINEAR EQUATIONS.

linearity – (1) *of a differential operator* is defined by the additive and multiplicative properties

$$L(x_1 + x_2) = L(x_1) + L(x_2) \quad \text{and} \quad L(ax) = a\,L(x).$$

Here x_1, x_2 are variables on which the operator L acts, and a is a constant.

linearity – (2) *of an equation* means that two solutions can be added, or a solution can be multiplied by a constant, and the result is also a solution.

linearity – (3) *of a system* linearity of the system S is defined by the following properties:

(a) If the separate inputs x_1, x_2 produce respective outputs y_1, y_2, then the combined input $x_1 + x_2$ produces an output $y_1 + y_2$;

(b) If input x produces output y, then input ax (where a is a constant) produces output ay.

linearization – a mathematical procedure that converts a set of inherently nonlinear equations into a set of linear equations, whose solutions describe small departures from a specified reference state (e.g. in classical acoustics, the equations of fluid motion are linearized to describe small perturbations in a stationary uniform medium).

linearly dependent – see LINEARLY INDEPENDENT SET.

linearly independent set – *of vectors or functions* a set with the property that no one member is a linear combination of the others. In the case of functions of a single variable x, linear independence requires that the WRONSKIAN DETERMINANT be non-zero for all x.

The opposite of linearly independent is *linearly dependent*, which means that at least one member of the set can be expressed as a linear combination of the remaining members.

linearly polarized – see POLARIZATION.

linearly stable – see STABLE SYSTEM.

linear momentum – a more specific term for MOMENTUM, used particularly where it is required to distinguish momentum from ANGULAR MOMENTUM. *Units* kg m s^{-1}.

linear system – any system in which the FREQUENCY RESPONSE FUNCTION connecting input and output signals is independent of amplitude.

linear time-invariant – describes a system whose input–output relation is linear and does not vary over time. Abbreviated as *LTI*.

linear viscous damping – see VISCOUS DAMPING COEFFICIENT.

liner – abbreviation for DUCT LINER.

line spectrum – a spectrum that consists of discrete lines. A line spectrum superimposed on a broadband background indicates a periodic signal with noise added.

listener envelopment – the sense of being surrounded by sound, especially in a concert hall. Listener envelopment is a component of SPATIAL IMPRESSION; it is currently considered to be related to the later reverberant sound received by the listener, and to increase with sound level.

lithotripsy – *in clinical practice* the use of focused transient acoustic pulses, or shock waves, to fragment kidney, gall, or salivary stones. Also known as *extracorporeal shockwave lithotripsy* (abbreviated *ESWL*).

liveness – the degree of perceived reverberation in a space, produced by the temporal smearing of sound that is associated with the reverberant field. The middle and upper parts of the audio frequency range are particularly important for liveness. Compare REVERBERANCE.

live room – a room in which the amount of sound absorption is small enough that the MEAN FREE PATH l_{av} is greater than $\sqrt{A}$, where A is the ROOM ABSORPTION. Compare DEAD ROOM.

Note: Since this definition is not intended to be precise, the mean free path may be taken as $l_{av} \approx 4V/S$ (S = boundary surface area; V = room volume).

Lloyd's mirror effect – *for propagation between a source and receiver near a plane boundary* the near-cancellation of direct and reflected fields that results from the boundary acting as a pressure-release surface in the grazing limit (**H LLOYD 1834**). In mathematical form, if the source and receiver are located at distances z_0, z from the boundary and their horizontal separation is r, the sound pressure at angular frequency ω varies asymptotically as

$$p \sim e^{-jkr}\frac{zz_0}{r^2}, \quad \text{for } r \gg kzz_0,$$

provided the pressure-release approximation is valid. Here k is the acoustic wavenumber ω/c, and c is the sound speed.

Note: In underwater acoustics, the water surface acts as an almost perfect pressure-release boundary at all angles.

L_N – *in environmental noise assessment* the level of A-weighted mean square sound pressure, with fast (F) time weighting, that is exceeded for N percent of a stated time period. For example, if an A-weighted level of 80 dB is exceeded for just 10 percent of a 24-hour period, the L_{10} value is 80 dB. Likewise if a level of 55 dB is exceeded for 90 percent of a 24-hour period, the L_{90} value is 55 dB. *Units* dB re (20 μPa)^2.

Note: L_N values are also known as ***percentile levels*** or ***exceedance levels***.

load impedance – *of a linear two-port system* the IMPEDANCE of a second linear system (the ***load***) that is connected to the output port of the first system.

localization – *in psychoacoustics* the perception by a listener that a sound is coming from a certain direction; the process of judging the direction of a source. In the most general sense of the term, localization includes judgments of distance as well as direction. Compare LATERALIZATION.

localized waves – *in ultrasonics* a class of pulsed wave fields, sometimes referred to as ***acoustic bullets***. Localized waves are characterized by confinement of the field energy to a transient "hot spot" that sweeps along the beam axis. The propagating focused pulse has a wide frequency bandwidth and remains more or less constant in size, both axially and laterally, over a significant axial region called the ***focus depth***. See also X WAVE.

Note: An alternative name for localized waves is ***transient Bessel beams***. However, unlike single-frequency BESSEL BEAMS, localized waves are non-separable in space and time: their generation requires a dynamic aperture, i.e. the region of excitation varies with time.

locally-reacting – describes a boundary that exhibits LOCAL REACTION. Also known as *point-reacting*.

local nonlinearity – *in a sound field of finite amplitude* the occurrence of local deviations from linear behaviour, with no tendency for the nonlinearity to accumulate along a propagation path. Compare CUMULATIVE NONLINEARITY.

Note: Finite-amplitude propagation of a plane progressive wave through a lossless uniform fluid involves both types of nonlinearity. Local nonlinearity is illustrated by the fact that the instantaneous acoustic pressure p and perturbation velocity u at a fixed point in space are not related by $u = p/\rho_0 c_0$ (as in a small-amplitude wave), but by

$$u = \int_{P_0}^{P_0+p} dP/\rho c \qquad (P = \text{absolute pressure}).$$

Here ρ is the density and c is the sound speed, both evaluated along an isentropic compression curve through P_0, and subscript 0 denotes undisturbed values.

local reaction – a property of a boundary, such that when an unsteady surface pressure distribution is applied (e.g. by an adjacent acoustic medium), the normal velocity at any point on the boundary depends only on the local pressure at that point, and not on the pressure elsewhere.

Note: In the case of a porous boundary, the normal velocity may refer either to the frame of the porous material, or to the FACE VELOCITY of the fluid medium. The term local reaction is ambiguous in this situation; it is necessary to specify which normal velocity is meant.

logarithmic decrement – *of a damped oscillatory signal* the quantity $\ln k$, where k is the factor by which successive peaks (of the same sign) are reduced in amplitude. For free oscillations with viscous damping, see under DAMPING CONSTANT. *Units* Np.

logarithmic frequency interval – *between two given frequencies* a generic term for the logarithm of the *frequency ratio* (f_2/f_1), where f_2 is the higher of the two frequencies and f_1 is the lower; sometimes abbreviated as FREQUENCY INTERVAL. The unit is an OCTAVE if the base of the logarithm is 2, and a DECADE if the base is 10. For example, an interval of two octaves corresponds to a factor of four in frequency. *Units* oct, dec.

logarithmic magnitude ratio – ◆ *of two complex numbers* the natural logarithm of the magnitude ratio of $\hat{y}$ and $\hat{x}$, given by

$$\mu = \ln|\hat{y}/\hat{x}| = \text{Re}\,\{\ln[\hat{y}/\hat{x}]\}.$$

Units Np.

Note (1): Complex numbers, can be used to denote single-frequency signals at the input and output ports of a linear two-port system; in this case $\hat{y}/\hat{x}$ represents the complex FREQUENCY RESPONSE FUNCTION of the system. The quantity μ, in this context, is sometimes called the **magnitude** of the system response.

Note (2): Compare the similar logarithmic definition for the PHASE, in radians, of $\hat{y}$ relative to $\hat{x}$:

$$\varphi = \arg[\hat{y}/\hat{x}] = \text{Im}\,\{\ln[\hat{y}/\hat{x}]\}.$$

log magnitude–phase plot – *for a linear system* a plot that shows the FREQUENCY RESPONSE FUNCTION of the system, $H(\omega)$, in phase and magnitude form over the angular frequency range $0 < \omega < \infty$. The system GAIN in decibels, $20\log_{10}|H(\omega)|$, is plotted against the PHASE RESPONSE, arg $H(\omega)$. Compare NYQUIST PLOT, BODE PLOT.

longitudinal quadrupole – an equivalent term for AXIAL QUADRUPOLE.

longitudinal stiffness – *of an elastic lumped element* the rate of change of force with linear extension. An equivalent term is SPRING CONSTANT. *Units* N m^{-1}.

longitudinal waves – waves in which the displacement of the medium at each point is normal to the local wavefront surface. Plane acoustic waves in fluids, and plane compressional waves in isotropic solids, are longitudinal (provided the medium is uniform on the scale of a wavelength). Axial waves in rods, and in-plane irrotational waves in isotropic plates, are only approximately so, but the description longitudinal is commonly applied. Compare TRANSVERSE WAVES.

longitudinal wave speed – the propagation speed of free LONGITUDINAL WAVES. In bulk media, the term is equivalent to COMPRESSIONAL WAVE SPEED. See also ROD LONGITUDINAL-WAVE SPEED, PLATE LONGITUDINAL-WAVE SPEED. *Units* m s^{-1}.

long-term shear modulus – see RELAXATION MODEL. *Units* Pa.

Lorentz transformation – a transformation of space–time coordinates in which the reference frame translates at a constant velocity, but the d'Alembertian operator remains unchanged. If the new reference frame moves at speed V in the x direction relative to the original frame, the transformed coordinates are

$$x' = \frac{1}{\beta}(x - Vt),\ y' = y,\ z' = z,\ t' = \frac{1}{\beta}(t - xV/c^2).$$

Here c is the wave speed appearing in the d'Alembertian, and $\beta = \sqrt{1 - M^2}$

where $M = V/c$. Originally discovered in the context of acoustics by Voigt (**W VOIGT 1887**), it was applied to electromagnetic waves by Lorentz (**H A LORENTZ 1904**), with c denoting the speed of light. The transformation was later found to have practical applications in aeroacoustics (**H G KÜSSNER 1940**).

Note: In situations where an acoustic source and the medium are in constant relative motion, it is often convenient to use a frame of reference that is fixed relative to the source, rather than relative to the medium. The Lorentz transformation permits this while retaining the wave operator in its original form (but in the new variables). However, a point source of density $S(t)\delta(x - Vt)\delta(y)\delta(z)$ in the original variables transforms into $\beta^{-1} S(\beta^{-1} t')\delta(x')\delta(y')\delta(z')$, i.e. all frequencies are raised by a factor β^{-1}. This can be avoided by adjusting the transformation: the final variables are

$$(x'', y'', z'', t'') = \frac{1}{\beta}(x', y', z', t'),$$

with the final result that the original wave equation

$$\Box^2 p = -S(t)\,\delta(x - Vt)\,\delta(y)\,\delta(z)$$

becomes

$$\Box''^2 p = -\frac{1}{\beta^2} S(t'')\,\delta(x'')\,\delta(y'')\,\delta(z'').$$

loss angle – *of a linear viscoelastic material* the frequency-dependent angle, δ, whose tangent is the ratio between the imaginary part and the real part of a specified COMPLEX MODULUS. In mathematical form, if the complex modulus is denoted by $E_1 + jE_2$,

$$\delta = \tan^{-1}\frac{E_2}{E_1} \qquad (e^{j\omega t} \text{ time factor}).$$

Units rad.

Note: Different loss angles are associated with the Young's modulus, the bulk modulus, and the shear modulus.

loss factor – *of a linear system driven at a single frequency in a particular mode* the ratio of the energy dissipated in one period of oscillation to $4\pi E_{kin}$, where E_{kin} is the vibrational kinetic energy of the system averaged over a cycle. If η denotes the loss factor, the time-average power dissipated under steady-state conditions is

$$W_{diss} = 2\omega\eta E_{kin}.$$

In general, the loss factor of a system in a given mode is dependent on frequency (but note the example below). See also RADIATION LOSS FACTOR, MODAL-AVERAGE LOSS FACTOR, COUPLING LOSS FACTOR. *Units* none.

Note: In a single degree of freedom system with VISCOUS DAMPING, the loss factor is independent of frequency; it may be expressed as

$\eta = R/\sqrt{km}$,

where R, k, m are constant parameters in the linear differential equation

$m\ddot{x} + R\dot{x} + kx = F\cos\omega t$

that describes the response $x(t)$ of the system to a sinusoidal driving force $F\cos\omega t$.

loss tangent – the tangent of the LOSS ANGLE; i.e. the ratio of the imaginary part to the real part, for a specified complex modulus. *Units* none.

loudness – (1) *in general usage* an observer's auditory impression of the strength of a sound.

loudness – (2) a standardized measure used to quantify LOUDNESS (1), based on the perceived ratio of the strength of a sound to that of a reference sound. See also LOUDNESS LEVEL. *Units* sone.

loudness level – for a given sound, the loudness level in phons is numerically equal to the sound pressure level of a reference sound that is judged by the average observer to match the given sound in loudness. The reference sound is a pure tone of frequency 1 kHz, in the form of a plane wave arriving from directly in front of the observer. There is a standardized relationship between the scale of loudness level in phons and the scale of loudness in sones. *Units* phon.

loudness recruitment – *in psychoacoustics* see RECRUITMENT.

loudspeaker – *for sound radiation in air* a transducer that accepts an electrical signal as input, and produces a radiated sound field as output, with minimal distortion over its design frequency range.

Love waves – *in a horizontally-stratified elastic medium* guided shear-wave modes that propagate horizontally, and in which all particles move in horizontal planes. In Love's original model (**A E H LOVE 1911**) a uniform layer with shear wave speed b_1 has one stress-free face, while the other face is bonded to a semi-infinite elastic medium that has shear wave speed b_2; this produces Love-wave phase speeds ranging from b_2 at low frequencies to b_1 at high frequencies.

low-pass filter – a FILTER designed to reject unwanted high-frequency components from a signal. If the filter rejects all frequencies above an upper limit f_c, then f_c is called the ***upper cutoff frequency*** of the filter.

LTI – *referring to a system* abbreviation for LINEAR TIME-INVARIANT.

lumped elements – idealized building-blocks used to represent the dynamic response of electrical or mechanical systems; the basic assumption is that the system component being represented has dimensions much less than the relevant wavelength. A lumped element can be either reactive (with no dissipation) or resistive (with no stored energy). Reactive elements are further categorized as springlike or masslike. The former store potential energy, the latter kinetic energy.

The examples below all relate to two-port lumped elements (see TWO-PORT SYSTEM). In each case, one variable – transmitted force, pressure, velocity, or volume velocity – is regarded as invariant through the element, as a result of the compactness assumption (i.e. element size $\ll$ wavelength). The defining property of the element is a FREQUENCY RESPONSE FUNCTION; in the examples listed below, it is formed as the ratio of the "difference variable" (e.g. for a mechanical compliance, the difference in displacement, Δx, across the element) to the "transmitted variable" whose value is invariant through the element.

Element name	Mechanical	Acoustical
COMPLIANCE	$\Delta x/F$ ($\Delta u/\dot{F}$)	$\Delta X/p$ ($\Delta U/\dot{p}$)
RESISTANCE	$\Delta F/u$	$\Delta p/U$
MASS, INERTANCE	$\Delta F/\dot{u}$	$\Delta p/\dot{U}$
DIFFERENTIAL IMPEDANCE	$\Delta F/u$	$\Delta p/U$
MOBILITY	$\Delta u/F$	$\Delta U/p$

Note: The first three examples describe simple basic elements for which the frequency response function is real and independent of frequency. The last two allow more general behaviour. The notation is:

F = force
p = pressure
X, U = volume displacement, volume velocity
x, u = displacement, velocity.

lumped-parameter system – a model system constructed from LUMPED ELEMENTS.

M

Mach angle – *associated with a weak stationary disturbance in a supersonic flow* the angle that MACH WAVES generated by the disturbance make with the direction of the incident flow. It is defined by

$$\theta = \sin^{-1}(1/M_{\text{rel}}),$$

where M_{rel} is the Mach number of the incident flow measured in a frame of reference in which the disturbance appears fixed. The name was apparently suggested by Prandtl in 1907, before the term MACH NUMBER was invented (see PRANDTL-MEYER FLOW). *Units* rad (but commonly expressed in degrees).

Mach cone – the cone that defines the region affected by pressure waves from an acoustic source, or a slender streamlined body, when the source or body is in steady supersonic motion relative to a uniform fluid. The generators of the Mach cone slope backwards from the source at the MACH ANGLE, θ, relative to the line of motion ($\sin\theta = 1/M$, where M is the Mach number of the source relative to the fluid). The first recorded experimental observations are due to Mach (**E MACH 1887**), but the phenomenon was predicted 40 years earlier by Doppler (**J C DOPPLER 1847**).

Note: Compare SONIC BOOM, the weak conical shock wave that is produced in the far field of a supersonic body when the near-field MACH WAVES have coalesced.

Mach number – (1) *of a body moving through a fluid* the velocity of the body relative to the fluid, divided by the ambient speed of sound; given this name by Ackeret in 1929 (**E MACH *1838–1916***). *Units* none.

Mach number – (2) *at a point in a steady flow* the ratio of the fluid velocity to the local speed of sound. Flow velocity vector components divided by the sound speed are called ***Mach number components***. *Units* none.

Mach number – (3) an abbreviation for ACOUSTIC MACH NUMBER. *Units* none.

Mach wave – *in steady supersonic flow* an infinitesimally weak wavefront of expansion or compression in a fluid, typically produced by steady supersonic flow past a slender body. Although conceptually identical to elementary acoustic disturbances (see WAVELET), Mach waves are associated with disturbances in an otherwise steady supersonic flow. They are produced either by supersonic relative motion of physical objects through a fluid, or otherwise by flow disturbances that move at supersonic speeds relative to the ambient medium (e.g. instability waves in a supersonic jet; see MACH WAVE RADIATION).

Mach wave radiation – *from a supersonic turbulent jet* a distinct component of the jet noise radiation field, characterized by a strong directional peak around the Mach angle θ defined by

$$\theta = \cos^{-1}(c_0/U_{\text{eddy}}).$$

Here θ is the radiation direction relative to the jet axis, c_0 is the ambient sound speed, and U_{eddy} is a typical eddy convection velocity, roughly equal to the flow velocity in the centre of the jet mixing region.

macrosonics – the science and technology of high-intensity sound or ultrasound as applied to industrial processes. Some typical applications are: mixing, degassing, agglomeration of particles, spray generation, and gas compression.

MAF – abbreviation for MINIMUM AUDIBLE FIELD. Usually expressed as a sound pressure level. *Units* dB re $(20\ \mu\text{Pa})^2$.

magnetoacoustic effect – the influence of magnetic fields on sound propagation, or attenuation, in certain materials.

magnetostriction – the tendency of a ferromagnetic material (e.g. nickel, iron) to expand or contract in a magnetic field.

magnitude – the complex number $z = x + jy$, where x and y are real, is said to have magnitude $|z| = \sqrt{x^2 + y^2}$. An equivalent term for $|z|$ is the **modulus**. See also PHASE AND MAGNITUDE REPRESENTATION.

magnitude and phase representation – *of a complex number* see PHASE AND MAGNITUDE REPRESENTATION.

magnitude response – *of a linear time-invariant system* an equivalent term for the AMPLITUDE RESPONSE FUNCTION of the system.

magnitude spectrum – *of a transient signal* the modulus of the signal FOURIER TRANSFORM, plotted as a continuous function of frequency. The signal may be either analogue (continuous-time) or digital (discrete-time); in the latter case the magnitude spectrum is the modulus of the DISCRETE-TIME FOURIER TRANSFORM, and is periodic.

Note: The magnitude spectrum is even with respect to frequency.

manikin – *in acoustics* a generic term for a HEAD-AND-TORSO SIMULATOR.

MAP – abbreviation for MINIMUM AUDIBLE PRESSURE. Usually expressed as a sound pressure level. *Units* dB re $(20\ \mu\text{Pa})^2$.

masked threshold – see ABSOLUTE THRESHOLD. *Units* dB re $(20\ \mu\text{Pa})^2$.

masking – (1) the process by which the threshold of hearing for one sound, called the *maskee*, is raised owing to the presence of another sound, called the *masker*. Masking may be either unintentional, or deliberately induced e.g. by ACOUSTIC WALLPAPER.

masking – (2) the amount of THRESHOLD SHIFT produced by MASKING (1), relative to the threshold of hearing in quiet conditions. *Units* dB.

mass – a conserved property of matter, responsible for both inertia and gravitational attraction. In Newtonian mechanics, the MOMENTUM of a particle is related to the velocity by a constant factor called the mass or *inertial mass* of the particle. *Units* kg.

mass–air–mass resonance – *of a double partition* a resonance associated with the stiffness of the air gap, combined with the mass-like impedances of the two individual partitions moving in opposite directions. Also known as *double-wall resonance*.

mass centre – (1) *of a system consisting of discrete mass elements* the point whose vector position is given by

$$\mathbf{r}_c = \frac{\sum m_i \mathbf{r}_i}{\sum m_i},$$

where m_i is the *i*th mass component and $\mathbf{r}_i$ is its position.

mass centre – (2) *of a continuous system* the point whose vector position is

$$\mathbf{r}_c = \frac{\int \rho(\mathbf{x})\mathbf{x}\,dV}{\int \rho(\mathbf{x})dV},$$

where $\rho(\mathbf{x})$ is the density of the material at position $\mathbf{x}$.

mass flux vector – *in continuum mechanics* the product of the local density of the material and the local velocity vector. In symbols,

$$\mathbf{m} = \rho \mathbf{u} \quad (\mathbf{u} = \text{local velocity}, \rho = \text{density}).$$

The component of **m** in a given direction gives the rate of mass flow per unit area, across a surface pointing in that direction. *Units* kg s^{-1} m^{-2}.

mass law – an approximate description of the sound transmission loss due to a partition, that applies when the inertia of the panel rather than its stiffness is the dominant effect. It states that for plane waves incident at a given angle, the transmission loss increases by 6 dB for each doubling of the mass per unit area, and by 6 dB for each doubling of the frequency. See also LIMP WALL.

master structure – *in the context of fuzzy structures* the part of a vibroacoustic system that is modelled by conventional deterministic means, such as finite element analysis.

matched field processing – a group of techniques used, in particular in ocean acoustics, to solve the INVERSE PROBLEM of determining the acoustic propagation environment from detailed sound field measurements. A parameterized model of the environment (including, for example, ocean sound speed profiles and wave speed profiles in the seabed) is used to predict the form of the sound field, and this field is then matched to measurements by adjusting the model parameters until an optimal fit is obtained.

matching – *of waves at a plane boundary* see COINCIDENCE.

material coordinates – an alternative term for LAGRANGIAN COORDINATES.

material derivative – *of a field variable in fluid or solid mechanics* the rate of change of the field variable with time, as measured in a frame of reference moving with the material. Equivalent terms are ***Lagrangian derivative*** and ***substantial derivative***.

The ***material derivative operator*** is often written as D/Dt. If q denotes the field variable, the mathematical definition of Dq/Dt is

$$\frac{Dq}{Dt} = \frac{\partial q}{\partial t} + (\mathbf{u} \cdot \nabla) q,$$

where $\mathbf{u}$ is the local material velocity and t denotes time.

matrix – an $m \times n$ matrix is a rectangular array of ***elements*** with m rows and n columns. In the example below, $\mathbf{A}$ is a 2×3 matrix with elements a_{ij} ($i = 1, 2$; $j = 1, 2, 3$):

$$\mathbf{A} = \begin{bmatrix} a_{11} & a_{12} & a_{13} \\ a_{21} & a_{22} & a_{23} \end{bmatrix}.$$

The elements of a matrix can be either scalars or matrices; see PARTITIONED MATRIX.

maximum frequency-weighted sound pressure level – *of a pressure signal with a time-varying envelope* see note under MAXIMUM SOUND PRESSURE LEVEL. Units dB re p_{ref}^2.

maximum length sequence (MLS) – a type of binary periodic sequence that has some similarities to random WHITE NOISE. One of its uses is as a pseudo-random input signal for linear system identification. Also known as a ***pseudorandom sequence***.

maximum length sequence diffuser – a reflecting plane surface containing a parallel array of slots or recesses, all of the same depth. The slot widths and spacings all belong to the basic sequence h, $2h$, $3h$, ..., etc. where the module h is chosen to be less than a quarter-wavelength of the sound to be diffused. The surface profile is periodic and forms a binary MAXIMUM LENGTH SEQUENCE, based on a single slot geometry that is either present or absent. See also SCHROEDER DIFFUSER.

maximum level – *with respect to time or position* the highest LEVEL reached by a specified quantity, either over a given time interval (if the quantity is non-stationary) or over a given region. Examples of the first type (temporal maximum) are: *maximum sound power level of a machine during its operating cycle*; *maximum A-weighted sound level at a given location during a 24-hour period*. An example of the second type (spatial maximum) is: *maximum sound pressure level in a room, under steady-state conditions*. Compare PEAK LEVEL. Symbol $L_{\max}$. *Units* dB.

Note (1): For non-stationary signals whose time scale is of the order of a second or less, the TIME WEIGHTING used to form a maximum level (with respect to time) must be specified; the standard fast or slow weightings may be indicated by adding a subscript F or S to the symbol above. Likewise the use of A or C frequency weighting may also be indicated by a subscript: for example, $L_{\mathrm{AF,max}}$ denotes a maximum level obtained with A frequency and F time weighting.

Note (2): The definition above accords with ANSI standard terminology, but not with IEC (which does not recognize maximum level and appears to adopt *peak level* both for maximum levels based on the time-weighted square of an oscillatory signal, and for peak levels based on the instantaneous signal magnitude). However, if one follows ANSI and reserves *peak* for instantaneous quantities, one can usefully distinguish between *maximum sound pressure level* and *peak sound pressure level*; see the respective entries for more information.

maximum sound pressure level – *of a sound with a time-varying envelope* the highest value of sound pressure level that occurs during a specified time interval, as measured with an instrument that performs a running average on $p^2(t)$, the square of the instantaneous sound pressure. The TIME WEIGHTING characteristic used in the running average must be specified. Compare PEAK SOUND PRESSURE LEVEL. *Units* dB re p_{ref}^2.

Note: If the sound pressure signal is FREQUENCY-WEIGHTED before squaring and averaging, the process above yields a ***maximum frequency-weighted sound pressure level***, sometimes abbreviated as ***maximum sound level***. Compare SOUND LEVEL.

Maxwell model – *of a linear viscoelastic material with time-independent proper-*

ties a stress-strain relationship in which the strain rate, $\dot{\varepsilon}$, at any instant is a linear function of the stress, σ, and its time derivative at that instant. In mathematical form,

$$\dot{\varepsilon} = a\sigma + b\dot{\sigma} \quad (a, b \text{ constants}),$$

(J C MAXWELL *1831–1879*). Compare VOIGT MODEL.

mean – an average value, defined either for a discrete set of numbers, or for a continuous variable. Three specific uses of the term are noted below.

(1) *For digital or sampled data:*

$$\langle x \rangle = \frac{1}{N} \sum_{i=1}^{N} x_i \qquad (\textit{arithmetic mean}).$$

This defines the mean of the N individual numbers x_i.

(2) *For a continuous variable:*

$$\langle x \rangle = \frac{1}{b-a} \int_a^b x(t)\,dt \qquad (\textit{mean with respect to } t).$$

Here the variable x is being averaged over the range $a < t < b$, and $\langle x \rangle$ denotes its mean value over that range.

(3) *For a random variable:*

$$\langle x \rangle = E\{x\} \qquad (\textit{expected value, population mean}).$$

See also EXPECTATION OPERATOR, WEIGHTED MEAN.

mean free path – *for sound in a room* a characteristic ray path length between successive reflections. The mean free path is defined as

$$l_{\text{av}} = \frac{1}{\kappa_{\text{av}}},$$

where κ_{av} is an average *collision rate* between a ray and the room boundaries: to obtain κ_{av}, follow a typical ray over many reflections, count the number of encounters with the boundaries, and divide by the total path length.

Note (1): If the room boundaries are diffusely-reflecting, then for any shape of room

$$\kappa_{\text{av}} = \frac{S}{4V} \qquad (S = \text{boundary surface area}; V = \text{room volume}),$$

which yields the standard mean free path estimate $l_{\text{av}} = 4V/S$. Units m.

Note (2): For a room with SPECULAR REFLECTION at its boundaries, three stages of averaging are required in order to define an average collision rate κ_{av} and hence a mean free path $l_{\text{av}} = \kappa_{\text{av}}^{-1}$:

(a) For a ray launched from a given position P in a given initial direction **i**, the collision rate is first averaged over many reflections to give the directed mean value $\kappa_{\text{dir}}(\mathbf{i}, P)$.

(b) Secondly, averaging κ_{dir} over all possible initial ray directions **i** gives the position-dependent average collision rate $\kappa_{\text{av}}(P)$. Its inverse defines a ***specular-reflection mean free path*** for rays beginning at the specific point P:

$$l_{\text{av}}(P) = \frac{1}{\kappa_{\text{av}}(P)} \quad \text{(specular boundary reflection)}.$$

(c) Thirdly, averaging $\kappa_{\text{av}}(P)$ with respect to source position P over the entire room volume gives a characteristic collision rate for the room as a whole, denoted by κ_{av}. This specular-reflection average may differ from the diffuse-reflection average $S/4V$, although for certain regular room shapes the two are identical.

Note (3): In the directional-averaging step (b) above, the average of the collision rate $\kappa_{\text{dir}}(\mathbf{i}, P)$ was taken with respect to initial ray direction **i**. Some authors have proposed instead taking the average of the inverse quantity, namely the directed mean path length $l_{\text{dir}}(\mathbf{i}, P) = 1/\kappa_{\text{dir}}(\mathbf{i}, P)$. This gives a different result for the specular-reflection mean free path $l_{\text{av}}(P)$.

mean square deviation – *of a signal from its mean* the MEAN of the squared difference $(x - \langle x \rangle)^2$, where x is the signal and $\langle x \rangle$ is the mean of the signal.

mean square value – (1) *of a random variable* the EXPECTED VALUE of the squared variable; i.e. $E\{X^2\}$, where X is the random variable.

mean square value – (2) *of a time-varying signal* the AVERAGE value of the squared signal, over a specified time interval; i.e. $\langle x^2(t) \rangle$, where $x(t)$ denotes the signal as a function of time.

mechanical admittance – *of a point-excited mechanical system* the complex ratio of velocity to applied force, at a single frequency. Equivalently, it is a frequency response function in which a velocity at one point is regarded as the output, and a force (at the same or a different point) is regarded as the input. Compare RECEPTANCE. See also MOBILITY, DRIVING-POINT ADMITTANCE. *Units* m s^{-1} N^{-1}.

mechanical impedance – *of a point-excited mechanical system* the complex ratio of applied force to velocity, at a single frequency. The reciprocal of MECHANICAL ADMITTANCE. *Units* N s m^{-1}.

Note (1): Equivalently, the mechanical impedance is a frequency response function in which a force at one point is regarded as the output, and a velocity (at the same or a different point) is regarded as the input.

Note (2): For related definitions, see DRIVING-POINT IMPEDANCE (1), DIRECT IMPEDANCE, CROSS IMPEDANCE.

mechanical index – *in biomedical ultrasonics* an indicator of the likelihood that CAVITATION will occur in a liquid, in response to a short-duration pulse of high-frequency sound. The mechanical index, abbreviated *MI*, for a pulse with peak negative pressure $\{P\}$ MPa and carrier frequency $\{F\}$ MHz is defined as the dimensionless quantity

$$MI = \frac{\{P\}}{\sqrt{\{F\}}};$$

the notation $\{P\}$, $\{F\}$ refers to the number of MPa and MHz respectively. *Units* none.

Note: To ensure that transient cavitation (bubble growth followed by violent collapse) will not occur in water or blood, the mechanical index must be kept below about 0.7. This figure is based on a typical carrier frequency in the range 1–10 MHz, and a pulse envelope lasting for at most a few cycles.

mechanical ohm – see SI MECHANICAL OHM.

mechanical radiation impedance – (1) *of a fluid-loaded rigid body, in translational oscillation at a single frequency along a specified axis* the complex ratio

$$Z_{m,\text{rad}} = F/u_{\text{body}}, \quad \text{(subscript } m \text{ for mechanical impedance)}.$$

Here u_{body} is the translational velocity of the body, and F is the force that the body exerts on the fluid (assumed to be in the same direction). *Units* N s m^{-1}.

Note: If the body shape is not symmetric, the fluid loading produces off-axis force components. The impedance presented by the fluid in this case is a matrix quantity (compare DRIVING-POINT IMPEDANCE MATRIX), and may be called the ***mechanical radiation-impedance matrix***. See also VIRTUAL MASS, which describes the fluid loading in the incompressible limit.

mechanical radiation impedance – (2) *of an opening or compact vibrating surface regarded as a lumped element* the product $A^2 Z_{m,\text{rad}}$, where A is the area of the opening and $Z_{m,\text{rad}}$ is the ACOUSTIC RADIATION IMPEDANCE. Also known as the ***radiation impedance*** of the opening or vibrating surface. *Units* N s m^{-1}.

Note: Compare PISTON RADIATION IMPEDANCE (defined as a mechanical impedance), which is a more precisely specified quantity.

mechanical resistance – the real part of a MECHANICAL IMPEDANCE. *Units* $N\,s\,m^{-1}$.

mechanical system – a SYSTEM in which forces, or stresses, provide the only means of transferring ENERGY.

mechanical wave propagation – a term that covers all types of mechanical wave motion in fluids, solids, and structures. Also known as *mechanical radiation*.

medium – a material continuum that supports acoustic or elastic wave propagation, usually in three dimensions.

membrane absorber – an acoustic ABSORBER formed by mounting a flexible impervious sheet or membrane in front of a rigid wall, often with sound-absorbing material placed in the intervening airspace. By using a lossy membrane of low bending stiffness, significant dissipation is achieved in the membrane itself. Compare MULTILAYER ABSORBER, PANEL ABSORBER, PERFORATED-PLATE ABSORBER.

Note (1): The elastic bending stiffness of the sheet in this type of absorber is typically too small to influence the fundamental resonance frequency, which is determined by the volume stiffness of the airspace and the inertia (mass per unit area) of the sheet. However, bending motion may account for a significant part of the total damping if the sheet material is lossy and not too thin (compare note (2) below).

Note (2): In some cases the membrane consists of a thin foil stretched over an edge support. The fundamental resonance frequency is much higher than for an unstretched membrane of the same material, since the stiffnesses of the membrane and the trapped air are additive; it can also be controlled, by altering the membrane tension. This type of membrane absorber is known as an *elastic foil absorber*.

membrane transverse-wave speed – *in a uniform membrane under tension* the propagation speed of free transverse waves of small amplitude, given by $\sqrt{T/m}$ where T is the tension per unit width in the membrane and m is the mass per unit area. *Units* $m\,s^{-1}$.

metric sabin – 1 m^2; the unit of ROOM ABSORPTION based on the metre as unit of length, alternatively called the *SI sabin*.

MI – *in biomedical ultrasonics* abbreviation for MECHANICAL INDEX. *Units* none.

microphone – a transducer that accepts an acoustic pressure signal as input, and produces an electrical signal as output, with minimal distortion over its design frequency range. Compare PRESSURE-GRADIENT MICROPHONE.

microphone flow noise – see FLOW NOISE (2).

midband frequency – *in proportional-bandwidth frequency analysis* the geometric mean of the upper and lower band limits, or filter cutoff frequencies, for a given frequency band. Also known as the ***geometric centre frequency***. *Units* Hz.

middle ear – an air-filled cavity, situated within the temporal bone of the skull, that connects the EXTERNAL EAR to the INNER EAR. The connecting ports are the tympanic membrane on the external ear side, and two delicate membranes called the oval window and round window on the inner ear side. The cavity is vented to the outside by a narrow duct (the ***Eustachian tube*** or ***auditory tube***), which opens into the pharynx and provides drainage and pressure equalization.

The OSSICLES connect the tympanic membrane mechanically to the oval window, providing an impedance-matched transmission path for vibrational signals between the external and inner ear. The acoustic driving force on the tympanic membrane is multiplied by a factor of around 10 at the oval window, as a result of the bone linkage system.

minimum audible field – *under either monaural or binaural listening conditions* the ABSOLUTE THRESHOLD for a listener exposed to a frontally-incident plane sound wave, expressed as a sound pressure level in the unobstructed sound field. "Unobstructed" means that the listener is absent, and the sound field is measured at the point where the centre of the listener's head would be. Abbreviated as ***MAF***. *Units* dB re $(20\ \mu\text{Pa})^2$.

minimum audible pressure – the ABSOLUTE THRESHOLD of the listener under test for a specified test signal, expressed as a sound pressure level at a defined point within the ear canal close to the eardrum. Abbreviated as ***MAP***. *Units* dB re $(20\ \mu\text{Pa})^2$.

Note: MAP is usually measured under conditions of monaural listening, with the stimulus delivered by earphone.

minimum level – *with respect to time or position* the lowest LEVEL reached by a specified quantity, either over a given time interval (if the quantity is non-stationary) or over a given region. *Symbol* $L_{\min}$. *Units* dB.

Note: Compare MAXIMUM LEVEL, where examples are given.

minimum phase system – (1) *in continuous time* a minimum phase system has all its poles and zeros in the left half of the complex S-PLANE (i.e. where the real part of s is negative). One of the important properties of a minimum phase system is that it has a stable inverse. Thus if one inverts the transfer function, the zeros become the poles of the inverted system, and these must be in the left half of the s-plane if the inverted system is to be stable.

Note (1): The transfer function $\tilde{H}(s)$ of any real system can be expressed in the form

$$\tilde{H}(s) = \tilde{H}_m(s)\, \tilde{H}_a(s),$$

where $|\tilde{H}_a(s)|$ is constant along the imaginary axis ($s = j\omega$). The factor $\tilde{H}_m(s)$ is the transfer function of a minimum-phase system as defined above, and $\tilde{H}_a(s)$ represents an **all-pass** system (gain independent of frequency) that contains all the pure delays present in $\tilde{H}(s)$.

Note (2): For a minimum phase system, the logarithm of the transfer function is analytic in the right half of the complex s-plane. The KRAMERS–KRONIG RELATIONS can therefore be applied to the function

$$\ln H_m(\omega) = \mu(\omega) + j\varphi(\omega) \qquad (\mu,\, \varphi \text{ real}).$$

where $H_m(\omega) = \tilde{H}_m(j\omega)$ is the system FREQUENCY RESPONSE FUNCTION. It follows that the phase-shift function $\varphi(\omega)$ of a minimum phase system can be deduced from the logarithmic magnitude $\mu(\omega)$, and vice versa.

Note (3): The description *minimum phase* refers to the fact that of all the possible phase-shift functions associated with a given magnitude response of a linear system, a minimum phase system has the smallest possible (negative) phase shift as a function of frequency.

minimum phase system – (2) *in discrete time* a minimum phase system has all its poles and zeros inside the unit circle in the complex Z-PLANE (i.e. where the modulus of z is less than unity). If one inverts the transfer function, the zeros become the poles of the inverted system; the zeros of the original system must therefore be inside the unit circle if the inverted system is to be stable.

mixed layer – *in ocean acoustics* see SURFACE DUCT.

mixing layer – *in fluid mechanics* a turbulent shear layer between two parallel streams of different velocity.

mnemonic interference – a form of noise MASKING that uses complex non-repetitive sounds, such as birdsong, to hide offensive sounds that have a simpler character, such as the hum of machinery.

mobility – (1) an equivalent term for MECHANICAL ADMITTANCE. *Units* m s^{-1} N^{-1}.

Note: The term *acoustic mobility* is likewise sometimes used as an equivalent for ACOUSTIC ADMITTANCE. There is scope for confusion between the use of mobility to mean a point admittance, as here, and the lumped-element use of the term mobility to mean a differential admittance, as in sense (2) below.

mobility – (2) *of a compliant lumped element* the complex ratio of two phasors, the first of which is the differential velocity across the element (or the angular velocity, or volume velocity), and the second of which is the force transmitted through the element (or the torque, or pressure). An alternative term is ***differential admittance***, by analogy with DIFFERENTIAL IMPEDANCE.

Note: The three types of mobility above are given the names ***mechanical mobility***, ***rotational mobility***, and ***acoustic mobility*** respectively. The essential feature of all three is that the "transmitted variable" is invariant through the element; see LUMPED ELEMENTS.

mobility analogy – see ELECTRICAL ANALOGIES.

mod – symbol for MODULO.

modal-average loss factor – *associated with the multimode response of a linear lightly-damped system driven by band-limited noise* the ratio of the time-average power dissipated in the vibrating system to $\omega E_{tot} \approx 2\omega E_{kin}$. Here ω is the midband angular frequency, E_{tot} is the total vibrational energy of the system, and E_{kin} is the mean kinetic energy component. If η_{av} denotes the modal-average loss factor, the dissipated power is given by

$$W_{diss} = \omega \eta_{av} E_{tot} \approx 2\omega \eta_{av} E_{kin}.$$

Units none.

modal bandwidth – a measure of the frequency bandwidth of the peak that occurs, at resonance, in the response function of a particular mode. It may be expressed either as a HALF-POWER BANDWIDTH, B_h, or as an ENERGY BANDWIDTH, $B_e = (\pi/2)B_h$. *Units* Hz.

modal characteristic impedance – *for a single mode propagating along an acoustic waveguide* the complex ratio of the local acoustic pressure to the particle velocity component in the propagation direction, measured at the same point. *Units* Pa s m^{-1}.

Note: If the waveguide contains a mean flow, the modal characteristic impedances are different for upstream and downstream propagation.

modal density – the average number of modal resonances per unit interval of frequency, for a particular system. If $N(f)$ denotes the asymptotic smoothed MODE COUNT at frequency f, then the corresponding modal density is the derivative $dN(f)/df$. Also known as ***asymptotic modal density***. *Units* Hz^{-1}.

modal energy – *in statistical energy analysis* an abbreviation for AVERAGE MODAL ENERGY. *Units* J.

modal impedance matrix – *at a given cross-section in a multimode waveguide* a square matrix of complex coefficients that relates a column vector of modal acoustic pressures **p** to a column vector of modal velocities **u**, as in the relations

$$\mathbf{p} = \mathbf{Z}\mathbf{u} \quad \text{or} \quad p_m = Z_{ml}u_l.$$

The diagonal terms are called *self-impedances* and the off-diagonal terms are *coupling impedances*. Compare TERMINATION IMPEDANCE MATRIX. *Units* Pa s m^{-1}.

modal loss factor – the LOSS FACTOR of an individual mode. *Units* none.

modal numbers – alternative term for MODE NUMBERS.

modal overlap – *for a multimode vibrating system* the expected number of system resonances occurring in the frequency band between $f - \frac{1}{2}B$ and $f + \frac{1}{2}B$. Here f is the frequency of interest, and B is the MODAL BANDWIDTH of a typical mode (expressed as an energy bandwidth). See also SCHROEDER FREQUENCY. *Units* none.

modal radiation impedance – *of a vibrating surface that radiates sound into an adjacent medium* the complex ratio of the surface pressure amplitude in one mode to the surface normal velocity amplitude in another mode, when the surface vibrates at a single frequency. In the general case, for an arbitrary pattern of surface vibration, the resulting surface pressure in mode i is given by

$$p_i = Z_{ij}\dot{w}_j, \quad \text{(summation over } j, \text{ including } j = i)$$

where $\dot{w}_j$ is the surface normal velocity in mode j and Z_{ij} are the components of the *modal radiation impedance matrix*. The diagonal elements ($j = i$) are called *self-impedances*. *Units* Pa s m^{-1}.

Note: Any convenient set of orthogonal BASIS FUNCTIONS can be used for the modal decomposition; for each set of basis functions there is a different impedance matrix Z_{ij}. Choosing the set of functions, or mode shapes, for which the matrix is diagonal leads to some simplification; these functions are called *radiation modes*.

modal radiation loss factor – *for a single resonant mode of a fluid-loaded structure* the part of the modal LOSS FACTOR that arises from radiation of sound energy into the surrounding fluid. See RADIATION LOSS FACTOR (1). *Units* none.

modal wavenumber – *for acoustic modes governed by the scalar Helmholtz equation* any of the wavenumbers κ_N for which κ_N^2 is a modal EIGENVALUE. The index N (N = 1, 2, 3, etc.) labels the various modes. *Units* m^{-1}.

Note: The modal wavenumber may be complex.

mode – (1) a spatial pattern of vibration, whose shape remains invariant as the vibration *either* evolves over time *or* propagates spatially. An example of the first definition is a natural acoustic mode in a room; an example of the second definition is an acoustic mode of propagation in a uniform waveguide. See NATURAL MODE, PROPAGATION COORDINATE.

mode – (2) a member of a set of linearly independent functions used as BASIS FUNCTIONS to describe a pattern of vibration, e.g. the normal displacement of a plate or the pressure field in an enclosure.

mode conversion – *for sound incident on an interface, or waveguide modes incident on a discontinuity* conversion from one type of propagation to another. For example, fluid-borne sound incident on a solid boundary is transmitted partly as shear waves and partly as compressional waves in the solid. Mode conversion can also occur on reflection, as when plane acoustic waves travelling in a rigid pipe are reflected at an open end.

mode count – *for an enclosed region with specified boundary conditions* the number of modes whose eigenvalues lie below a given value, or (alternatively) whose resonance frequencies lie below a given frequency. For example, the acoustic modes obtained by solving the Helmholtz equation $(\nabla^2 + \kappa^2)p = 0$ for the pressure p in a bounded region R, with either Neumann or Dirichlet boundary conditions on the boundary ∂R, have discrete eigenvalues κ^2; the number of such modes with κ less than k may be estimated in the Neumann case as

$$N(k) \approx \frac{Vk^3}{6\pi^2} + \frac{Sk^2}{16\pi} \quad \text{(3D region; volume } V\text{, surface area } S\text{)};$$

$$N(k) \approx \frac{Ak^2}{4\pi} + \frac{Lk}{4\pi} \quad \text{(2D region; area } A\text{, perimeter } L\text{)}.$$

The relative error tends to zero for large N. For the Dirichlet boundary condition, i.e. $p = 0$ on ∂R, the sign of the second term in both expressions above is negative. *Units* none.

mode numbers – a sequential numbering scheme that provides a means of labelling individual MODES.

mode ray angle – *in a waveguide with non-absorbing or non-leaky boundaries* the angle of propagation measured from the waveguide axis. The mode ray angle θ is defined for a given mode by $\sin \theta = 1/\xi$, where ξ is the mode CUTOFF RATIO. *Units* rad.

modulation – the systematic variation of a periodic carrier signal by a lower-frequency signal. Assuming the carrier signal is proportional to $\cos 2\pi f_0 t$ (where f_0 is the carrier frequency), the general form of the modulated signal is

$$y(t) = A(t) \cos [2\pi f_0 t + \theta(t)].$$

The amplitude $A(t)$ and phase angle $\theta(t)$ are both slowly-varying functions of time (as measured on the time scale of the carrier signal), and their time dependence may be used to encode a message signal.

Three particular cases are:

(1) *Amplitude modulation*: Here $A(t) = A_0[1 + m(t)]$, while $\theta(t)$ is held constant. The effect is to shape the envelope of the carrier according to the input signal $m(t)$, provided the latter has amplitude less than 1. See also MODULATION INDEX.

(2) *Phase modulation*: Here $A(t)$ is held constant, and $\theta(t)$ fluctuates slowly in proportion to an input signal.

(3) *Frequency modulation*: As case (2), except that the derivative $\dot{\theta}(t)$ fluctuates slowly in proportion to an input signal.

modulation index – *of an amplitude-modulated signal* a measure of the depth of modulation imposed on the carrier signal. A signal of the form

$$x(t) = [1 + m(t)] A \cos \omega t \qquad (t = \text{time}),$$

where $m(t)$ is a slowly-varying function on the time scale of the carrier frequency ω, is said to have a modulation index of $\max |m(t)|$. *Units* none.

modulation transfer function (MTF) – *for a specified carrier signal* a measure of a transmission channel's ability to convey information via amplitude modulation. Band-filtered PINK NOISE is fed into the channel at the input, with sinusoidal modulation of the input signal power:

$$I_{\text{in}}(t) = A(1 + \cos 2\pi F t), \qquad F = \text{modulation frequency}.$$

Under steady-state conditions, the output signal power will be of the form

$$I_{\text{out}}(t) = B[(1 + m \cos 2\pi F(t - \tau)], \qquad \tau = \text{const}.$$

The factor m is the output power modulation index corresponding to unit input power modulation. It depends on the modulation frequency F; the function $m(F)$ defines the modulation transfer function of the transmission channel, for the given noise band.

Note (1): The MTF for sound transmission in a room, measured by applying modulation frequencies in the range 0.63–12.5 Hz to octave bands of noise in the speech frequency range, has been found to correlate with speech intelligibility. See SPEECH TRANSMISSION INDEX.

Note (2): If $\hat{h}(t)$ is the normalized impulse response of the transmission channel, defined such that $\int_0^\infty \hat{h}^2(t)dt = 1$, then the MTF is the modulus of the Fourier-transformed squared impulse response:

$$m(F) = \left| \int_0^\infty e^{-j2\pi Ft} \hat{h}^2(t) dt \right|.$$

Omitting the modulus signs gives the *complex modulation transfer function*.

modulator – an equivalent term for ACTIVE TRANSDUCER.

modulo – two numbers are congruent modulo N (meaning to the modulus N) if their difference is an integer multiple of N. The relative phase, φ, of two sinusoidal signals is usually given modulo 2π, meaning that the value of φ is brought within the range $-\pi < \varphi \leq \pi$ (or alternatively $0 \leq \varphi < 2\pi$) by subtracting a multiple of 2π.

modulus – (1) *of a complex number* the real positive quantity $|z| = \sqrt{x^2 + y^2}$, where $x = \operatorname{Re} z$ and $y = \operatorname{Im} z$ are the real and imaginary parts of the complex number z.

modulus – (2) a generic name for any elastic coefficient appearing in an equation that expresses HOOKE'S LAW. Examples are the SHEAR MODULUS and YOUNG'S MODULUS of an isotropic solid. An ideal fluid is characterized by its BULK MODULUS, and has zero shear modulus.

modulus – (3) see MODULO.

molar gas constant – the GAS CONSTANT for one MOLE of any gas; also known as the *universal gas constant*. Its value is 8.314 51 J mol^{-1} K^{-1}.

molar mass – the mass per MOLE of any substance. *Units* kg mol^{-1}.

mole – the SI base unit for the amount of substance in a system; it is defined as the amount that contains the same number of molecules (or atoms, or ions) as there are atoms in 0.012 kg of carbon-12. *Unit symbol* mol.

Note (1): The conversion factor from moles to number of particles is called the AVOGADRO CONSTANT, with symbol N_A.

Note (2): For undissociated gases, the molecule is generally understood to be the basic particle; the *number of molecules per mole of gas* is therefore N_A.

molecular relaxation – the process of energy redistribution at the molecular level in a gas, among the various degrees of freedom (translational, rotational, and vibrational), that occurs as local equilibrium is regained following a rapid disturbance. The rate at which a particular energy component E approaches its equilibrium value E_{eq} is $|E - E_{eq}|/\tau$, where τ is a constant called the RELAXATION TIME.

molecular weight – see RELATIVE MOLAR MASS.

moment – (1) *of a distribution* a weighted integral (for a continuous distribution), or weighted sum (for a discrete distribution), in which the weighting factor is x^k. Here x is the distribution variable, and $k = 0, 1, 2$, etc. is the *order* of the moment. Mathematically, the order-k moment of either a continuous or a discrete distribution is defined by

$$M_k = \int_{-\infty}^{\infty} f(x)\, x^k\, dx \quad \text{(continuous distribution)};$$

$$M_k = \sum_i n_i\, x_i^k \quad \text{(discrete distribution)}.$$

In the first equation above, $f(x)$ is the DENSITY FUNCTION of the distribution. In the second equation, the sum is taken over the N paired values (n_i, x_i) that form the discrete FREQUENCY DISTRIBUTION, with $i = 1, 2, \ldots, N$.

Note (1): The term *moment*, with no order specified, is generally understood as referring to $n = 1$.

Note (2): In physical applications, x is commonly a cartesian coordinate, and the equations above yield ***moments about the x axis***. Alternatively, x may be replaced in the equations above by a radius r in polar coordinates, giving ***polar moments*** about the origin.

Note (3): In statistics, the moments of a PROBABILITY DENSITY FUNCTION yield information about the statistics of the random variable concerned. For example, M_1 is the variable's expected value or MEAN, M_2 is its MEAN SQUARE VALUE, and so on. See POPULATION MOMENTS, SAMPLE MOMENTS.

moment – (2) *of a force* the torque or turning effect produced by the force. Equivalent definitions in scalar and vector form are given below.

(1) The moment of a force about a specified axis is a scalar, equal to the torque exerted about that axis. It equals FL, where F is the force magnitude and L is the minimum distance that separates the line of action of the force from the given axis.

(2) The moment of a force $\mathbf{F}$ about a point A is a vector $\mathbf{G} = (\mathbf{r} - \mathbf{r}_A) \times \mathbf{F}$. Here $\mathbf{r}$ is the position vector of any point on the line of action of $\mathbf{F}$ (e.g. the point of application of the force); $\mathbf{r}_A$ is the position vector of A. The

component of **G** in a specified direction gives the torque exerted about an axis through A, parallel to that direction.

Units N m.

momentum – *of a system* a vector quantity given by the vector sum $\Sigma m_i \mathbf{v}_i$, where m_i is the mass of an individual particle and $\mathbf{v}_i$ is its velocity relative to an INERTIAL FRAME OF REFERENCE; the sum extends over the total mass m_{tot} of the system. An equivalent term is **linear momentum**. *Units* kg m s^{-1}.

Note (1): The conservation law for linear momentum states that the momentum, **M**, of a closed system changes with time according to

$$\frac{d\mathbf{M}}{dt} = \mathbf{F}$$

where **F** is the total force applied to the system.

Note (2): If the MASS CENTRE of the system has velocity $\mathbf{v}_c$, the total momentum of the system is $\mathbf{M} = m_{\text{tot}} \mathbf{v}_c$.

monaural – using, or used with, only one ear.

monochromatic – single-frequency; e.g. ~ *signal*, ~ *sound field*. Equivalent to HARMONIC (2).

monophonic – using only one channel between sound source and listener. Monaural listening (i.e. with only one ear) is monophonic. Likewise a single-microphone recording system is monophonic.

monopole – *used as a noun* abbreviation for POINT MONOPOLE.

monopole source – a sound generation mechanism that drives the surrounding medium by a process equivalent to a time-varying volume displacement. In the COMPACT limit, the sound power output from a monopole source varies as c^{-1}, where c is the sound speed in the radiating medium.

monopole strength – an equivalent term for ACOUSTIC SOURCE STRENGTH.

monostatic – adjective used to describe an ACTIVE SONAR system in which the transmitter and receiver are at the same location.

motion response – *of a vibration isolator excited by a force input* the amplitude of the output displacement under single-frequency force excitation, normalized by the isolator deflection under a static load equal to the input force amplitude. Compare TRANSMISSIBILITY. *Units* none.

motion transmissibility – *of a vibration isolator excited by a displacement input* the ratio of output to input displacement amplitude, under single-frequency excitation. Compare RELATIVE TRANSMISSIBILITY. *Units* none.

muddy – *in auditorium acoustics* see entry for CLARITY.

muffler – an acoustic isolator or absorber designed for insertion in a duct system. An alternative term is DUCT SILENCER.

multi-degree-of-freedom system – a dynamical system that requires at least two variables, and their time derivatives, to define its state at any instant. See DEGREES OF FREEDOM, GENERALIZED COORDINATES.

multilayer absorber – a sound absorber formed by positioning two or more layers of sound-absorbing material, of differing properties and possibly separated by resistive sheets or screens, in front of a rigid surface. The layers may also be separated by airspaces.

multimode scattering matrix – (1) ◆ *for a waveguide termination* a square matrix of complex coefficients that relates a vector of modal reflected pressures, $\mathbf{p}^-$, to a vector of modal incident pressures $\mathbf{p}^+$. In mathematical form, it is the matrix $\mathbf{S}$ in the equation

$$\mathbf{p}^- = \mathbf{S}\mathbf{p}^+.$$

An equivalent term is REFLECTION COEFFICIENT MATRIX. *Units* none.

Note: In principle, the vectors $\mathbf{p}^+$, $\mathbf{p}^-$ and the matrix $\mathbf{S}$ are of infinite size (since the number of waveguide modes is infinite). In practice, the modal series would be truncated to limit the number of evanescent modes.

multimode scattering matrix – (2) ◆ *for a junction connecting two waveguides* a 2×2 SCATTERING MATRIX each of whose elements is a square matrix of modal coefficients. In mathematical form, it is the partitioned matrix defined by

$$\mathbf{S} = \begin{bmatrix} \mathbf{R}^{(1)} & \mathbf{T}^{(2)} \\ \mathbf{T}^{(1)} & \mathbf{R}^{(2)} \end{bmatrix}$$

where $\mathbf{R}^{(1)}$ is the pressure REFLECTION COEFFICIENT MATRIX for modes incident from side (1), $\mathbf{T}^{(1)}$ is the corresponding pressure TRANSMISSION COEFFICIENT MATRIX, and so on. Thus on side (1), the vector of modal scattered pressures $\mathbf{p}_1^{out}$ (outgoing with respect to the junction) is related to the two modal incident pressure vectors by

$$\mathbf{p}_1^{out} = \mathbf{R}^{(1)}\mathbf{p}_1^{in} + \mathbf{T}^{(2)}\mathbf{p}_2^{in}; \quad \text{likewise}$$

$$\mathbf{p}_2^{out} = \mathbf{T}^{(1)}\mathbf{p}_1^{in} + \mathbf{R}^{(2)}\mathbf{p}_2^{in}.$$

The four modal coefficient matrices have the form

$\mathbf{R}^{(1)} = [R_{ml}^{(1)}], \quad \mathbf{T}^{(1)} = [T_{ml}^{(1)}],$ etc.

where index l denotes the incident mode and m denotes the scattered mode. Compare MULTIPORT SINGLE-MODE SCATTERING MATRIX. *Units* none.

multipath – *in underwater acoustics* one of the ray paths involved in MULTIPATH PROPAGATION; hence *multipath arrival*.

multipath propagation – *in underwater acoustics* propagation due to the combined effect of many ray paths.

multiple pure tone noise – *of a supersonic fan rotor* an equivalent term for BUZZ-SAW NOISE.

multipole – *used as a noun* abbreviation for POINT MULTIPOLE.

multipole order – see ORDER (12).

multipole source – a multipole source of order n is a sound generation mechanism that drives the surrounding medium by a process equivalent to order-n SPHERICAL HARMONIC vibration of a spherical boundary. In the COMPACT limit, the sound power output from such a source varies as $c^{-(2n+1)}$, where c is the sound speed in the surrounding medium.

multiport admittance matrix – ◆ *the inverse of a* MULTIPORT IMPEDANCE MATRIX.

multiport impedance matrix – ◆ *for a network with two or more ports, or any extended system with inputs and responses at two or more discrete points* a square matrix of complex coefficients that relates a vector of output quantities **y** to a vector of input quantities **x**, as in the relations

$\mathbf{y} = \mathbf{Z}\mathbf{x}$ or $y_j = Z_{ji} x_i.$

Here the elements of **x** are the inputs at each port, and the elements of **y** are the outputs at each port.

Note: The word *impedance* is used if **x** represents currents, velocities, or volume velocities and **y** represents voltages, forces, or pressures. In the opposite situation (e.g. with **x** representing pressures and **y** representing volume velocities), **Z** would be called a *multiport admittance matrix*.

multiport single-mode scattering matrix – ◆ *for an N-port junction connecting two or more single-mode waveguides* an $N \times N$ matrix of complex coefficients that relates a column vector of scattered (or outgoing-wave) pressures at the N ports, $\mathbf{p}^{out}$, to a column vector of incident or incoming-wave pressures, $\mathbf{p}^{in}$, as in the relations

$$\mathbf{p}^{out} = \mathbf{S}\mathbf{p}^{in} \quad \text{or} \quad p_m^{out} = S_{ml}p_l^{in}.$$

Units none.

Note: The term *scattering matrix*, in the single-mode sense used here, is standard terminology for waveguide junctions in electrical engineering. The only reason for introducing the "multiport" qualifier is to avoid confusion with *multimode* scattering, as in MULTIMODE SCATTERING MATRIX.

multivariate – *in statistics* relating to more than two random variables, as opposed to UNIVARIATE (one variable) or BIVARIATE (two variables). See JOINT PROBABILITY DENSITY FUNCTION.

mutual impedance – *of elements in a radiating array* the complex acoustic pressure that a single active element produces at the location of another element, divided by the normal velocity of the active element. If one defines an *array impedance matrix* in which the (i, j) term is the complex ratio of the pressure at the ith element to the velocity of the jth element, the mutual impedances are the off-diagonal terms. *Units* Pa s m^{-1}.

Note: A more precise term for the ratio defined above would be *mutual specific impedance*. Likewise a *mutual acoustic impedance* would refer to the complex ratio of pressure to volume velocity. Compare TRANSFER IMPEDANCE.

N

n – symbol for the prefix nano, denoting a submultiple of 10^{-9} in the SI system of units. Thus 1 nm denotes 10^{-9} m (1 nanometre).

narrow-band random noise – RANDOM NOISE whose BANDWIDTH (2) is small compared with other characteristic frequencies.

natural boundary conditions – *for a second-order differential equation* conditions imposed on the first derivative of the solution, at points on the boundary. Compare NEUMANN BOUNDARY CONDITIONS. Other names sometimes used are ***dynamic*** or ***non-essential*** boundary conditions.

Note: For a higher-order differential equation of order $2m$ (where m is an integer), natural boundary conditions are those which involve the derivatives of the solution from order m up to order $(2m - 1)$. Compare ESSENTIAL BOUNDARY CONDITIONS.

natural frequency – *of a given mode of a system* the frequency of FREE OSCILLATION. If the oscillations are exponentially damped, the ***damped natural frequency*** is defined as $1/T_d$, where T_d is the period between successive zero crossings in the same direction. See also COMPLEX NATURAL FREQUENCY. *Units* Hz.

Note: A lumped-parameter system with N degrees of freedom has N natural frequencies, corresponding to its N modes of free oscillation (i.e. the ***natural modes***). A continuous system has an infinite number of natural frequencies, although the number in a given frequency band is finite.

natural mode of vibration – *of a linear system* a spatial pattern of free vibration. Each natural mode oscillates at a single frequency after all external forcing is removed. In systems with damping the amplitude of free vibration decays exponentially; see COMPLEX NATURAL FREQUENCY.

Navier–Stokes equations – *in fluid mechanics* the equations of viscous fluid motion (**C-L-M NAVIER 1822; G G STOKES 1845**).

near field – *of a finite-size source, radiating at a given frequency, under free-field conditions* the region surrounding the source within which FAR-FIELD conditions do not apply. See also GEOMETRIC NEAR FIELD, HYDRODYNAMIC NEAR FIELD, FRESNEL NEAR FIELD.

neck – a narrow tube or passage connecting two regions of larger cross-section, as in a HELMHOLTZ RESONATOR. Fluid flowing from one region to the other accelerates to a relatively high velocity in the neck, compared with its velocity on either side.

neper – a logarithmic unit of amplitude ratio, using base e. If two sinusoidal quantities of the same frequency and same type (e.g. the acoustic pressures at two points in a single-frequency sound field) have amplitudes p_1 and p_2, their logarithmic amplitude ratio is $\ln(p_2/p_1)$ nepers. Thus one neper corresponds to a factor-e ratio in amplitude. Based on the Latin name Neperus (**J NAPIER *1550–1617***). *Unit symbol* Np.

Neumann boundary condition – *applied to solutions of a differential equation* a boundary condition that specifies the normal derivative (normal gradient) of the solution at every point on the boundary: for example, the normal gradient is required to equal zero (**C G NEUMANN 1877**).

Neumann function – the cylinder function of the second kind, denoted by $Y_\mu(z)$ or $N_\mu(z)$ where z is the argument of the function and μ is the order (**C G NEUMANN 1867**). See CYLINDER FUNCTIONS.

newtonian fluid – see VISCOSITY (**I NEWTON 1686**). Both capital (Newtonian) and lower-case spellings are used.

NIHL – abbreviation for NOISE-INDUCED HEARING LOSS. *Units* dB.

NIL – abbreviation for NOISE IMMISSION LEVEL. *Units* dB re E_{ref}; $E_{\text{ref}} = (1740/8)$ times the E_{ref} used for NOISE EXPOSURE LEVEL.

node – a point of minimum amplitude in a one-dimensional standing wave field (e.g. *pressure* ~ *in a standing-wave tube*; also *velocity* ~ *on a transversely vibrating string*). A **nodal line** is a line of minimum amplitude in a two-dimensional field, and a **nodal surface** is a surface of minimum amplitude in a three-dimensional field. Compare ANTINODE.

noise – (1) an undesired or extraneous signal.

noise – (2) *in an acoustical context* undesired or extraneous sound.

noise – (3) *in a signal-processing context* an abbreviation for RANDOM NOISE (2).

noise-equivalent bandwidth – *of a filter* the same as ***equivalent noise bandwidth***. See ENERGY BANDWIDTH (1). *Units* Hz.

noise exposure level – the A-weighted SOUND EXPOSURE within a given time period (usually 24 hours), expressed in dB relative to 1.15×10^{-5} Pa^2 s. Compare SOUND EXPOSURE LEVEL, which is the same quantity expressed relative to a different reference value. Symbol $L_{\text{EX,8h}}$. *Units* dB re E_{ref}; $E_{\text{ref}} = 1.15 \times 10^{-5}$ Pa^2 s.

Note: The subscript 8 h in $L_{EX,8h}$ refers to the fact that in the case of a constant noise that lasts for 8 h, the noise exposure level is numerically equal to the A-weighted sound pressure level. See also DAILY PERSONAL NOISE EXPOSURE, which is an identical measure, but is specifically restricted to occupational exposure over a typical 24-hour period.

noise immission level (NIL) – a measure of cumulative A-weighted SOUND EXPOSURE during a person's lifetime, based on a modification of the NOISE EXPOSURE LEVEL concept. The reference sound exposure is chosen in such a way that after working for N years in an environment with a constant sound level L_A, a person who works a typical 1740 hours per year will have a cumulated noise immission level given by

$$NIL = L_A + 10 \log_{10} N,$$

assuming the person's leisure-time exposure and previously-cumulated work exposure are negligible. The equation above implies a reference sound exposure of $E_{ref} = 2.5 \times 10^{-3}$ Pa² s, i.e. (1740/8) times the E_{ref} used for noise exposure level. *Units* dB re E_{ref}.

noise-induced hearing loss (NIHL) – cumulative hearing loss associated with repeated exposure to noise, principally due to irreversible changes in the ORGAN OF CORTI. The typical audiometric pattern is a maximum hearing loss (sometimes a notch) in the region 3–6 kHz. Hearing loss implies a comparison of the affected person's hearing threshold with that of an appropriate control; such a control (or notional person) is taken to be someone differing from the affected person only in respect of their exposure to noise. *Units* dB.

Note: The term normally excludes sudden damage and its associated hearing loss, e.g. due to an explosion (ACOUSTIC TRAUMA) or rapid change of barometric pressure (BAROTRAUMA).

noise level – (1) the level of NOISE (2). *Units* dB re p_{ref}^2.

noise level – (2) abbreviation for the SPECTRUM LEVEL of NOISE (2). *Units* dB re p_{ref}^2 Hz^{-1}.

noise reduction (NR) – (1) *of a silencer or other element in a sound transmission path* the reduction in sound pressure level, in a given frequency band, between a point just ahead of the silencer (the side nearer the source) and a point just behind the silencer (the side further from the source). *Units* dB.

noise reduction (NR) – (2) *of a common dividing wall or partition between two rooms* the LEVEL DIFFERENCE (2) between the rooms, in a given frequency band, when the field in both rooms is approximately diffuse. *Units* dB.

Note: For two rooms separated by a partition, the NR depends on the product, $\tau_{stat}S_w$, of the STATISTICAL SOUND POWER TRANSMISSION COEFFICIENT of the partition and its area S_w; it also depends on the room absorption of the receiving room, A_r. According to the diffuse-field hypothesis,

$$\mathrm{NR} = 10\log_{10}\left(1 + \frac{A_r}{\tau_{stat}S_w}\right).$$

Thus the noise reduction between a pair of rooms depends on which room contains the source (or sources) of sound.

non-causal response – *of a system* an equivalent term for ACAUSAL RESPONSE

non-compact – *at a given frequency* having a spatial extent comparable with the acoustic wavelength, or greater. The opposite of COMPACT.

non-deterministic signal – an equivalent term for RANDOM SIGNAL; the opposite of a DETERMINISTIC SIGNAL.

non-homogeneous differential equation – a DIFFERENTIAL EQUATION in which one or more forcing terms appear that do not contain the dependent variable. For example, the following differential equation in $u(x)$ is non-homogeneous because of the $F(x)$ term, which is independent of u:

$u'' + a(x)u' + b(x)u = F(x).$

The description ***inhomogeneous*** is sometimes used in place of non-homogeneous.

non-homogeneous set of linear equations – a set of M equations in N unknown variables $(x_1, x_2, \ldots, x_N)$, with a non-zero right-hand side appearing in at least one of the equations. Such a set of equations may be written as

Ax = b,

in matrix notation. Here **A** is an $M \times N$ matrix of known coefficients, and **x** is a column vector. The elements of the column vector **b** are the right-hand sides of each of the M equations; by definition at least one of them is non-zero.

Whether or not the set of equations has an exact solution depends on the relative sizes of M and N, as follows:

(a) For $M > N$ the solution is ***overdetermined***. A LEAST-SQUARES solution can be found, but no exact solution exists in general.

(b) For $M = N$ there is a unique solution for **x**, provided the determinant of **A** is non-zero.

(c) If $M < N$, then at least $N - M$ of the unknown variables are **undetermined**.

nonlinear acoustics – the branch of acoustics that deals with finite-amplitude sound fields. Compare LINEAR ACOUSTICS.

nonlinear acoustic saturation – see SATURATION.

nonlinear instability – see under STABILITY.

nonlinearity – a departure from LINEARITY.

nonlinearity parameter – *for a fluid under isentropic compression or expansion* the ratio B/A, where A, B are coefficients in a Taylor series expansion of the isentropic pressure-density relation about a reference state (P_0, ρ_0):

$$P - P_0 = A\zeta + \frac{B}{2!}\zeta^2 + \frac{C}{3!}\zeta^3 + \ldots, \qquad \left(\zeta = \frac{\rho - \rho_0}{\rho_0}\right).$$

Here P is the fluid pressure, ρ is the density, and ζ is the relative increase in density. For more information on the coefficients (A, B, etc.) see ISENTROPIC EQUATION OF STATE IN SERIES FORM. Compare COEFFICIENT OF NONLINEARITY. *Units* none.

non-singular – a SQUARE MATRIX is non-singular if its determinant is not equal to zero. An equivalent term is **invertible**. See INVERSE MATRIX.

nonstationary signal – a signal whose statistical properties vary with respect to time.

non-white noise – a stationary random signal whose AUTOSPECTRAL DENSITY or power spectrum falls off at a constant rate, in dB per octave, as the frequency increases. See also PINK NOISE, BROWN NOISE.

normal – *to a smooth surface* a line that cuts the surface at right angles.

normal distribution – an alternative term for GAUSSIAN DISTRIBUTION, especially for the UNIVARIATE version with a PROBABILITY DENSITY FUNCTION given by

$$f(x) = \frac{1}{\sqrt{2\pi\sigma^2}} \exp\left[\frac{-(x-\mu)^2}{2\sigma^2}\right].$$

The function $f(x)$ above has a first moment about the origin of μ, and a second moment of $(\mu^2 + \sigma^2)$; the corresponding moments about the mean are 0 and σ^2.

normal impedance – abbreviation for the SPECIFIC ACOUSTIC IMPEDANCE of a boundary in the normal direction. *Units* Pa s m^{-1}.

normal incidence – describes plane waves arriving at a plane boundary with an ANGLE OF INCIDENCE equal to zero (i.e. the propagation direction is normal to the boundary).

normalization constant – *for an eigenfunction* the constant N_m defined by $V^{-1} \int \psi_m^2 \, dV = N_m^2$. Here ψ_m is the mth in a sequence of orthogonal eigenfunctions or BASIS FUNCTIONS, defined over a finite region of volume V, and the integral extends over the entire region.

normal-mode field – *in underwater acoustics* the component of a radially-propagating sound field, in a stratified model of the ocean, that consists of shallow-angle trapped rays. Specifically, the normal-mode field corresponds to GRAZING ANGLES smaller than the CRITICAL ANGLE FOR TRANSMISSION, so that there is near-total reflection at the boundaries of the waveguide. An equivalent term is ***trapped modes***. Compare EVANESCENT MODE, VIRTUAL MODE.

normal modes – (1) *for any linear system* a set of EIGENFUNCTIONS of the governing equation that (a) satisfy chosen boundary conditions (not necessarily the actual boundary conditions), and (b) are ORTHOGONAL. Such functions can be used as a basis for representing forced vibrations of the system at any desired frequency (not necessarily an eigenfrequency).

normal modes – (2) the orthogonal EIGENFUNCTIONS that describe the modes of FREE OSCILLATION in an undamped system.

normal modes – (3) *in a waveguide* orthogonal eigenfunctions of the waveguide cross-section that satisfy the wall boundary conditions, and can be used to represent propagation of sound along the waveguide at any frequency. See also WAVEGUIDE MODE.

normal specific acoustic impedance – *at a boundary excited by a single-frequency sound field* the SPECIFIC ACOUSTIC IMPEDANCE $z = p/u$ evaluated at the boundary. Here p is the local acoustic pressure, and u represents the normal particle velocity directed into the boundary. *Units* Pa s m^{-1}.

Note: If the boundary consists of a rigid porous or perforated surface, u is interpreted as the ***face velocity***, i.e. the volume velocity per unit area entering the surface.

normal vector – *at a point on a smooth surface* a UNIT VECTOR normal to the surface. Alternative terms are ***unit normal vector*** or ***surface normal vector***.

Note: The word *vector* is sometimes dropped from all these terms, where there is no risk of ambiguity.

normal velocity transmission coefficient – *for single-frequency plane waves incident on a plane boundary or two-fluid interface* the ratio

$$T_u = u_{\text{trans}}/u_{\text{inc}},$$

where u_{trans} and u_{inc} represent the complex amplitudes of the transmitted and incident particle velocity normal to the boundary, measured in the same direction. The value of T_u generally depends on the frequency as well as the angle of incidence. *Units* none.

noy – a unit of noisiness used to quantify the annoyance potential of complex sounds. A one-third octave band of noise with a sound pressure level of 40 dB re $(20\ \mu\text{Pa})^2$, centred on 1 kHz, has a noisiness of 1 noy, and a PERCEIVED NOISE LEVEL of 40 PNdB. Compare SONE.

NPE – a lossless nonlinear PARABOLIC EQUATION that describes time-domain propagation of almost-axial waves through a uniform medium. The NPE, or *nonlinear propagation equation*, follows a wave packet or transient sound pulse as it propagates, by introducing a co-moving frame of reference: if z is the axial direction, the new coordinates are

$$x' = x,\ y' = y,\ z' = z - c_0 t \qquad (c_0 = \text{undisturbed sound speed}).$$

The evolution of the spatial pressure waveform with time, t, is described in these coordinates by

$$\frac{\partial p}{\partial t} = -\frac{c_0}{2}\int_{-\infty}^{z'} \nabla_{\perp}^2 p(t,x',y',\zeta)d\zeta - \frac{\beta}{2\rho_0 c_0}\frac{\partial p^2}{\partial z'},$$

where $p(t, x', y', z')$ is the acoustic pressure, ρ_0 is the ambient density, and β is the COEFFICIENT OF NONLINEARITY. The Laplacian $\nabla_{\perp}^2$ operates in the plane normal to the beam axis, and accounts for beam diffraction in that plane.

null matrix – a MATRIX whose elements are all zeros. An alternative term is *zero matrix*.

N wave – (1) a transient waveform having the shape of a letter N; it consists of two successive step functions separated by a downwards ramp. An N-wave signal of duration T and peak value A, beginning at time $t = t_0$, is defined mathematically by

$$x(t) = \begin{cases} A\left(1 - \dfrac{2t'}{T}\right), & 0 < t' < T; \\ 0, & t' < 0 \text{ and } t' > T. \end{cases}$$

Here $t' = t - t_0$ is the time measured from the onset of the N wave.

N wave – (2) *in nonlinear acoustics* one of the two possible asymptotic forms taken by transient propagating pressure pulses at large distances, but prior to the OLD-AGE REGION; see SATURATION, SAWTOOTH WAVE (2), WEAK-SHOCK THEORY.

Note: Pressure waves of approximately N-wave shape occur in the far field of supersonic aircraft (see SONIC BOOM) and of airborne explosions. However, turbulence in the lower levels of the atmosphere, and ground reflection effects, modify the waveform received near the ground.

Nyquist frequency – *for a signal sampled at finite intervals* one-half of the SAMPLING FREQUENCY. Frequency components in the original signal that are below the Nyquist frequency are correctly represented in the sampled signal; all other frequencies are ALIASED. *Units* Hz.

Nyquist plot – *for a linear system* a polar plot of the FREQUENCY RESPONSE FUNCTION of the system, $H(\omega)$, in the complex plane over the angular frequency range $-\infty < \omega < +\infty$. The imaginary part, $\operatorname{Im} H(\omega)$, is plotted against the real part, $\operatorname{Re} H(\omega)$. Compare LOG MAGNITUDE–PHASE PLOT, BODE PLOT.

Nyquist rate – an obsolete term for SAMPLING FREQUENCY. *Units* Hz.

Nyquist theorem – see SAMPLING THEOREM.

O

O, o – order symbols. See ORDER (1).

OASPL – abbreviation for *overall sound pressure level*. See OVERALL LEVEL. *Units* dB re p_{ref}^2.

objective – *used as an adjective* refers to a measurement in which human judgement is not involved; the opposite of SUBJECTIVE.

oblique mode – *of a rectangular room with rigid boundaries* a standing-wave acoustic field in which the spatial distribution of pressure, p, is of the form

$$p \propto \cos\frac{l\pi x}{L_x}\cos\frac{m\pi y}{L_y}\cos\frac{n\pi z}{L_z} \qquad (x, y, z = \text{cartesian coordinates})$$

with none of the integers l, m, n equal to zero. Here the room boundaries are at $x = 0, L_x$; $y = 0, L_y$; $z = 0, L_z$. The name "oblique mode" comes from the fact that a single-frequency field having the spatial pattern described above is equivalent to a set of eight plane progressive waves, all travelling obliquely to the coordinate axes, with wavenumber components

$$k_x = \pm\frac{l\pi}{L_x}, \quad k_y = \pm\frac{m\pi}{L_y}, \quad k_z = \pm\frac{n\pi}{L_z}.$$

oblique mode count – *in a hard-walled rectangular room* the number of OBLIQUE MODES whose EIGENVALUES κ^2, given by

$$\kappa^2 = \left(\frac{l\pi}{L_x}\right)^2 + \left(\frac{m\pi}{L_y}\right)^2 + \left(\frac{n\pi}{L_z}\right)^2 \qquad (l, m, n \text{ integers}),$$

are less than a specified value k^2. Equivalently, the oblique mode count is the number of such modes with modal wavenumber κ less than k. If the oblique mode count is denoted by $N_{\text{obl}}(k)$, a smoothed asymptotic formula for a room of volume V and surface area S is

$$N_{\text{obl}} \approx \frac{Vk^3}{6\pi^2} - \frac{Sk^2}{16\pi} + \frac{L_{\text{tot}}k}{4\pi}, \qquad (N_{\text{obl}} \gg 1)$$

where $L_{\text{tot}} = (L_x + L_y + L_z)$ is the sum of the three room dimensions. *Units* none.

OBS – abbreviation for OCEAN BOTTOM SEISMOMETER.

occluded-ear simulator – *in audiology* an EAR SIMULATOR that simulates the inner part of the ear canal. It provides a transfer impedance between an insert and the eardrum similar to that of the human ear.

ocean bottom seismometer – an instrument (e.g. a GEOPHONE) for measuring particle displacement or acceleration at the sea bed. Compare SEISMOMETER.

octant – one of the eight equal regions into which a set of three orthogonal planes divides three-dimensional space. An example of an octant is the region ($x > 0$, $y > 0$, $z > 0$), where x, y, z are rectangular coordinates.

octave – (1) a unit of LOGARITHMIC FREQUENCY INTERVAL. The interval in octaves between a lower frequency f_1 and a higher frequency f_2 is $\log_2(f_2/f_1)$. *Unit symbol* oct.

octave – (2) abbreviation for OCTAVE BAND. See also ONE-THIRD OCTAVE.

octave band, 1/1 octave band – a frequency band whose upper and lower limits are a factor of 2 (one octave) apart. See also ONE-THIRD OCTAVE BAND.

odd function – a function $f(x)$ with the property that $f(-x)$ equals $-f(x)$. See also PARITY.

old-age region – *in nonlinear acoustics* the region sufficiently far from a sound source that no further nonlinear distortion occurs, because the pressure amplitude has fallen too low – either because of nonlinear dissipation in shocks, or because of wavefront spreading (see AGE VARIABLE). The transition to old age occurs at the point where the tendency for waveform steepening becomes insignificant, in comparison with the tendency of dissipative processes to smooth the waveform.

omnidirectional response – *of a transducer to incident sound* a response that does not vary with the direction of the incident sound, or whose variation with direction is designed to be as small as possible.

one-dimensional – confined to one dimension, or dependent on only one coordinate. Compare ONE-DIMENSIONAL PROPAGATION.

one-dimensional propagation – propagation of a fixed wave pattern with respect to a single spatial coordinate, as in a SPHERICAL PROGRESSIVE WAVE. Plane and cylindrical progressive waves, and modal propagation in a uniform waveguide, are further examples.

Note: One-dimensional propagation is limited to SEPARABLE SOLUTIONS of the wave equation; the basic idea is that the spatial dependence of the field transverse to the propagation direction can be separated out, as a factor independent of the PROPAGATION COORDINATE and time.

one-tenth decade, 1/10 decade – (1) a factor of $10^{1/10}$; (2) the LOGARITHMIC FREQUENCY INTERVAL between a lower frequency f_1 and a higher frequency f_2, when f_2/f_1 equals $10^{1/10}$; (3) abbreviation for a frequency band whose upper and lower limits are a factor $10^{1/10}$ apart.

one-third octave, 1/3 octave – (1) the LOGARITHMIC FREQUENCY INTERVAL between a lower frequency f_1 and a higher frequency f_2, when f_2/f_1 equals $2^{1/3}$; (2) abbreviation for a frequency band whose upper and lower limits are a factor $2^{1/3}$ apart.

Note: A frequency ratio of 10 one-third octaves is close to one DECADE (within 0.8 per cent). The difference between bandwidths of *one-tenth decade* and *one-third octave* is small enough to be ignored in many practical situations, and spectral analysis in contiguous bands one-tenth decade wide is routinely described (for historical reasons) as ***one-third-octave analysis***.

one-third octave band, 1/3-octave band – a frequency band whose upper and lower limits are a factor of $2^{1/3}$ (i.e. a one-third octave) apart. The distinction between $2^{1/3}$ and $10^{1/10}$ is commonly ignored; see the previous and following entries.

one-third-octave exact midband frequencies – a sequence of standard (base-ten) midband frequencies, at 1000 Hz and at one-tenth-decade intervals on either side, as used for spectral analysis in one-tenth decade or one-third octave bands. The centre frequency of band N is $10^{N/10}$ Hz. *Units* Hz.

one-third-octave nominal midband frequencies – the sequence of rounded frequencies that includes 100, 125, 160, 200, 250, 315, 400, 500, 630, 800 and 1000 Hz, extended by powers of 10 in either direction. *Units* Hz.

open system – a system whose boundaries permit the flow of matter into or out of the system.

operator – *in mathematics* a generalization of the concept of a FUNCTION. Whereas a function maps a number into another number, or a set of numbers into another set of numbers, an operator can also map a function into another function. An example is the differential operator, D or d/dx:

$$Df(x) = \frac{d}{dx} f(x) = f'(x).$$

Here the function $f(x)$ is mapped into its derivative $f'(x)$.

opposite phase – see ANTIPHASE.

optoacoustic excitation – the production of pressure disturbances in a fluid (gas or liquid) by absorption of a modulated light beam, e.g. from a laser. Compare PHOTOACOUSTIC EXCITATION.

order – (1) *of an asymptotic approximation that becomes exact as a parameter* ε *tends to zero* a measure of how fast the error in the approximation approaches zero, based on comparison with a suitable function of ε. Two symbols are used for this purpose, big-*O* and little-*o*, with the following meanings:

$f(\varepsilon) = o[g(\varepsilon)]$ as $\varepsilon \to 0$ means $\lim_{\varepsilon \to 0} \left| \dfrac{f(\varepsilon)}{g(\varepsilon)} \right| = 0$;

$f(\varepsilon) = O[g(\varepsilon)]$ as $\varepsilon \to 0$ means $\lim_{\varepsilon \to 0} \left| \dfrac{f(\varepsilon)}{g(\varepsilon)} \right| < \infty$.

For example, as $\varepsilon \to 0$, the statements

$\sin \varepsilon = o(1)$ and $\sin \varepsilon = O(\varepsilon)$

are both true: the ratio of $\sin \varepsilon$ to 1 is *zero* in the limit, and the ratio of $\sin \varepsilon$ to ε is *bounded* (it equals 1). In the same limit, the error in approximating $\sin \varepsilon$ by ε is $O(\varepsilon^3)$, although one could also express it as $o(\varepsilon^2)$.

order – (2) *of a derivative* the number of times a function is differentiated.

order – (3) *of a determinant* the number of rows or columns.

order – (4) *of a polynomial* the largest exponent involved; also known as the *degree*. See POLYNOMIAL.

order – (5) *of a spherical Bessel function, or a spherical surface harmonic* see SPHERICAL BESSEL FUNCTIONS, SPHERICAL HARMONICS.

order – (6) *of a cylindrical Bessel function* see CYLINDER FUNCTIONS.

order – (7) *of the associated Legendre functions* the azimuthal order, corresponding to the upper index μ in the P_ν^μ and Q_ν^μ notation.

order – (8) *of a moment* see MOMENTS, DENSITY FUNCTION.

order – (9) *of an FFT* the number of points on which the transform is based. See FAST FOURIER TRANSFORM.

order – (10) *of a tensor* in three dimensions, a TENSOR of order 1 is a 3-component vector; in N dimensions, a tensor of order k has a total of N^k components. The terms **rank** and order are equivalent in this context.

order – (11) *of a frequency component* the ratio of a particular HARMONIC component in a periodic signal to the fundamental frequency.

order – (12) *of an acoustic source distribution* the number of divergence operations associated with the distribution. Thus source distributions $\mathcal{F}$ of order 0, 1, 2 are described respectively by

$$\mathcal{F} = F, \quad -\frac{\partial F_i}{\partial x_i}, \quad \frac{\partial^2 F_{ij}}{\partial x_i \partial x_j};$$

the general expression for order n is

$$\mathcal{F} = \frac{(-\partial)^n F_{ijk\ldots}}{\partial x_i \, \partial x_j \, \partial x_{k\ldots}}.$$

Respective names for the first three distributions are ***monopole***, ***dipole***, and ***quadrupole***; an order-3 source distribution would be called ***octupole***.

Note: The quantity F_i is called a dipole distribution or ***dipole density*** (the term ***dipole source distribution*** refers to the associated source density given by $-\partial F_i/\partial x_i$). Similarly, F_{ij} is a quadrupole distribution or ***quadrupole density***. See also SOURCE MOMENTS, MULTIPOLE SOURCE.

ordinary differential equation – a DIFFERENTIAL EQUATION in which only one independent variable appears.

organ of Corti – a complex structure associated with the hearing mechanism, located in the COCHLEA; it includes the BASILAR MEMBRANE and attached HAIR CELLS.

orthogonal coordinate system – a coordinate system whose coordinate surfaces intersect at right angles: for example RECTANGULAR COORDINATES fall in this category, because the three coordinate surfaces $x = a$, $y = b$, $z = c$ (with a, b, c constants) are orthogonal.

orthogonal eigenfunctions – see ORTHOGONAL FUNCTIONS, including the note.

orthogonal functions – the functions ψ_1, ψ_2 defined over a region R are orthogonal if their product integrates to zero over the region:

$$\int_R \psi_1 \psi_2 \, dV = 0.$$

A more general definition allows the integrand to contain a WEIGHTING FUNCTION $w(\mathbf{x})$, that varies over the region R according to position $\mathbf{x}$. The generalized orthogonality statement is

$$\int_R \psi_1(\mathbf{x}) \, \psi_2(\mathbf{x}) \, w(\mathbf{x}) \, dV = 0.$$

Note: The orthogonality statement above holds for any two different EIGENFUNCTIONS ψ_1, ψ_2 of the linear differential equation

$$L\,u(\mathbf{x}) + \lambda\,w(\mathbf{x})\,u(\mathbf{x}) = 0,$$

provided the operator L is SELF-ADJOINT (for example, $L = \nabla^2$) and the unknown function $u(\mathbf{x})$ satisfies appropriate ORTHOGONALITY BOUNDARY CONDITIONS on the boundary of R. As an example, the HELMHOLTZ EQUATION

$$\nabla^2 p + k^2 p = 0$$

has orthogonal eigenfunctions, provided the boundary of R is LOCALLY-REACTING.

orthogonality boundary conditions – *for a linear differential equation of specified form* a set of BOUNDARY CONDITIONS under which the EIGENFUNCTIONS of the differential equation are guaranteed to be ORTHOGONAL over the region concerned. For example, many single-frequency fields in acoustics are described by equations of the form

$$\mathrm{div}\{a(\mathbf{x})\,\mathrm{grad}\,u\} + b(\mathbf{x})\,u + \lambda\,w(\mathbf{x})\,u = 0;$$

here λ (a constant with respect to position $\mathbf{x}$) is a parameter that can be varied, and u is an unknown function of $\mathbf{x}$. The orthogonality boundary condition for this class of equations is

$$\frac{\partial u}{\partial n} = A u \qquad (A = \text{constant for all } \mathbf{x} \text{ and } \lambda),$$

where $\partial u/\partial n$ is the normal derivative of u on the boundary.

orthogonal matrix – a SQUARE MATRIX $\mathbf{A}$ is said to be orthogonal if

$$\mathbf{A}^T \mathbf{A} = \mathbf{I},$$

where $\mathbf{I}$ is the IDENTITY MATRIX of the same size as $\mathbf{A}$, and $\mathbf{A}^T$ is the TRANSPOSE of $\mathbf{A}$. Equivalently, if the elements of $\mathbf{A}$ are denoted by a_{ij}, orthogonality means that

$$a_{ij}\,a_{ik} = \delta_{jk};$$

here δ_{jk} is the KRONECKER DELTA, and the repeated index i implies summation.

Note: An orthogonal matrix has a determinant equal to ± 1, and an INVERSE equal to its TRANSPOSE: $\mathbf{A}^{-1} = \mathbf{A}^T$.

orthogonal random variables – two RANDOM VARIABLES are said to be orthogonal if their product has an EXPECTED VALUE of zero. An equivalent term for orthogonal in this sense is UNCORRELATED.

Note: If two zero-mean random variables are STATISTICALLY INDEPENDENT, they are necessarily orthogonal. However, the reverse statement (that orthogonality implies statistical independence) is true only for Gaussian signals. See note (1) under GAUSSIAN DISTRIBUTION.

orthogonal vectors – VECTORS with mutually perpendicular directions.

orthonormal functions – ORTHOGONAL FUNCTIONS that have been normalized, usually so that the mean square value of each function is unity over the region concerned. To achieve this, each function is divided by an appropriate NORMALIZATION CONSTANT.

orthonormal vectors – UNIT VECTORS with mutually perpendicular directions.

oscillation – a time-dependent SIGNAL; also, a time-dependent fluctuation in some physical quantity such as pressure, temperature, or displacement. The term *spatial oscillation* is used to describe similar fluctuations with respect to position, rather than time. See also WAVE.

ossicles – the chain of tiny articulated bones in the tympanic cavity, named *malleus*, *incus* and *stapes* ("hammer", "anvil" and "stirrup" respectively), that connects the TYMPANIC MEMBRANE to the oval window of the COCHLEA, and thence to the perilymph within the cochlea.

otoacoustic emission – a vibrational disturbance, originating in the COCHLEA, that is observable as sound in the external ear canal. Otoacoustic emissions are associated with motile behaviour of the outer hair cells; they may be spontaneous or evoked by an externally applied acoustic stimulus. Otoacoustic emissions evoked by a transient acoustic stimulus (e.g. a click) occur with a time delay of a few ms, and have been referred to informally as *cochlear echoes*.

otoadmittance tests – *in audiology* objective tests that involve measuring the acoustic admittance of the ear, usually with the aid of a probe inserted into the ear canal. Principal among these are TYMPANOMETRY and ACOUSTIC REFLEX TESTS.

otology – study of the ear, its anatomy and diseases.

ototoxic – poisonous to the auditory nerve.

ototoxic drugs – drugs used to suppress or reduce activity in the INNER EAR.

outer ear – alternative term for EXTERNAL EAR.

outgoing waves – *with respect to a given surface surrounding a source* waves leaving the source that propagate across the surface in the outward direction.

Note: Sound radiation from any object into a FREE FIELD consists of outgoing waves, in the sense that the entire exterior sound field can be synthesized from elementary outgoing waves whose centre is located within the radiating object. Three-dimensional outgoing-wave fields are represented in terms of SPHERICAL WAVES, and two dimensional fields in terms of CYLINDRICAL WAVES.

output – (1) *from a physical or analogue system* a physical response obtained by applying an excitation to such a system: e.g. *output acceleration*. Also, a continuous signal representing the response of an ANALOGUE SYSTEM. Compare RESPONSE.

output – (2) *from a digital system* a sequence of numbers obtained from a DIGITAL SYSTEM in response to an input sequence.

overall – including all frequency components; with no frequency weighting applied.

overall level – the LEVEL of a signal to which no filtering or frequency weighting has been applied. *Units* dB.

overdamped – describes a system that returns smoothly to its equilibrium position after a displacement, without overshooting. See also DAMPING RATIO, CRITICAL DAMPING.

oversampling – *in analogue-to-digital conversion* use of a sampling frequency that exceeds the minimum required to prevent ALIASING. See SAMPLING THEOREM.

overtones – *of an oscillatory system* natural frequencies of the system that lie above the FUNDAMENTAL FREQUENCY (2). The overtones may or may not be integer multiples of the fundamental frequency.

P

p – symbol for the prefix pico, denoting a submultiple of 10^{-12} in the SI system of units. Thus 1 pW denotes 10^{-12} W (1 picowatt).

Pacinian corpuscles – lamellar nerve endings, approximately 1–2 mm in size, that are distributed in the human body as exteroceptors (e.g. on the palms of the hands and soles of the feet), as interoceptors (in the viscera), and as proprioceptors (e.g. in some tendons and ligaments). As exteroceptors they provide vibrotactile sensitivity to surface acceleration, particularly in the range 20–200 Hz; the larger the area of skin excited, the lower the acceleration threshold for detection (**F PACINI *1812–1883***).

PACS – abbreviation for the Physics and Astronomy Classification Scheme of the American Institute of Physics.

Paley–Wiener criterion – *for time-invariant linear systems* a limitation placed on the frequency dependence of the FREQUENCY RESPONSE FUNCTION magnitude, if the system is to be CAUSAL and therefore physically realizable. According to the Paley–Wiener criterion, causality requires

$$\int_{-\infty}^{\infty} \frac{|\ln|H(\omega)||}{a^2 + \omega^2} d\omega < \infty \quad (a = \text{const.}).$$

Here $H(\omega)$ is the system frequency response function, assumed to be SQUARE-INTEGRABLE with respect to frequency:

$$\int_{-\infty}^{\infty} |H(\omega)|^2 d\omega < \infty.$$

Note: A practical implication of the criterion is that no realizable non-periodic filter can have a perfect stop band (i.e. with infinite transmission loss).

panel absorber – an acoustic ABSORBER formed by mounting a flexible panel in front of a rigid wall, often with sound-absorbing material placed in the intervening airspace. Compliant lossy elements bridging the gap between panel and wall can also be used to increase the damping. Compare MULTI-LAYER ABSORBER, MEMBRANE ABSORBER, PERFORATED-PLATE ABSORBER.

parabolic equation (PE) – *for near-axial or narrow-beamwidth wave propagation* an approximate version of the HELMHOLTZ EQUATION, used in acoustics to describe the one-way propagation of sound in a non-uniform medium. All parabolic equations involve some kind of small-angle or paraxial approximation, and they generally treat the nonuniformity of the medium as a small parameter. See also NPE.

Note (1): A standard example that illustrates the application of PEs to acoustics is the problem of azimuthally-symmetric sound propagation in a medium of uniform density, whose sound speed varies with range and depth. The exact propagation equation in cylindrical coordinates (r, ϕ, z) is the 2D Helmholtz equation

$$\frac{\partial^2 p}{\partial r^2} + \frac{1}{r}\frac{\partial p}{\partial r} + \frac{\partial^2 p}{\partial z^2} + k_0^2 n^2 p = 0,$$

where p is the sound pressure, k_0 is a reference acoustic wavenumber, and $n(r, z)$ is a refractive index. To obtain a PE version of this equation, an outgoing-wave solution is assumed in the form

$$p(r, z) = \psi(r, z) H_0^{(2)}(k_0 r) \qquad (e^{j\omega t} \text{ time factor}).$$

The paraxial approximation

$$\left|\frac{\partial^2 \psi}{\partial r^2}\right| \ll 2k_0 \left|\frac{\partial \psi}{\partial r}\right|$$

is then introduced. The result is the standard parabolic equation

$$2jk_0 \frac{\partial \psi}{\partial r} \approx \frac{\partial^2 \psi}{\partial z^2} + k_0^2(n^2 - 1)\psi;$$

but various improved versions exist, capable of handling wider angles of propagation without serious phase errors.

Note (2): Although parabolic equations are only approximate descriptions of sound propagation, their solutions satisfy the RECIPROCITY PRINCIPLE of linear acoustics exactly.

parallel-baffle silencer – an equivalent term for SPLITTER SILENCER.

parameter of nonlinearity – *for a fluid in isentropic compression or expansion* an alternative term for NONLINEARITY PARAMETER; usually denoted by the symbol B/A. See ISENTROPIC EQUATION OF STATE IN SERIES FORM. *Units* none.

parametric array – a device for generating directional low-frequency sound, based on nonlinear acoustic interaction in the near field of a high-frequency radiating transducer. The transducer is driven simultaneously at two closely-spaced frequencies, f_1 and f_2; it produces a beam of primary sound that attenuates significantly within the RAYLEIGH DISTANCE based on the primary frequencies. Nonlinearity in the medium leads to virtual acoustic sources at the difference frequency $|f_1 - f_2|$; compare KZK EQUATION. Because the axial extent of these sources can greatly exceed the sound wavelength at their radiating frequency, the parametric array is able to act as a highly directional radiator of low-frequency sound, with a HALF-POWER BEAMWIDTH almost independent of the transducer radius.

Note: A disadvantage of the parametric array is the low efficiency with which it converts sound power at the primary frequencies, f_1 and f_2, into sound power at the difference frequency.

parametric excitation – *of a system* any type of excitation that is mathematically equivalent to forcing some parameter of the system to vary with time. An example of parametric excitation is a straight rod that is driven by an unsteady axial load applied at one end, with the other end held fixed. The rod responds axially in the normal manner, but nonlinear coupling causes an additional vibrational response in the transverse direction.

Note (1): The following equation governs the transverse response of a straight uniform rod, of mass per unit length m and bending stiffness D, to an axial load $P(t)$:

$$m\frac{\partial^2 w}{\partial t^2} + P(t)\frac{\partial^2 w}{\partial x^2} + D\frac{\partial^4 w}{\partial x^4} = 0.$$

Here $w(x, t)$ is the transverse displacement of the rod, expressed as a function of axial position x and time t. Note that the axial driving force $P(t)$ appears as a coefficient in the equation for w, rather than as a separate forcing term.

Note (2): Parametric excitation can be responsible for large responses to excitation frequencies that are not natural frequencies of the system; see PARAMETRIC RESONANCE.

parametric resonance – resonance of a system under PARAMETRIC EXCITATION. Typically, parametric resonance occurs when the forcing frequency is a SUBHARMONIC of one of the system natural frequencies.

parity – the property of being odd or even. Parity can be assigned to integers, to mathematical functions, and to quantities that are expressed in terms of spatial coordinates. Examples are given below.

(1) *Odd and even integers*: 1 is odd, 2 is even, and so on.

(2) *Parity value of integers*: The parity of an integer n is the sum of its binary digits, expressed modulo 2.

(3) *Odd and even functions*: The function $\sin x$ is called odd, since $\sin(-x)$ is the negative of $\sin x$; on the other hand, $\cos(-x)$ equals $\cos x$, so the function $\cos x$ is even.

(4) *Odd and even parity with respect to inversion of coordinates*: Suppose the coordinate system x, y, z is inverted (i.e. all points are reflected through the origin). The spatial coordinates (x, y, z) of a given point become $(-x, -y, -z)$; they therefore have odd parity. The distance of the point from the origin, however, has even parity. Properties of the

coordinates that are even in this sense are said to have **space reflection symmetry**.

Parseval relation, Parseval's formula – *for integral transforms* a type of energy relation connecting the cross-product of two functions with the cross-product of their transforms. If the original functions $f(x)$, $g(x)$ transform into $f(\kappa)$, $G(\kappa)$ then the typical form of such relations is

$$\int_{D(x)} x f^*(x) g(x) \, dx = \int_{D(\kappa)} \kappa \, F^*(\kappa) \, G(\kappa) \, d\kappa,$$

where the integrals extend over the corresponding domains, $D(x)$ and $D(\kappa)$, over which the functions and their transforms are defined. For examples, see the entries for FOURIER TRANSFORM, HILBERT TRANSFORM, and HANKEL TRANSFORM. Parseval relations also connect continuous functions with their discrete representations in terms of a complete set of eigenfunctions.

Parseval's theorem – (1) *for periodic signals* the statement that the average POWER of a periodic signal equals the sum of the average powers of its individual harmonic components. In mathematical terms, Parseval's theorem may be stated as

$$\frac{1}{T} \int_{\langle T \rangle} |x(t)|^2 dt = \sum_{k=-\infty}^{\infty} |c_k|^2;$$

here t denotes time, T is the period of the signal $x(t)$, and c_k are its complex Fourier coefficients (see FOURIER ANALYSIS). The notation $\int_{\langle T \rangle}$ denotes integration over one complete period of the signal, starting at any point.

The discrete-time version of this result relates the mean square value of a periodic sequence to the spectral coefficients in a DISCRETE FOURIER SERIES representation:

$$\frac{1}{N} \sum_{n=\langle N \rangle} |\tilde{x}[n]|^2 = \sum_{k=\langle N \rangle} |c_k|^2.$$

Here the periodic sequence $\tilde{x}[n]$ has period N, and c_k are its spectral coefficients.

Parseval's theorem – (2) *for signals of finite duration* the statement that the ENERGY of a finite-duration signal is given by integrating the square of its FOURIER TRANSFORM magnitude over all frequencies. In mathematical terms,

$$\int_{-\infty}^{\infty} |x(t)|^2 dt = \int_{-\infty}^{\infty} |X(f)|^2 df$$

where t denotes time, f denotes frequency and $X(f)$ is the FOURIER TRANSFORM of the signal $x(t)$.

The discrete-time version of this result is

$$\sum_{n=-\infty}^{\infty} |x[n]|^2 = \frac{1}{2\pi} \int_{\langle 2\pi \rangle} |X(\Omega)|^2 d\Omega,$$

where Ω denotes angular frequency in the digital domain and $X(\Omega)$ is the DISCRETE-TIME FOURIER TRANSFORM of the finite sequence $x[n]$. The notation $\int_{\langle 2\pi \rangle}$ denotes integration over one complete period in Ω (an interval of 2π), starting at any point.

partial differential equation – any equation for an unknown function that involves partial derivatives with respect to more than one variable. An example is LAPLACE'S EQUATION in two dimensions:

$$\frac{\partial^2 u}{\partial x^2} + \frac{\partial^2 u}{\partial y^2} = 0.$$

partially coherent – see COHERENT.

partially-diffuse field – a sound field that approximates to a DIFFUSE FIELD in some respects but differs in others: for example, a large number of uncorrelated plane progressive waves, whose distribution of intensity with respect to direction is non-uniform.

partially-diffuse reflection – a reflection process that is neither specular nor completely diffuse. Compare SPECULAR REFLECTION, DIFFUSE REFLECTION.

particle – a point mass, or a vanishingly small element of a continuous medium.

particle displacement – *in a continuous medium* the displacement of a material particle from its equilibrium position, e.g. as a result of acoustic excitation. *Units* m.

particle velocity – *in a continuous medium* the velocity of a material particle. In acoustics, the term is used to describe velocity fluctuations associated with a sound field; an equivalent term is ***acoustic particle velocity***. *Units* m s^{-1}.

particular integral – any particular solution of a linear NON-HOMOGENEOUS DIFFERENTIAL EQUATION, as opposed to a general solution containing arbitrary constants. Supplementing a particular integral with a general form of the HOMOGENEOUS SOLUTION provides the general solution to the original equation. A particular integral used in this way is not required to satisfy boundary or initial conditions, since such conditions apply only to the complete solution.

partitioned matrix – a matrix whose elements are also matrices. For example,

$$\mathbf{A} = \begin{bmatrix} a_{11} & a_{12} \\ a_{21} & a_{22} \end{bmatrix}$$

represents a partitioned matrix if each of the elements a_{ij} is itself a matrix, with elements $(a_{ij})_{mn}$. The matrices $[a_{ij}]$ into which the larger matrix $\mathbf{A}$ is partitioned are called ***submatrices***.

Note: Compatibility of dimensions requires that the submatrices in each row all have the same number of rows, and similarly the submatrices in each column must have the same number of columns.

passband – *of a filter* the frequency range over which the TRANSMISSION LOSS is close to zero (analogue filter), or the frequency response function of the filter has a magnitude close to unity (digital filter).

passive – having no source of energy. A *passive impedance* has a non-negative real part.

passive sonar – (1) the detection and location of underwater objects, by listening underwater and processing the acoustic signals received.

passive sonar – (2) equipment used for PASSIVE SONAR (1).

pdf – abbreviation for PROBABILITY DENSITY FUNCTION.

peak – *of an oscillatory signal* the peak value of a time-varying real signal $x(t)$ is the largest instantaneous value of $|x(t)|$ that occurs within a specified time interval. Symbol x_{peak}. Compare PEAK SOUND PRESSURE.

peak action level – *in noise at work regulations* a single-event PEAK SOUND PRESSURE, currently set at 200 Pa, at which either an employer is required to take remedial measures, or an employee is required to wear hearing protection. See also ACTION LEVEL. *Units* Pa.

peak frequency-weighted sound pressure level – see PEAK SOUND PRESSURE LEVEL. *Units* dB re p_{ref}^2.

peak level – *of an oscillatory signal over a specified time interval* the level of the squared peak signal, defined by

$$L_{\text{pk}} = 10 \log_{10}(x_{\text{peak}}^2/x_{\text{ref}}^2) = 20 \log_{10}(x_{\text{peak}}/x_{\text{ref}}).$$

Here x_{peak} is the PEAK value of the real oscillatory signal $x(t)$. See also PEAK SOUND PRESSURE LEVEL. *Units* dB re x_{ref}^2.

Note: The definition above accords with ANSI standard terminology, but see note (2) under MAXIMUM LEVEL.

peak sound pressure – *at a given point in a transient or nonstationary sound field* the largest instantaneous SOUND PRESSURE magnitude (based on either positive or negative pressure) that occurs during a given time interval. *Units* Pa.

Note: The *peak negative sound pressure* is the magnitude of the largest negative pressure occurring in the given interval. Likewise, the *peak positive sound pressure* is the largest instantaneous positive pressure.

peak sound pressure level – *of a sound over a specified time interval* the squared peak magnitude of the instantaneous sound pressure, positive or negative, expressed as a LEVEL. Compare PEAK SOUND PRESSURE.

For example, a transient pressure pulse in air, with a maximum pressure excursion of either $+100$ Pa or -100 Pa, has a peak sound pressure level of 134 dB re $(20\ \mu\text{Pa})^2$. Peak sound pressure level is distinguished from MAXIMUM SOUND PRESSURE LEVEL; see note (2) under MAXIMUM LEVEL. *Units* dB re p_{ref}^2.

Note: If the sound pressure signal is FREQUENCY-WEIGHTED before finding the peak magnitude, the result of the process described above is a *peak frequency-weighted sound pressure level*.

peak-to-peak value – *of a signal or oscillatory quantity* the maximum range over which the quantity oscillates. The peak-to-peak value of a sinusoidal oscillation equals twice the AMPLITUDE.

Pekeris waveguide – *in underwater acoustics* an idealized model of the ocean, used in theoretical studies (**C L PEKERIS 1948**). It consists of a water layer of uniform depth, sound speed, and density, with a pressure-release boundary at its upper surface, and a uniform fluid half-space beneath. The properties of the lower fluid are chosen to match the density and compressional wave speed of the sea bed.

penetration depth – *for oscillatory flow near a boundary* see THERMAL PENETRATION DEPTH, VISCOUS PENETRATION DEPTH. *Units* m.

perceived noise level – a FREQUENCY-WEIGHTED sound pressure level used to predict the annoyance potential of complex sounds. It is calculated from the sound pressure levels in 24 one-third octave bands ranging from 50 Hz to 10 kHz. See also D-WEIGHTING. *Units* dB re $(20\ \mu\text{Pa})^2$.

percentage syllable articulation – *in a given speech sample presented to a listener* the percentage of syllables that are heard correctly. The listener records each syllable he/she hears, and the result is compared with what the speaker said. *Units* %.

percentile levels – *in environmental noise assessment* the level of A-weighted mean square sound pressure, with fast (F) time weighting, that is exceeded for N percent of the time interval considered. For examples, see the entry under L_N. Symbol $L_{AN,T}$, where T denotes the time interval (e.g. 24 h). *Units* dB re $(20\ \mu\text{Pa})^2$.

perfect gas – a term used for an IDEAL GAS whose specific heats (C_p at constant pressure and C_v at constant volume) do not vary with temperature.

Note: Since the difference between C_p and C_v for an ideal gas equals the GAS CONSTANT, an equivalent statement is that the specific-heat ratio is independent of temperature.

perforated-plate absorber, perforated-panel absorber – an acoustic ABSORBER formed by mounting a rigid perforated plate in front of a rigid wall, often with sound-absorbing material placed in the intervening airspace. If the perforations in the plate are sufficiently small (i.e. their diameter is less than the thickness of the acoustic boundary layer, at the frequencies of interest), viscothermal fluid damping in the holes provides sufficient absorption without the addition of sound-absorbing material.

Note: This type of absorber is particularly effective at low audio frequencies, typically between 20 Hz and 300 Hz. Its frequency bandwidth can be increased by using a double-layer construction; see MULTILAYER ABSORBER.

perilymph – the fluid that fills the outer two channels of the cochlea and the vestibular organ; see INNER EAR. Compare ENDOLYMPH.

period – the smallest interval (of time or distance) over which a periodic oscillation repeats. In acoustics, the word period (without qualification) normally implies the *time period*, and the term *spatial period* is used for spatial waveforms. The *period of damped oscillations* is the interval between successive zero crossings in the same direction.

periodic function – a FUNCTION whose value remains unchanged when the argument is increased by an integer multiple of the period; e.g. $\sin x$, $\tan x$ are periodic functions with periods 2π and π respectively.

periodic oscillation – a steady-state oscillation with respect to time, whose time history consists of repeated identical cycles. The terms *periodic signal*, *periodic waveform* are similarly defined.

Note: An oscillation can also be *spatially periodic*, in which case distance replaces time as the dependent variable.

peripheral auditory system – the outer and middle ear, from the pinna to the oval window of the cochlea. This is the air-filled part of the system, in contrast to the fluid-filled cochlea and vestibular organ.

permanent threshold shift – *in psychoacoustics* the component of THRESHOLD SHIFT that shows no recovery with time after the apparent cause has been removed. *Units* dB.

permeability – *for slow viscous flow through a porous solid* a parameter related to pore geometry; see DARCY'S LAW. *Units* m^2.

persistent threshold shift – *in psychoacoustics* a THRESHOLD SHIFT that shows gradual recovery with time after the apparent cause has been removed, but may last for several days or weeks. *Units* dB.

ph – denotes the phase function,

ph $z \equiv \arg z$,

where z is a complex number. See ARGUMENT, PHASE AND MAGNITUDE REPRESENTATION.

phase – (1) *of a complex number* the argument of the complex number; for example, $re^{j\varphi}$ (r, φ real) has phase φ. An equivalent term is **angle**. See also PHASE AND MAGNITUDE REPRESENTATION. *Units* rad.

phase – (2) *of a sinusoidal signal* abbreviation for the PHASE ANGLE of the signal. *Units* rad.

Note: The term phase is sometimes used to refer to the INSTANTANEOUS PHASE of a sinusoidal signal, which is a time-dependent quantity. The meaning is usually clear from the context; an example of phase used in the instantaneous sense would be the statement "angular frequency is the rate of change of phase with time".

phase – (3) *of a linear time-invariant system* abbreviation for the PHASE SHIFT produced by the system between input and output. When the phase shift is considered as a function of frequency, it is called the PHASE-SHIFT FUNCTION of the system. See also ABSOLUTE PHASE.

phase – (4) one of the three states of matter (solid, liquid or gas) in which a pure substance such as water may exist.

phase advance – a PHASE SHIFT introduced by a system, between its input and output, that causes the output waveform at a given frequency to lead the input waveform. *Units* rad.

Note: If the output at angular frequency ω leads the input by a time shift

$\tau(\omega)$, the phase factor between output and input is $e^{j\omega\tau}$, corresponding to a phase advance of $\omega\tau$ radians. Note that the phase shift changes sign from positive to negative frequencies.

phase and magnitude representation – representation of the complex number $z = x + jy$ in polar form, as $re^{j\varphi}$.

The real number $r = |z| = \sqrt{x^2 + y^2}$ is called the *modulus* of z; alternative names are *magnitude* or *absolute value*. The phase angle φ is called the *argument*, *phase* or *angle* of z, commonly written $\arg z$ (for alternative notations, see ANGLE). Its value is uniquely determined, to within a multiple of 2π, by the conditions

$$\cos\varphi = x/|z|, \quad \sin\varphi = y/|z|.$$

Note (1): Alternative conventions may be used for the range of φ, e.g. $-\pi/2 \leq \varphi < 3\pi/2$ or $0 \leq \varphi < 2\pi$. The former is particularly convenient when z represents an impedance: the real part of z is then constrained by physical considerations to be non-negative, and one can achieve this by specifying $-\pi/2 \leq \varphi < \pi/2$. See STRICTLY POSITIVE-REAL FUNCTION.

Note (2): An algorithm that places φ in the range $0 \leq \varphi < 2\pi$ is

$$\varphi = \pi - (\pi - \alpha)\operatorname{sgn} y, \quad 0 \leq \varphi < 2\pi.$$

Here α is the principal value of $\cos^{-1}(x/|z|)$, defined such that $0 \leq \alpha < \pi$. An alternative algorithm that places φ in the range $-\pi/2 \leq \varphi < 3\pi/2$ is

$$\varphi = \beta + (1 - \operatorname{sgn} x)\frac{\pi}{2}, \quad -\frac{\pi}{2} \leq \varphi < \frac{3\pi}{2},$$

where β is the principal value of $\tan^{-1}(y/x)$ defined such that $-\pi/2 \leq \beta < \pi/2$. See INVERSE CIRCULAR FUNCTIONS.

phase angle – *of a sinusoidal signal* if a single-frequency signal, with amplitude A and angular frequency ω, is represented by

$$x(t) = A\cos(\omega t + \theta), \quad (t = \text{time}),$$

then the real quantity θ is called the phase angle of the signal. Equivalent terms are *phase* and *initial phase*. Units rad.

Note: Compare INSTANTANEOUS PHASE, which refers to the time-dependent quantity $(\omega t + \theta)$.

phase angle of the cross-spectral density – the phase angle defined by

$$\varphi_{xy}(f) = \arg[C_{xy}(f) + jQ_{xy}(f)]$$

$$= \tan^{-1}\frac{Q_{xy}(f)}{C_{xy}(f)}, \quad -\frac{\pi}{2} < \varphi_{xy}(f) < \frac{\pi}{2}$$

where $C_{xy}(f)$ and $Q_{xy}(f)$ are the real and imaginary parts of the CROSS-

SPECTRAL DENSITY. When $Q_{xy}(f)$ is positive, the signal $y(t)$ leads $x(t)$ at frequency f by a phase angle between 0 and $\pi/2$; see PHASE LEAD. *Units* rad.

phase change – (1) *of a signal* the amount by which the phase (2) changes, for example between two points in a sound field. The term can also refer to a phase change of π, i.e. a change of sign. *Units* rad.

phase change – (2) *of a material* transformation from one state of matter (solid, liquid or gas) to another.

phase-closure principle – *for a one-dimensional bounded system* states that the condition for the system to have a natural frequency is that the total phase change of a single-frequency wave, after travelling one complete circuit round the system, equals $2n\pi$. Here n is an integer, and the total phase change includes the phase change due to reflection of the wave at each end.

Note: The principle can be stated in the following generalized form, for stable systems without damping. If the complex amplitude of a wave increases by a factor $e^{(\mu + j\varphi)}$ on each round trip, where μ and φ are real functions of frequency, the condition for a natural frequency to occur is

$$\mu = 0, \quad \varphi = 2n\pi \quad (n \text{ integer}).$$

phase coefficient – *for a single-frequency progressive wave* the real part of the PROPAGATION WAVENUMBER. Equivalently, it is the imaginary part of the PROPAGATION COEFFICIENT. *Units* rad m^{-1}.

phase conjugation – *of an acoustic field* the construction, from the original field, of a field that (over a specified region) is a time-reversed replica of the original, apart from an amplitude scaling factor. Time reversal of a finite-length signal is equivalent to taking the complex conjugate in the frequency domain; hence the name phase conjugation.

phase constant – an alternative term for PHASE COEFFICIENT. *Units* rad m^{-1}.

Note: The 1994 ANSI standard on acoustical terminology recognizes *phase constant* (with the prefix "acoustic"), but not *phase coefficient*; the 1994 IEC standard recognizes phase coefficient (also with the prefix "acoustic") but not phase constant.

phase delay – a PHASE SHIFT introduced by a system, between its input and output, that causes the output waveform at a given frequency to lag the input waveform. *Units* rad.

phase difference – *between two sinusoidal signals* the magnitude of their RELATIVE PHASE. *Units* rad.

phase distortion – *of a linear time-invariant system* refers to distortion of the output waveform that is caused by nonlinear dependence of the system PHASE-SHIFT FUNCTION on frequency.

phase factor – *between the input and output of a linear time-invariant system* if the frequency response function of the system, at angular frequency ω, is written in polar form as

$$H(\omega) = |H(\omega)|e^{j\varphi(\omega)},$$

then the unit-magnitude complex factor $e^{j\varphi(\omega)}$ is called the phase factor of the system. Compare GAIN FACTOR; also PROPAGATION FACTOR, TIME FACTOR. *Units* none.

phase function – *for a single-frequency sound field* the INSTANTANEOUS PHASE $\varphi(\mathbf{x}, t)$, regarded as a function of position $\mathbf{x}$ and time t.

phase lag – *between two sinusoidal signals at the same frequency* if the phase angle of one signal is less than the phase angle of the other by a positive amount α (less than π), the first signal *lags* the second by α; it is said to have a phase lag of α. *Units* rad.

phase lead – *between two sinusoidal signals at the same frequency* if the phase angle of one signal is greater than the phase angle of the other by a positive amount α (less than π), the first signal is said to have a phase lead of α. *Units* rad.

phase modulation – see MODULATION (2).

phase plane – *for a single-degree-of-freedom system* the plane defined by coordinates $(u, \dot{u})$, where $u(t)$ is a variable that describes the system response (e.g. the displacement from equilibrium).

phase response – *of a linear time-invariant system* an equivalent term for the PHASE-SHIFT FUNCTION of the system.

phase reversal – (1) *of a sinusoidal signal* addition of π to the PHASE ANGLE, or (equivalently) multiplication by -1.

phase reversal – (2) *in digital signal processing* a reversal in the sign of the PHASE assigned to the reconstructed signal, caused by undersampling of the original signal.

phase shift – *of a linear time-invariant system* the PHASE of the system FREQUENCY RESPONSE FUNCTION. *Units* rad.

Note: Suppose a linear system is driven by a sinusoidal input. The system output will not necessarily be in phase with the input. The RELATIVE PHASE of the output signal, compared with the input, is the phase shift at the driving frequency. It will generally be a function of frequency: see KRAMERS–KRONIG RELATIONS.

phase-shift function – *of a linear time-invariant system* the UNWRAPPED PHASE of the system FREQUENCY RESPONSE FUNCTION, presented as a continuous function of frequency. If the frequency response function at angular frequency ω is written in polar form as

$$H(\omega) = |H(\omega)|e^{j\varphi(\omega)},$$

the phase-shift function corresponds to the real quantity $\varphi(\omega)$. Also known (with or without phase unwrapping) as the *phase response* of the system. *Units* rad.

Note: Some authors use the opposite sign convention, with $e^{j\varphi(\omega)}$ replaced by $e^{-j\varphi(\omega)}$ in the equation above.

phase spectrum – (1) *of a periodic signal* the set of discrete Fourier harmonic phases, plotted against frequency. See FOURIER ANALYSIS, COMPLEX FOURIER AMPLITUDES, SPECTRAL COEFFICIENTS.

phase spectrum – (2) *of a transient signal* the phase of the signal FOURIER TRANSFORM, plotted as a continuous function of frequency.

Note: In both cases above, if the signal is real it follows that the phase spectrum is odd with respect to frequency.

phase speed – (1) *for single-frequency free wave propagation in a given medium* the rate of advance of a constant-phase surface. For plane waves of angular frequency ω, the phase speed in the medium is given by

$$v_{\text{ph}} = \omega/\text{Re}[k(\omega)],$$

where $k(\omega)$ is the CHARACTERISTIC WAVENUMBER. *Units* m s^{-1}.

Note: If the medium is anisotropic, or has a mean flow velocity, the phase speed is a function of the WAVENORMAL DIRECTION of the plane waves.

phase speed – (2) *along a specified axis, for a single-frequency progressive wave field* the speed traced out along the axis by a wavefront. The phase speed along the x axis of an obliquely-propagating plane progressive wave, for example, is related to the wavenumber component k_x in the x direction by

$$v_{\text{ph}}^{(x)} = \omega/k_x \quad \text{(at angular frequency } \omega\text{)}.$$

The phase speed in this sense is the TRACE VELOCITY of wavefronts along the specified axis. *Units* m s^{-1}.

phase unwrapping – the phase shift between input and output of a linear system is indeterminate within an integer multiple of 2π. However, by requiring the phase shift to be a continuous function of frequency, a graph of phase shift versus frequency for the system can be obtained in which phase relationships between different frequencies no longer contain an arbitrary component. This process is called phase unwrapping.

phase velocity – *of plane progressive waves at a single frequency* the velocity $\mathbf{n}\omega/k$, where $\mathbf{n}$ is the unit vector in the propagation direction and k is the wavenumber. *Units* m s^{-1}.

Note: Although the phase velocity as defined above has both magnitude and direction, it does not have the essential vector property of being resolvable into components: the phase speed in a specified direction is *not* equal to the component of the phase velocity in that direction. Compare SLOWNESS VECTOR.

phasor – a complex quantity proportional to $e^{j\omega t}$ (or to $e^{-i\omega t}$), whose real part represents a time-harmonic signal or physical variable. Here t is time and ω is the ANGULAR FREQUENCY. In the COMPLEX PLANE, phasors are represented by rotating vectors.

phon – the unit of LOUDNESS LEVEL. Under reference listening conditions, the loudness level of a 1 kHz pure tone is numerically equal to its SOUND PRESSURE LEVEL; this relationship does not apply to sounds in general. *Unit symbol* phon. Not an SI unit.

phonon – a quantum of vibrational energy in a solid. Elastic waves of frequency f carry energy in quanta of amount hf, where h is Planck's constant.

phonophoresis – literally "sonic transport". A biomedical term that refers to the accelerated diffusion of substances through the skin under the action of ultrasound. Particle velocities of order 0.1 m s^{-1} and frequencies of a few MHz are found to be effective for this purpose.

photoacoustic excitation – refers to generation of sound by the absorption of a modulated light beam, for example at a solid surface; compare OPTOACOUSTIC EXCITATION. The terms are often used interchangeably.

photoacoustic spectroscopy – the use of selective optical absorption to detect the presence of certain molecules in air, or in a gas mixture. Modulated laser light (or infrared radiation) is used to irradiate a gas sample in a rigid-walled cavity. The laser is tuned to the absorption band of the target molecules, and modulated at a frequency that produces resonant acoustic excitation. The strength of the acoustic response is then a measure of the target gas content within the cavity.

physiological acoustics – the branch of acoustics that deals with the auditory system, in humans or in animals, from a physiological viewpoint. Covers all aspects of function, from the biochemical and cellular levels to cochlear mechanics and electrophysiology; it includes the effects of noise and trauma on the auditory system.

pink noise – a stationary random signal whose power spectrum falls off at a constant rate of 3 dB per octave, as the frequency increases. See also WHITE NOISE, BROWN NOISE.

pinna – the part of the EAR that projects outside the head; the ear flap.

piston – *in acoustics* a rigidly oscillating surface that drives an acoustic medium.

piston radiation impedance – the radiation impedance of a vibrating flat rigid piston, or of a plane opening driven by a normal velocity that is uniform over the opening area. The piston radiation impedance may be expressed in one of three ways: (a) as a *mechanical impedance*, $F/\dot{w}$, i.e the complex ratio of the force, F, exerted by the piston to the piston velocity $\dot{w}$; (b) as an average *specific acoustic impedance*, $F/(A\dot{w})$; or (c) as an average *acoustic impedance*, $F/(A^2\dot{w})$. In the last two expressions, A is the area of the piston face or the opening.

pitch – the attribute of auditory sensation that is related primarily to frequency.

Plancherel's equality – an alternative name for PARSEVAL'S THEOREM, which relates the mean square value of a periodic function to the Fourier coefficients of the function (**M PLANCHEREL 1910**).

Plancherel's theorem – another name for the CONVOLUTION THEOREM.

plane progressive wave – a PLANE WAVE whose wavefronts propagate in one direction. The direction of advance (i.e normal to the wavefronts) is called the wavenormal direction. In mathematical terms, the acoustic pressure in a plane progressive wave travelling in the x direction has the general form

$$p = f(t - x/c) \quad (t = \text{time}; c = \text{propagation speed}),$$

provided the medium is uniform and lossless; here f is any function of the indicated argument. In a medium with attenuation and dispersion, plane progressive waves at angular frequency ω are described by the complex pressure

$$p = Ae^{j(\omega t - kx)} \quad (k = \text{characteristic wavenumber}),$$

where A is a constant. See also PHASE VELOCITY.

plane wave field – *in acoustics* a sound field in which the instantaneous field variables are uniform over any plane normal to a specified axis. A plane wave field in a uniform medium can be expressed as a sum of forward and backward waves, which are PLANE PROGRESSIVE WAVES travelling in opposite directions parallel to the axis.

plane-wave mode – *in a uniform fluid-filled waveguide with rigid walls* the lowest-order acoustic mode of propagation, in which the pressure is uniform over the cross-section of the waveguide. Compare QUASI-PLANE WAVE MODE.

plate absorber – an alternative term for PANEL ABSORBER. See also PERFORATED-PLATE ABSORBER.

plate bending-wave speed – the speed at which free bending waves travel along a uniform thin plate; it is proportional to $\sqrt{fh}$, where f is frequency and h is the plate thickness. See LAMB WAVES for further information. *Units* m s^{-1}.

plate longitudinal-wave speed – the speed at which free inplane longitudinal waves, of wavelength large compared with the plate thickness, travel along a uniform thin plate; it is a property of the plate material, and does not depend on either the frequency or the plate thickness. See LAMB WAVES for further information. *Units* m s^{-1}.

plate waves – an equivalent term for LAMB WAVES.

PNL – abbreviation for PERCEIVED NOISE LEVEL. *Units* dB re $(20\ \mu\text{Pa})^2$.

PNLT – abbreviation for TONE-CORRECTED PERCEIVED NOISE LEVEL. *Units* dB re $(20\ \mu\text{Pa})^2$.

point admittance – the inverse of a POINT IMPEDANCE; otherwise known as *driving-point admittance*.

point dipole – an idealized source of vanishingly small size whose acoustic source strength is zero, but which has a finite DIPOLE MOMENT that determines the sound field. A point dipole can be constructed from two POINT MONOPOLES with equal and opposite strengths (say S and $-S$), by allowing the monopole separation d to tend to zero while maintaining the product Sd constant. Also known as a *dipole* or *acoustic dipole*.

point impedance – *for any linear system* an alternative term for DRIVING-POINT IMPEDANCE, in the general sense of an impedance-type frequency response function based on an input and output measured at the same point in the system. Another alternative term is *point input impedance*.

point impedance matrix – *for a point-excited mechanical system* abbreviation for DRIVING-POINT IMPEDANCE MATRIX.

point monopole – an ACOUSTIC SOURCE DENSITY that is concentrated at a single point in space, with the mathematical form

$$\mathcal{F}(\mathbf{x}, t) = S(t)\delta(\mathbf{x} - \mathbf{x}_0).$$

Here δ denotes the DIRAC DELTA FUNCTION, $S(t)$ is the acoustic source strength, t is time, and the point monopole is located at position $\mathbf{x} = \mathbf{x}_0$.

point multipole – an idealized MULTIPOLE SOURCE that is concentrated at a single point. The corresponding ACOUSTIC SOURCE DENSITY has the mathematical form

$$\mathcal{F}(\mathbf{x}, t) = S^{(n)}_{ijk\ldots}(t) \frac{(-\partial)^n \delta(\mathbf{x} - \mathbf{x}_0)}{\partial x_i \partial x_j \partial x_k \ldots}, \quad \text{(multipole of order } n\text{)}.$$

Here δ denotes the DIRAC DELTA FUNCTION, $S^{(n)}_{ijk\ldots}(t)$ are the components of the ***multipole strength***, t is time, and the point multipole is located at position $\mathbf{x} = \mathbf{x}_0$.

point of constant phase – see INSTANTANEOUS PHASE (1).

point quadrupole – an idealized point source whose net monopole and dipole moments are both zero, so that the sound field is determined by the source QUADRUPOLE MOMENT. See also AXIAL QUADRUPOLE, LATERAL QUADRUPOLE.

point-reacting – an equivalent term for LOCALLY-REACTING.

point reciprocity – *in linear acoustics* see RECIPROCITY PRINCIPLE.

point source – (1) a monopole sound source of vanishingly small size, i.e. a POINT MONOPOLE.

point source – (2) informal abbreviation for a POINT MULTIPOLE source of any order (monopole, dipole, quadrupole, etc.).

Poisson's ratio – *for a solid* a measure of transverse contraction under tensile stress. It is defined as the ratio of the transverse contraction of a sample to its longitudinal extension. See also COMPLEX MODULUS, AUXETIC MATERIAL.

Poisson sum formula – relates the sum of a repetitive sequence of pulses to the sampled Fourier transform of a single pulse. If $y(t)$ denotes a typical pulse in the time domain, and $Y(f)$ is its Fourier transform, then

$$\sum_{n=-\infty}^{\infty} y(t+n\Delta) = \frac{1}{\Delta} \sum_{n=-\infty}^{\infty} Y(n/\Delta) E^n.$$

Here Δ is the displacement of successive pulses along the time axis, and E stands for $\exp j(2\pi/\Delta)t$. A special case is an infinite sequence of delta functions, for which the formula gives

$$\sum_{n=-\infty}^{\infty} \delta(t+n\Delta) = \frac{1}{\Delta} \sum_{n=-\infty}^{\infty} E^n.$$

Note: A similar result holds with time and frequency interchanged. If Δ is reinterpreted as the frequency displacement between successive "pulses" in the frequency domain, then their sum to infinity may be expressed as

$$\sum_{n=-\infty}^{\infty} Y(f+n\Delta) = \frac{1}{\Delta} \sum_{n=-\infty}^{\infty} y(n/\Delta) E^{-n}.$$

where E now stands for $\exp j(2\pi/\Delta)f$.

polar angle – *in spherical polar coordinates* the angle θ that measures rotation away from the axis direction. *Units* rad (but commonly expressed in degrees).

polar coordinates – a set of coordinates (r, θ) in a plane, based on radial distance r from the origin and polar angle θ.

polar correlation technique – *for source location* a remote sensing technique for determining the axial distribution of sound power output from a linear source array. The name polar correlation refers to the use of a circular-arc microphone array (i.e. a ***polar array***) centred on the source region, with the array processing procedure based on cross spectra between pairs of microphone signals. Individual array elements are assumed to all have the same directionality, but there is no restriction on the degree of mutual coherence between elements.

polar form – a complex number is said to be in polar form if it is expressed as $re^{j\varphi}$, where r, φ are real numbers. See also PHASE AND MAGNITUDE REPRESENTATION.

polarity – the sign of a real number. In acoustics, the term can be applied to any real signal or physical quantity that can change sign. Two real signals are said to have ***opposite polarity*** when one is a negative constant times the other.

polarity reversal – *of a signal, not necessarily sinusoidal or periodic* multiplication by -1.

polarization – *of a transverse wave* orientation of the wave's POLARIZATION VECTOR in a particular direction. The wave is said to be ***linearly polarized*** if the polarization vector points in a fixed direction. If the polarization vector rotates in the transverse plane at a constant rate of once per cycle, the wave is called ***circularly polarized***, with positive **helicity** if the rotation is clockwise (as viewed by an observer looking in the propagation direction).

polarization vector – *for plane-progressive transverse waves* a unit vector that is normal to the propagation direction, and parallel to the transverse displacement produced by the wave.

polar moment – see note (2) under MOMENT (1).

poles – (1) *of a function of a complex variable* the values of the complex argument z at which the function is singular. Thus a function with a pole at z_0 is undefined at $z = z_0$. However, an ANALYTIC FUNCTION of z can be expanded in the neighbourhood of z_0 as a convergent series in inverse powers of $z - z_0$. More precisely, a function $f(z)$ with a pole at z_0 can be expanded in the series form

$$f(z) = a_0 + a_1(z - z_0) + a_2(z - z_0)^2 + \ldots + \frac{b_1}{z - z_0} + \frac{b_2}{(z - z_0)^2} + \ldots,$$

convergent within the region R defined by $0 < |z - z_0| < r_{max}$, provided $f(z)$ is analytic in R.

poles – (2) *of a linear time-invariant system* the points in the S-PLANE where the system transfer function becomes singular.

polynomial – any expression of the form

$$a_0 + a_1 x + a_2 x^2 + \ldots + a_n x^n \qquad (n \geq 1; a_n \neq 0)$$

is called a polynomial in x, of degree n.

polytropic exponent – *for a gas* the exponent n in a pressure-volume relation of the form

$$PV^n = \text{const.} \qquad \text{(for constant } n\text{)}.$$

Here P is the absolute pressure, and V is the specific volume of the gas. A ***polytropic process*** is a compression or expansion process that is described by the equation above. Compare ISENTROPIC EXPONENT. *Units* none.

Note: The equation $PV^n = $ const. describes either isothermal or isentropic processes for an IDEAL GAS, by putting $n = 1$ and $n = \gamma$ respectively (where γ is the ratio of specific heats).

polytropic gas – *mainly in physics* a term sometimes used for an IDEAL GAS.

population – *in statistics* the set of all possible observations, or values of a RANDOM VARIABLE, from which SAMPLES may be taken in order to estimate the statistical properties of the whole set.

population moments – *in statistics* the kth population moment of a RANDOM VARIABLE X is defined as the EXPECTED VALUE of X^k, denoted here by M_k:

$$M_k = E\{X^k\}, \quad (k = 0, 1, 2, \text{etc.}).$$

The quantity M_k is often called simply the kth moment of X; compare MOMENT.

Note: For CONTINUOUS RANDOM VARIABLES, the population moments of X are expressible in terms of the PROBABILITY DENSITY FUNCTION $f(x)$ as

$$M_k = \int_{-\infty}^{\infty} x^k f(x)\, dx.$$

For QUANTIZED VARIABLES the corresponding expression is

$$M_k = \sum_i x_i^k P(X = x_i),$$

where x_i ($i = 1, 2, 3$, etc.) are the quantized values that the random variable X may take, and $P(A)$ is the PROBABILITY that statement A is true.

population statistics – the statistical properties of an entire POPULATION, as opposed to those based on a subset or SAMPLE. The latter are called SAMPLE STATISTICS.

pore – a cavity or channel in a cellular or fibrous material.

porosity – *of a porous solid* the fraction of the total volume of the material that is occupied by interconnecting open spaces or voids through which fluid can flow. *Units* none.

porous absorber – a SOUND ABSORBER constructed from porous material, and provided with a porous or perforated outer surface that allows the external fluid medium to penetrate into the pores of the material.

porous sound-absorbing material – a porous solid material with interconnecting voids and channels through which fluid can flow. The skeleton may be either rigid or flexible.

port – (1) an entrance or exit for a network; a point in a system at which an input can be specified, or where an output is measured. Each port in a linear network or system acts as either an input or output port; two systems can be coupled by connecting a port of one system to a port of the other. In electrical circuit terms, a pair of terminals constitutes a port.

Note: Generally two quantities, called *state variables*, are chosen to characterize the physical interaction between the system and its surroundings at each of the ports. Examples are the pressure and volume velocity at the neck opening of a Helmholtz resonator, regarded as an acoustic lumped-element system with a single port.

port – (2) an additional opening in a loudspeaker cabinet, usually designed so as to improve the low-frequency performance of the speaker.

potential – a scalar field Φ, whose spatial gradient $\nabla\Phi$ defines an IRROTATIONAL vector field. See CONSERVATIVE FORCE FIELD.

potential energy – *of a system* a quantity whose decrease gives the work done on a system either by localized CONSERVATIVE FORCES, or by a CONSERVATIVE FORCE FIELD such as gravity. The total work done on the system by such forces, when the system moves from one position to another, can be expressed as $-\Delta V$ where V is the associated potential energy of the system. *Units* J.

potential energy density – see ACOUSTIC ENERGY DENSITY. *Units* J m^{-3}.

power – (1) *in acoustics* an abbreviation for SOUND POWER. *Units* W.

power – (2) *in physics and applied mechanics* the rate of ENERGY transfer across a specified boundary, as a result of mechanical forces. In its general sense, power includes all non-thermal energy transfer mechanisms; for example the power input to a system may include both electrical and mechanical contributions. *Units* W.

power – (3) *of a time-stationary signal or ergodic process* the average of the squared signal, given by $\langle |x|^2(t) \rangle$, where $x(t)$ is the signal (possibly complex) and angle brackets $\langle \ldots \rangle$ denote the time averaging operation. The *ac power* is defined as the power in the ac component of the signal; if $\bar{x}$ is the dc component or mean value of $x(t)$, the ac power equals $\langle |x|^2(t) \rangle - \bar{x}^2$. The *dc power* is defined as the power in the dc component; it equals $\bar{x}^2$. See also AUTOSPECTRAL DENSITY.

power density – *in a three-dimensional wave field* a general term for the time-average power crossing a specified surface per unit area; an equivalent term is *energy flux density*. Specific terms used in particular disciplines are *acoustic intensity* (acoustics) and *radiant flux density* (optics). Units W m^{-2}.

Note: In a two-dimensional wave field (e.g. ocean surface waves) the power density is a line density rather than an area density, and has units W m^{-1}.

power spectral density – *of a real time-stationary continuous signal* an equivalent term for AUTOSPECTRAL DENSITY.

power spectral matrix – *of a vector of real time-stationary continuous signals* an equivalent term for AUTOSPECTRAL MATRIX.

power transmission factor – *for structural or acoustic waves in a one-dimensional system* the ratio

$$\alpha_t = W_{\text{trans}}/W_{\text{inc}},$$

where W_{trans} is the net transmitted power and W_{inc} is the forward-wave (or incident) power. *Units* none.

Pr – symbol for PRANDTL NUMBER; also written Pr (**L PRANDTL** *1875–1953*). *Units* none.

Prandtl–Glauert transformation – *for steady compressible flow* a transformation of spatial coordinates that converts the small-perturbation equation for *steady* pressure disturbances in a uniform subsonic stream to the corresponding incompressible equation. The transformed coordinates are

$$x' = \frac{1}{\beta}x, \; y' = y, \; z' = z, \quad \text{with } \beta = \sqrt{1 - M^2};$$

here the x axis is in the flow direction and M is the flow Mach number (**H GLAUERT, L PRANDTL, J ACKERET** *1928*). Compare the LORENTZ TRANSFORMATION, which uses the same spatial coordinate stretching.

Prandtl–Meyer expansion fan – *in steady supersonic flow past a two-dimensional convex corner* a family of MACH WAVES, radiating out from the corner into the flow, that turn the flow so that it follows the boundary.

Prandtl–Meyer flow – supersonic steady isentropic flow with curved streamlines, as produced near a boundary that curves in the streamwise direction: in normal fluids, curvature into the flow causes compression, and curvature away from the flow causes expansion. The continuous change in flow direction along each streamline is associated with MACH WAVES that intersect the streamlines obliquely. An equivalent term is ***simple isentropic compression*** (or expansion, as appropriate). (**L PRANDTL, T MEYER** *1907–8*).

Prandtl–Meyer function – *for simple isentropic compression or expansion in supersonic flow* the function $v(M)$ defined by

$$v(M) = \int_1^M \frac{(M^2 - 1)dM}{1 + (\beta - 1)M^2} \qquad (M = \text{flow Mach number}).$$

Here β is the COEFFICIENT OF NONLINEARITY of the fluid, equal to $(\gamma + 1)/2$ for an ideal gas with constant specific-heat ratio γ. The Prandtl–Meyer function relates the initial and final Mach numbers M_1 and M_2 to the streamline turning angle $\Delta\theta$, as follows:

$v(M_2) = v(M_1) - |\Delta\theta|$ for compressive turning,

$v(M_2) = v(M_1) + |\Delta\theta|$ for expansive turning.

Prandtl number – *of a fluid* the ratio of the momentum diffusivity to the thermal diffusivity. In symbols, the Prandtl number is given by

$$\Pr = \frac{C_p \mu}{\kappa}$$

where μ is the viscosity of the fluid, κ is the thermal conductivity, and C_p is the constant-pressure specific heat. In a fluid of high Prandtl number, velocity differences are smoothed out by molecular diffusion more rapidly than temperature differences. *Units* none.

precedence effect – *in psychoacoustics* when two short-duration sounds are heard in rapid succession, the human auditory system appears to suppress the second sound (under certain conditions discussed below) and give precedence to the first, for purposes of LOCALIZATION. The pair of sounds is perceived as a single sound coming from the direction of the first source; the phenomenon is called the precedence effect.

Note: For the precedence effect to operate, there has to be a minimum separation of about 1 ms between the arrival of the two sounds. There is also an upper limiting separation, but its value depends on the nature of the sounds (simple or complex) and on whether the pair of signals is heard once or repeatedly: a typical range for this maximum separation is from 5 ms to about 40 ms. Finally, if the lagging sound is too loud relative to the first sound (more than about 10–15 dB higher in level), the precedence effect is overridden.

pre-envelope signal – an equivalent term for ANALYTIC SIGNAL.

presbyacusis – name given to the progressive hearing impairment that accompanies ageing, in the absence of other identifiable causes. An alternative spelling is ***presbycusis***.

presence – a possible alternative term for INTIMACY in concert halls. An intimate hall has presence.

pressure – *at a point in a continuous medium* the average value, with respect to direction of the surface, of the normal force per unit area that the medium on one side of an imaginary surface applies to the medium on the other side. In a fluid at rest, no directional averaging is required as the normal stress is the same in all directions. See also TOTAL PRESSURE, STAGNATION PRESSURE, STATIC PRESSURE, DYNAMIC PRESSURE. *Units* Pa.

Note: The positive normal direction is defined *into* the medium on which the force is measured. Thus a positive pressure tends to compress the medium.

pressure coefficient – *in fluid flow* the increase in static pressure between a reference position and the measurement position, normalized by the DYNAMIC PRESSURE at the reference position. In equation form, the pressure coefficient C_p is given by

$$C_p = \frac{P - P_{\text{ref}}}{q_{\text{ref}}}, \quad \text{(subscript ref} = \text{value at the reference point)}$$

where P is static pressure and q is dynamic pressure. *Units* none.

pressure–density relation – *for a fluid* a relation between pressure and density that holds under defined conditions. For example, the *isentropic* ~ holds for reversible adiabatic compression or expansion; it may be written as $P = P(\rho)$, indicating that there is a one-to-one relation between pressure P and density ρ under these conditions (in which the specific entropy remains constant). The differential form of the isentropic pressure–density relation, $dP = c^2 d\rho$, is one of the basic equations of linear acoustics.

Note (1): The pressure–density relation for a given fluid element is not necessarily the same as that at a fixed point in space. In the isentropic example above, the pressure P and density ρ refer to a fluid element.

Note (2): To obtain a one-to-one $P(\rho)$ relation, some additional constraint is always necessary. The reason is that the equilibrium pressure in a fluid is a function of two thermodynamic coordinates. For example, except along a phase boundary – where $P = P(T)$ – the pressure may be expressed as $P(\rho, T)$ where T is the temperature of the fluid.

pressure gain factor – *of an acoustic transducer that produces a single-frequency focused beam* an equivalent term for AMPLITUDE FOCAL GAIN. Compare GAIN FACTOR. *Units* none.

pressure-gradient microphone – a transducer whose output is proportional to the acoustic pressure gradient at the position of the active element, rather than to the acoustic pressure as in a normal microphone.

pressure reflection coefficient – *for single-frequency incident waves* abbreviation for SOUND PRESSURE REFLECTION COEFFICIENT. It is the complex ratio of reflected to incident sound pressure, at a specified position in a standing-wave field. An equivalent term is COMPLEX PRESSURE REFLECTION COEFFICIENT. *Units* none.

pressure reflection factor – an alternative term for COMPLEX PRESSURE REFLECTION COEFFICIENT. *Units* none.

pressure-release – describes a boundary on which the acoustic pressure is zero at all times. Reflection of underwater sound at an air–water interface is often approximated by a pressure-release boundary condition.

pressure–residual intensity index – *of an instrument for measuring sound intensity* a measure of the dynamic range of the instrument, defined by

$$\delta_{pI_0} = L_p - L_{I_0}.$$

Here L_p and L_{I_0} are the indicated SOUND PRESSURE LEVEL and INTENSITY LEVEL at the measurement point, when the intensity probe is oriented to measure in a null direction. Although the intensity in this direction is zero (by definition), there will appear to be a finite intensity I_0 owing to limitations of the instrument. Also known as the ***residual intensity index***. *Units* dB re $p_{\text{ref}}^2/I_{\text{ref}}$.

Note (1): The definition above assumes that measurements are made in air at standard ambient conditions.

Note (2): The residual intensity index at a given frequency is influenced, in general, by the spatial properties of the sound field (e.g. standing wave versus travelling wave). However, the dependence is weak at low frequencies (specifically, where the sound wavelength is large compared with the probe dimensions), and in the limit, δ_{pI_0} for a given instrument becomes a function of frequency only.

pressure-squared integral – *for transient sound signals* the integral with respect to time of the instantaneous squared sound pressure, $p^2(t)$. In equation form, it is given by

$$E = \int p^2(t)dt.$$

Compare SOUND EXPOSURE, which is a similar quantity but with FREQUENCY WEIGHTING applied to the pressure signal before squaring. *Units* Pa^2 s.

pressure transmission coefficient – *for single-frequency incident waves* abbreviation for SOUND PRESSURE TRANSMISSION COEFFICIENT. It is the complex ratio of transmitted to incident sound pressure, defined either for plane waves incident at a given angle on a uniform plane boundary, or for plane waves in a one-dimensional system (e.g. a pipe or waveguide) being transmitted between specified points in the system. An equivalent term is COMPLEX PRESSURE TRANSMISSION COEFFICIENT. *Units* none.

prevalence – *of a particular disease or disability* the number of people affected at any given time in a specified population, usually normalized to a standard population size of 10^5. Also known as ***prevalence rate***.

Pridmore-Brown equation – *for sound propagation in a parallel shear flow* see note (3) under LILLEY'S EQUATION.

principle of ~, principle of acoustic ~ – see ~.

prism – a three-dimensional object, formed by translating a plane polygon (the cross-section) in a direction normal to the plane (the axial direction). The cross-section need not be regular.

Note: The object defined above is called by mathematicians a ***right prism***, to distinguish it from the more general case where the end faces are not orthogonal to the axis.

probability – *of an event* a measure of the likelihood of the event, defined in such a way that:

(1) If A and B are mutually exclusive events, the probability that at least one of them will occur is the sum of the separate probabilities of A and B. In symbolic form,

$$P(A \cup B) = P(A) + P(B),$$

where $P(X)$ is the probability of event X.

(2) The probabilities of A (the event occurring) and of $A^\sim$ (the event not occurring) add to 1:

$$P(A) + P(A^\sim) = 1.$$

Units none.

probability density function (pdf) – *for a single random variable* the function $f(x)$ whose integral between any two values x_1 and x_2 gives the PROBABILITY that the variable will lie between those limits. In mathematical form,

$$P(x_1 < X < x_2) = \int_{x_1}^{x_2} f(x)\, dx.$$

Here X denotes the random variable, and $P(A)$ is the probability that statement A is true. See also DENSITY FUNCTION.

Note: The probability density function $f(x)$ is the derivative of the CUMULATIVE DISTRIBUTION FUNCTION, defined as $P(X \leq x)$.

probability distribution function – see CUMULATIVE DISTRIBUTION FUNCTION.

progressive wave – a wave that propagates in one direction (or parallel to a given axis), in contrast to a STANDING WAVE which has both forward and backward travelling components. See PLANE PROGRESSIVE WAVE, AXIALLY PROGRESSIVE WAVE. In addition, the term can refer to outgoing or incoming waves in spherical or cylindrical coordinates: see CYLINDRICAL PROGRESSIVE WAVE, SPHERICAL PROGRESSIVE WAVE.

projector – see SONAR PROJECTOR.

Prony series – *for spectral analysis* a representation of a finite-length data record in terms of a finite number of complex frequency components (**BARON DE PRONY *1755–1839***). In contrast to the Fourier series approach, the frequencies in a Prony series are not determined in advance but arise from a fit to the data.

For uniformly-sampled discrete data, this approach leads to a technique (***Prony method***) for fitting the data with a finite series of exponentials. The order-K model, with K complex frequencies, is defined by the sequence $\hat{x}[n]$:

$$\hat{x}[n] = \sum_{k=1}^{K} a_k \, e^{j\omega_k n \Delta} \quad (\Delta = \text{sampling interval}, n = \text{sample number}).$$

This model is fitted to the measured data $x[n]$ by solving two systems of linear equations and finding the roots of a polynomial: the process yields values for the complex amplitude coefficients a_k and the complex angular frequencies ω_k.

Note: The solution thus obtained is not optimal in a least squares sense, and the accuracy of ω_k (especially its imaginary part) is often sensitive to noise. Nevertheless the technique is widely used for high-resolution spectral analysis of short data lengths.

propagating wave – alternative term for PROGRESSIVE WAVE.

propagation – (1) transmission of a signal in a wavelike manner.

propagation – (2) the progressive advance of a WAVE through space, either in a specified direction (as in a waveguide) or normal to the WAVEFRONT surface.

propagation coefficient – *of a single-frequency progressive wave system* the quantity $\gamma = jk$, where k is the complex wavenumber in the propagation direction and $j = \sqrt{-1}$. The PROPAGATION FACTOR is then $\exp(j\omega t - \gamma x)$, for propagation in the x direction at angular frequency ω. *Units* (real part) Np m^{-1}; (imaginary part) rad m^{-1}.

Note: The real and imaginary parts of γ are the ATTENUATION COEFFICIENT and the PHASE COEFFICIENT respectively.

propagation constant – an alternative term for PROPAGATION COEFFICIENT. *Units* (real part) Np m^{-1}; (imaginary part) rad m^{-1}.

Note: The 1994 ANSI standard on acoustical terminology recognizes *propagation constant* (with the prefix "acoustic"), but not *propagation coefficient*; the 1994 IEC standard recognizes propagation coefficient (with the prefix "sound") but not propagation constant.

propagation coordinate – refers to the free coordinate in a separable-variables solution of the homogeneous wave equation. It may be a spatial coordinate (as in waveguide mode propagation), or time (as in freely-decaying room modes).

propagation factor – *of a single-frequency progressive wave system* the expression $\exp j(\omega t - kx)$ that contains the dependence of the wave variable on time, t, and the PROPAGATION COORDINATE, x. Here ω is the angular frequency, and the coefficient k is the PROPAGATION WAVENUMBER of the progressive wave system. *Units* none.

Note: An alternative way of writing the propagation factor is $\exp(j\omega t - \gamma x)$. Here γ is the PROPAGATION COEFFICIENT, related to the propagation wavenumber by $\gamma = jk$.

propagation loss – *in underwater acoustics* an equivalent term for TRANSMISSION LOSS (3). *Units* dB re 1 m^2.

propagation speed – *of free progressive waves* the speed of advance of WAVEFRONTS. For waves in a dispersive medium, there is in general no single clearly-defined propagation speed; see PHASE SPEED, GROUP VELOCITY. *Units* m s^{-1}.

propagation vector – *of single-frequency plane progressive waves in a lossless medium* the vector $\mathbf{k} = k\mathbf{n}$, where $\mathbf{n}$ is the unit vector in the WAVENORMAL direction, and k is the propagation wavenumber or CHARACTERISTIC WAVENUMBER. *Units* rad m^{-1}.

Note: The term propagation vector is recognized in both acoustics and optics. In acoustics, the alternative term WAVENUMBER VECTOR is commonly used for this quantity.

propagation wavenumber – *of single-frequency progressive waves* the phase change per unit distance along the propagation coordinate axis. Also known as CHARACTERISTIC WAVENUMBER. *Units* (real part) rad m^{-1}; (imaginary part) Np m^{-1}.

Note: Since a given type of wave usually has two possible directions of propagation (e.g forward and backward plane acoustic waves in a pipe), propagation wavenumbers usually come in pairs.

proportional-bandwidth frequency analysis – the partitioning of signal energy or power into CONTIGUOUS BANDS whose width is proportional to the MIDBAND FREQUENCY.

proprioceptor – a type of sensory receptor located in the muscles, tendons, or joints that transmits information about the position and movement of parts of the body.

PSD – abbreviation for *power spectral density*. For a definition, see AUTOSPECTRAL DENSITY.

pseudo-coincidence – ◆ *between an acoustic waveguide mode and any of the waveguide walls* COINCIDENCE between the axial phase speed of free bending waves in the wall of the waveguide, and the phase speed at which the acoustic mode would propagate if the walls were rigid. The prefix "pseudo" signifies that the fields do not match in the transverse direction.

Note: The pseudo-coincidence effect is significant for coupling between structural and acoustic waveguide modes, under conditions of light fluid loading (e.g. noise breakout from ventilation ducts).

pseudoplastic – *in viscoelasticity* an equivalent term for SHEAR-THINNING.

pseudo-random numbers – numbers that are generated by a deterministic algorithm, but nevertheless pass statistical tests for randomness. The algorithm is called a *random number generator*. See RANDOM NUMBER SEQUENCE.

pseudo-Rayleigh waves – *at a plane solid–fluid interface* an alternative term for LEAKY RAYLEIGH WAVES. The waves are damped by sound radiation into the fluid.

pseudosound – *in subsonic turbulent flow* local pressure fluctuations that are essentially incompressible, with typical magnitude $\frac{1}{2}\rho u'^2$ where u' is the rms turbulent velocity fluctuation.

pseudo-Stoneley waves – *at a plane solid–solid interface* damped interface waves analogous to STONELEY WAVES; unlike Stoneley waves, they radiate energy away from the interface in the form of shear waves.

psychoacoustics – the science of human and animal psychological responses to sound. The term is broadly equivalent to PSYCHOLOGICAL ACOUSTICS.

psychological acoustics – the branch of acoustics that deals with phenomenological aspects of sound perception, in humans or in animals. Covers hearing thresholds, loudness, masking, localization, and binaural hearing; also ageing effects, deafness, hearing aids, and hearing protection. Often understood to include sensory perception of vibration and motion, at infrasonic or audio frequencies.

PTS – abbreviation for PERMANENT THRESHOLD SHIFT. *Units* dB.

public address system – a system that uses amplified sound to provide public information, usually to listeners who are preoccupied with other tasks. The sound may be either live or pre-recorded. Compare SOUND REINFORCEMENT.

pulsatance – alternative term for ANGULAR FREQUENCY; now virtually obsolete. *Units* rad s^{-1}.

pulse – a transient signal of finite duration, usually short compared with other time scales of interest.

pulse bandwidth – the BANDWIDTH (2) of a pulse. *Units* Hz.

pure imaginary – a complex number is called pure imaginary if it has zero real part.

pure standing wave – a one-dimensional STANDING WAVE that has equal wave amplitudes in the forward and backward propagation directions.

pure tone – a single-frequency acoustic signal; see also TONE.

pure-tone audiometry – a technique for determining a person's hearing threshold levels for pure tones by behavioural means (i.e. involving a voluntary response from the subject). Sound may be applied by earphone or as an incident field in an audiometric test room (***air-conduction audiometry***); alternatively a pure-tone vibrational input may be applied to the skull by a bone vibrator (***bone-conduction audiometry***).

Note: The descriptions *air-conduction* and *bone-conduction* indicate the type of transducer employed, rather than the exclusive pathway of sound transmission to the cochlea.

pure-tone diffuse sound field – see DIFFUSE FIELD.

P waves – *in a solid medium* an equivalent term for COMPRESSIONAL WAVES. Compare S WAVES.

PWL – abbreviation for SOUND POWER LEVEL. *Units* dB re 1 pW.

Q

Q – abbreviation for Q-FACTOR. *Units* none.

***Q*-factor** – *for a peak in the frequency response of a lightly-damped linear system* a measure of the sharpness of the peak. It is defined for a given resonant mode as

$$Q = \frac{f_\text{res}}{B_h},$$

where f_res is the resonance frequency and B_h is the HALF-POWER BANDWIDTH of the resonance. *Units* none.

Note: The Q-factor is the reciprocal of the system LOSS FACTOR.

quadratic residue diffuser – a reflecting plane surface containing a parallel array of slots or recesses all of the same width, with slot depths that vary according to a mathematical sequence based on a prime number. Reflected sound is diffused in the plane perpendicular to the slots, with the effective frequency range covering a bandwidth of several octaves. Diffusers with a two-dimensional array of different-depth wells (rather than slots) are also possible. Quadratic residue diffusers have been used in a few concert halls and are popular in recording studio control rooms. See also SCHROEDER DIFFUSER.

quadrature – two sinusoidal signals of the same frequency are in quadrature if there is a phase difference of $\pi/2$ (or 90°) between them.

quadrature component – *of a single-frequency signal relative to a reference signal at the same frequency* the component of the signal that leads or lags the reference signal by $\pi/2$ in phase.

quadrature function – *of a time-domain signal* the quadrature function of any signal $x(t)$ is its HILBERT TRANSFORM, $x_H(t)$. The name arises from the fact that in the frequency domain, positive-frequency components of $x_H(t)$ are phase-delayed by $\pi/2$ relative to the original signal (based on the $e^{j\omega t}$ convention as used in the definition of FOURIER TRANSFORM). See also ANALYTIC SIGNAL.

Note: The opposite process, i.e. advancing positive frequency components of $x(t)$ by $\pi/2$ and delaying negative frequencies, yields the inverse Hilbert transform $\text{IHT}[x(t)] = -x_H(t)$.

quadrature spectral density – an expanded version of the term QUAD-SPECTRAL DENSITY.

quadrupole density – see ORDER (12).

quadrupole moment – the second spatial moment of an ACOUSTIC SOURCE DENSITY; also known as the *quadrupole-moment tensor*. For a quadrupole constructed from two equal and opposite POINT DIPOLES of strength $+\mathbf{D}$ and $-\mathbf{D}$, with the $+\mathbf{D}$ dipole displaced by $\mathbf{d}$ relative to $-\mathbf{D}$, the cartesian components of the quadrupole-moment tensor are $\frac{1}{2}(D_j d_i + D_i d_j)$. See QUADRUPOLE-ORDER SOURCE DISTRIBUTION.

quadrupole-order source distribution – a sound source distribution describable by an ACOUSTIC SOURCE DENSITY $\mathcal{F}$ whose zeroth and first spatial MOMENTS with respect to position are both zero, i.e.

$S = \int \mathcal{F} d^3\mathbf{y} = 0$ (zero source strength);

$D_i = \int y_i \mathcal{F} d^3\mathbf{y} = 0$ (zero dipole moment),

but whose second moment – with components Q_{ij} defined below – is non-zero:

$Q_{ij} = \frac{1}{2} \int y_i y_j \mathcal{F} d^3\mathbf{y}$ (non-zero quadrupole moment).

Here y_i, y_j are cartesian components ($i, j = 1, 2, 3$) of the position vector $\mathbf{y}$, $d^3\mathbf{y}$ denotes a volume element, and Q_{ij} are the cartesian components of the *quadrupole-moment tensor*.

Note: The far-field radiation of such a source distribution can be represented in the COMPACT limit by placing four equal-magnitude POINT SOURCES at the vertices of a parallelogram, with cyclically alternating strengths (S, $-S$, S, $-S$), so as to replicate the Q_{ij} values of the actual source. See also POINT QUADRUPOLE.

quadrupole source – a sound generation mechanism that drives the surrounding medium by a process equivalent to a time-varying applied stress field. In the COMPACT limit, the sound power output from a quadrupole source varies as c^{-5}, where c is the sound speed in the radiating medium.

quad-spectral density – *of two real signals that are jointly stationary with respect to time* the imaginary part of the CROSS-SPECTRAL DENSITY. The quad-spectral density at frequency f is related to the CROSS-CORRELATION FUNCTION by

$$Q_{xy}(f) = \int_0^\infty [R_{xy}(-\tau) - R_{xy}(\tau)] \sin 2\pi f \tau \, d\tau,$$

where $x(t)$ and $y(t)$ represent the two signals and $R_{xy}(\tau)$ is their cross-correlation function. See also PHASE ANGLE OF THE CROSS-SPECTRAL DENSITY.

Note: The quad-spectral density is normally given as a single-sided spectrum,

$$\bar{Q}_{xy}(f) = 2Q_{xy}(f) \qquad (f > 0).$$

quad-spectrum – an equivalent term for QUAD-SPECTRAL DENSITY.

quality factor – (1) an alternative name for Q-FACTOR. *Units* none.

quality factor – (2) *for wave propagation in viscoelastic media* the ratio of the real part to the imaginary part of the appropriate complex modulus. Specifically, in a material with complex shear modulus $G = G' + jG''$, the *quality factor for shear waves* is

$$Q_s = \frac{G'}{G''}, \qquad (G'' > 0 \text{ for positive frequencies}).$$

Likewise in a material with complex bulk modulus $B = B' + jB''$, the *quality factor for compressional waves* is

$$Q_d = \frac{G' + (4/3)B'}{G'' + (4/3)B''}, \qquad (B'' > 0 \text{ for positive frequencies}).$$

In a lossless medium, both Q_s and Q_d tend to infinity. *Units* none.

Note (1): Both Q_s and Q_d are functions of frequency. An equivalent interpretation is

$$\frac{1}{Q} = \frac{\alpha \lambda}{\pi} \qquad \text{(for either shear or compressional waves),}$$

where α is the plane-wave ATTENUATION COEFFICIENT and λ is the wavelength.

Note (2): In terms of energy, the quality factor is 2π times the ratio w_{max}/w_{diss}, where w_{max} is the peak energy density in the medium due to the wave, and w_{diss} is the energy dissipated per unit volume over one cycle.

quantization error – numerical inaccuracy caused by representing numbers in quantized form, e.g. by using a limited word length in a computer. An equivalent term is *round-off error*.

quantized data – numerical DATA that is limited to discrete values, or rounded to a fixed number of digits. The term *discrete data* is a technically correct equivalent, but is liable to cause confusion with discrete SAMPLING.

quantized variable – any numerical variable that is limited to certain discrete values.

quartz wind – an obsolete term for the ACOUSTIC STREAMING that occurs when a beam of ultrasound is absorbed in a fluid. See SONIC WIND.

quasi-plane wave mode – *in a uniform fluid-filled waveguide with almost-hard walls* the lowest-order acoustic mode of propagation, in which the pressure is almost uniform over the cross-section of the waveguide. Compare PLANE-WAVE MODE.

Note: A quasi-plane wave mode also exists in a rigid-walled tube containing a viscous fluid. See LARGE-TUBE ATTENUATION COEFFICIENT.

quasi-stationary – a random signal $x(t)$ whose statistics vary slowly on the typical timescale of the signal itself is called quasi-stationary. By selecting time-segments that are short compared with the timescale of the non-stationarity, but still long compared with the typical timescale of $x(t)$, sufficient averaging can be achieved to produce local statistical estimates. See AVERAGE, example (3).

R

radian frequency – an alternative term for ANGULAR FREQUENCY. *Units* rad s^{-1}.

radiation – *in acoustics* outgoing waves from a surface or object. The radiation may either be scattered (i.e. produced by passive interaction of the surface with incident sound or other disturbances), or direct (i.e. produced by some external agency driving the surface).

radiation condition – see SOMMERFELD RADIATION CONDITION.

radiation damping – DAMPING (1) of a structure caused by sound power radiation from the structure into a surrounding fluid. See also RADIATION LOSS FACTOR.

radiation efficiency – a nondimensional measure of the SOUND POWER radiated by a vibrating surface into an adjacent fluid. In symbols, the radiation efficiency σ is defined as W/W_0, where W is the power actually radiated, and W_0 is the power that would be radiated by the same surface vibration under ρc loading:

$$W_0 = \rho c \int \langle \dot{w}^2 \rangle \, dS.$$

In the expression above, ρ and c are the fluid density and sound speed, $\langle \dot{w}^2 \rangle$ is the local mean square velocity of the vibrating surface, and the integral extends over the whole vibrating region. *Units* none.

Note: The radiation efficiency may be specified for a single frequency, or as an average value over a frequency band (e.g. one-third octave): its value depends on the surface vibration pattern, the geometry of the surface, and the size of the surface in relation to the sound wavelength. It is not limited to values less than 1, and for this reason the alternative terms *radiation factor* or *radiation ratio* are often preferred.

radiation factor – an equivalent term for RADIATION EFFICIENCY. *Units* none.

radiation force – see ACOUSTIC RADIATION FORCE. *Units* N.

radiation impedance – a generic term for the impedance presented to a vibrating surface by the adjacent acoustic medium. An equivalent term is *fluid loading impedance*. For rigid vibrating surfaces, see PISTON RADIATION IMPEDANCE, MECHANICAL RADIATION IMPEDANCE. For acoustically compact openings and surfaces (i.e. small compared with the wavelength), see ACOUSTIC RADIATION IMPEDANCE, CONDUCTIVITY. For surfaces with arbitrary normal velocity distributions, see MODAL RADIATION IMPEDANCE, SPECIFIC RADIATION IMPEDANCE, WAVE IMPEDANCE.

radiation impedance matrix – see MODAL RADIATION IMPEDANCE.

radiation index – $10 \log_{10} \sigma$, where σ is the radiation factor or RADIATION EFFICIENCY. *Units* dB.

Note: The relation of radiation index to radiation factor is the same as that of directivity index to directivity factor.

radiation loss factor – (1) *for a single resonant mode of a fluid-loaded structure* the ratio of the energy radiated as sound in one period of oscillation to $4\pi E_{kin}$, where E_{kin} is the mean vibrational kinetic energy of the structure. It follows that at frequencies close to resonance, the sound power radiated from the structural mode concerned is

$$W_{rad} = 2\omega \eta_{rad} E_{kin}, \quad (\omega = \text{angular frequency}),$$

where η_{rad} denotes the radiation loss factor for that mode. See also LOSS FACTOR. *Units* none.

radiation loss factor – (2) *of a fluid-loaded structure driven by band-limited noise* the contribution to the MODAL-AVERAGE LOSS FACTOR that arises from radiation of acoustic energy into the surrounding fluid. If η_{rad} denotes the multimode radiation loss factor defined in this way, the radiated sound power is given by

$$W_{rad} = \omega \eta_{rad} E_{tot} \approx 2\omega \eta_{rad} E_{kin};$$

here ω is the mid-band angular frequency, E_{tot} is the total vibrational energy of the system, and E_{kin} is the mean kinetic energy component. *Units* none.

radiation modes – *of a fluid-loaded vibrating surface* see MODAL RADIATION IMPEDANCE.

radiation pressure – abbreviation for ACOUSTIC RADIATION PRESSURE. *Units* Pa.

radiation ratio – an equivalent term for radiation factor or RADIATION EFFICIENCY. *Units* none.

radiation stress tensor – abbreviation for ACOUSTIC RADIATION STRESS TENSOR. *Units* Pa.

radius of gyration – if the zeroth and second moments of a DENSITY FUNCTION are denoted by M_0 and M_2, the radius of gyration k is defined by

$$k^2 = M_2/M_0.$$

Applied to the bending stiffness of rods, beams, or plates, the term refers to the quantity k defined by

$$k^2 = \frac{1}{A_{\text{tot}}} \int A(y) y^2 dy,$$

where $A(y)$ is the area distribution of the cross-section with respect to distance y from the neutral axis, and A_{tot} is the total cross-sectional area.

ramp function – a function $x(t)$, defined over a time interval $t_1 < t < t_2$, such that $x(t)$ increases or decreases linearly with t. In equation form,

$$x(t) = a(t - t_0), \qquad t_1 < t < t_2;$$

the constants a and t_0 represent the slope and the zero-crossing time respectively.

random incidence – *at a point on a plane boundary* describes an incident sound field that, if represented as a sum of plane progressive waves, would have its mean square pressure uniformly divided among the whole hemisphere of incident directions. If the total sound power incident on the boundary per unit area is denoted by Φ, then the part due to waves arriving with incidence angles between θ and $\theta + d\theta$ is

$$d\Phi = (2 \sin \theta \cos \theta \, d\theta)\Phi.$$

Here Φ, sometimes called the IRRADIATION STRENGTH, represents the total normal intensity due to the incident field.

Note: Reflection of a random-incidence field at a rigid plane boundary produces a combined field at the surface that is not DIFFUSE, even though the intensity is the same in all directions, because the incident and reflected waves are coherent.

random-incidence absorption coefficient – an equivalent term for STATISTICAL ABSORPTION COEFFICIENT. *Units* none.

random-incidence sound power transmission coefficient – an equivalent term for STATISTICAL SOUND POWER TRANSMISSION COEFFICIENT. *Units* none.

random noise – (1) an undesired or extraneous RANDOM SIGNAL.

random noise – (2) an alternative term for RANDOM SIGNAL. It usually implies that the signal has a smooth or featureless AUTOSPECTRAL DENSITY as a function of frequency, unless qualified as in NARROW-BAND ∼. See WHITE NOISE, PINK NOISE, COLOURED NOISE.

random number generator – see PSEUDO-RANDOM NUMBERS.

random number sequence – a sequence of decimal digits (0–9), or binary digits (0, 1), such that each digit is equally likely to occur at any stage; there is therefore no way of predicting the next digit.

random process – a function (usually of time) that takes on a definite waveform each time a chance experiment is performed, but that cannot be predicted in advance. See also STOCHASTIC PROCESS.

random signal – a SIGNAL that cannot be predicted exactly, although its statistical properties can be described. Thus a signal $x(t)$ is called random (as opposed to DETERMINISTIC) when its values for $t > t_1$ cannot be predicted with certainty based on the waveform for $t < t_1$.

random variable – a single variable that takes a definite numerical value each time a chance experiment is performed, but whose next value cannot be predicted in advance. A *continuous random variable* takes all possible values over an interval, finite or infinite. A *discrete random variable* is limited to a discrete set of values (i.e. it is quantized).

random waveform – an equivalent term for RANDOM SIGNAL.

range – (1) *in underwater or atmospheric acoustics* an abbreviation for HORIZONTAL RANGE. The term is used in sonar for the horizontal distance between a sonar system and its target. *Units* m.

range – (2) *in sonar* a measurement range or test range, often a loch or fjord, used for testing sonar equipment. Also known as a *sonar range* (but compare DETECTION RANGE).

range-normalized pressure – *in underwater acoustics* the far-field acoustic pressure received from a sound source or scattering target, normalized to unit distance from the source or scatterer. Echo-free transmission conditions are assumed, and the field is corrected for losses in the medium. *Units* Pa @ 1 m.

Note: The range-normalized pressure is used in defining the source level and source spectrum level of steady-state underwater sound sources, and the energy source level and energy source spectrum level of transient sources. Compare SOURCE LEVEL, ENERGY SOURCE LEVEL.

range rate – *in sonar* the rate of increase of range with time, between a sonar system and its target. *Units* m s^{-1}.

rarefaction wave – *in a fluid or solid* a progressive wave, or wavefront, that causes expansion of the medium rather than compression. Compare COMPRESSION WAVE.

RASTI (RApid Speech Transmission Index) – an objective measure of speech INTELLIGIBILITY, based on a simplified SPEECH TRANSMISSION INDEX scheme. Originally developed in order to allow objective measurement of speech intelligibility using portable instruments. *Units* none.

rate of decay – see DECAY RATE. *Units* dB s^{-1}.

rate of strain – the time derivative of STRAIN. *Units* s^{-1}.

rational function – a ratio of two POLYNOMIALS.

rational number – a number that can be expressed as the ratio of two integers.

ray – *used as a noun* abbreviation for RAY PATH.

ray acoustics – an alternative term for GEOMETRICAL ACOUSTICS. In the ray acoustics approximation, sound travels between any two points along lines (rays), which are analogous to rays in optics. See also FERMAT'S PRINCIPLE.

ray direction – *at any point on a ray path* the local tangent to the RAY PATH. It is expressed mathematically by the unit vector

$$\mathbf{m} \equiv \frac{\mathbf{v}_{ray}}{|\mathbf{v}_{ray}|},$$

where $\mathbf{v}_{ray}$ is the RAY VELOCITY vector.

Note: The ray and WAVENORMAL directions are not the same in general. They coincide for sound waves in an isotropic medium at rest, but not for waves in an ANISOTROPIC solid. The two directions are also different when sound waves in a fluid are superimposed on a mean flow.

rayl – see SI RAYL. The original rayl was defined as a cgs unit of specific acoustic impedance, equal to 1 dyn s cm^{-3}.

Rayleigh distance – *for high-frequency radiation from a coherent source region of finite extent* the distance l_{trans}^2/λ beyond which far-field behaviour is approached. The expression given applies to free-field radiation of wavelength λ from a source region of maximum dimension l_{trans}, where l_{trans} is measured in a plane transverse to the radiation direction. Compare FRAUNHOFER REGION. *Units* m.

Note: It is assumed that $l_{trans} > \lambda$. Otherwise, the GEOMETRIC NEAR FIELD extends out to the Rayleigh distance (at least) and the Rayleigh distance loses its significance.

Rayleigh integral – *for the sound field in a half-space driven by a vibrating plane boundary* a representation of the field in terms of a surface integral, taken over the infinite plane boundary S_∞, of the normal velocity multiplied by the inverse distance $1/r$. The acoustic pressure p is given by

$$p(\mathbf{x}, t) = \frac{\rho}{2\pi} \int_{S_\infty} \frac{u_n(\mathbf{y}, t - r/c)}{r} dS(\mathbf{y}).$$

Here u_n is the normal velocity on the boundary (positive into the fluid); $r = |\mathbf{x} - \mathbf{y}|$ is the distance between the field point $\mathbf{x}$ and the source point $\mathbf{y}$ on the boundary; $t - r/c$ is the retarded time; and ρ, c are the fluid density and sound speed.

Note: The time-domain version of the Rayleigh integral given above is restricted to lossless fluids. This limitation can be removed by transforming into the frequency domain.

Rayleigh radiation pressure – see ACOUSTIC RADIATION PRESSURE. *Units* Pa.

Rayleigh–Ritz method – a powerful numerical technique for obtaining natural frequencies and mode shapes in acoustical or mechanical vibrating systems; also known as the Ritz method (**W RITZ 1908**). The method provides approximate solutions to problems that can be stated in terms of a VARIATIONAL PRINCIPLE. The dependent variables are expressed as sums of the form

$$y = \sum_{i=1}^{n} a_i \psi_i,$$

where the a_i are unknown coefficients and the ψ_i are basis functions (suitably differentiable) that satisfy the ESSENTIAL BOUNDARY CONDITIONS, and are therefore suitable for approximating the solution. Maximizing or minimizing the value of the functional, denoted by J, implies the necessary conditions

$$\frac{\partial J}{\partial a_1} = 0, \quad \frac{\partial J}{\partial a_2} = 0, \ldots, \quad \frac{\partial J}{\partial a_n} = 0,$$

which yield equations to determine the coefficients a_i.

Rayleigh's criterion – *in thermoacoustics* a necessary condition for the onset of instability, in an irrotational flow at low Mach number that contains a compact heat source of time-varying output $Q(t)$. It states that oscillations can become unstable when the heat input is in phase with the local acoustic pressure $p(t)$, at any given frequency (**LORD RAYLEIGH 1878**). A more precise statement applicable to nonlinear oscillations is that instability requires

$$\oint p(t) Q(t) dt > 0 \qquad \text{(integral over one cycle)}.$$

Rayleigh wave – a surface wave that propagates along the free boundary of a solid half-space, i.e. with a vacuum next to the solid surface (**LORD**

RAYLEIGH 1885). Rayleigh waves are nondispersive, in contrast to guided waves in a slab of finite thickness (see LAMB WAVES). Compare also LEAKY RAYLEIGH WAVE, SCHOLTE WAVE, STONELEY WAVE.

ray parameter – *in a horizontally-stratified medium* a ray-path invariant, equal to the inverse of the horizontal phase speed. In a stationary medium with sound speed variation $c(z)$, where z is the vertical coordinate, the ray parameter is given by

$$a = \frac{\cos \theta(z)}{c(z)}.$$

Here $\theta(z)$ is the ray GRAZING ANGLE. See also SNELL'S LAW. *Units* s m^{-1}.

ray path – *between two points in a time-invariant medium* a line in space joining the two points, whose direction is everywhere parallel to the direction of energy propagation as defined in the GEOMETRICAL ACOUSTICS limit. Equivalently – provided the medium is lossless – the ray path is determined by FERMAT'S PRINCIPLE. Compare RAY TRAJECTORY.

ray theory – *in acoustics* an alternative term for GEOMETRICAL ACOUSTICS.

ray trajectory – *for an element of an acoustic wavefront* the trajectory traced out by a point that at every instant moves with the local RAY VELOCITY. In a time-varying medium, different wavefronts will take different routes, in general, between the same fixed points.

ray tube – *in geometrical acoustics* a tube whose boundaries are formed by neighbouring individual RAYS. An ***infinitesimal ray tube*** has cross-sectional dimensions that are vanishingly small, compared with both the radius of curvature of the ray trajectory and the distance from the source.

ray-tube area – the cross-sectional area of a RAY TUBE, measured in the plane normal to the local RAY DIRECTION. *Units* m^2.

ray velocity – *in geometrical acoustics* the TRACE VELOCITY (2) of a local wavefront along the ray path. For acoustic waves in a moving fluid, the ray velocity vector is the vector sum of the propagation velocity relative to the fluid, denoted by $c\mathbf{n}$ (where c represents the local sound speed and $\mathbf{n}$ is the unit WAVENORMAL vector), and the background motion of the fluid, denoted by $\mathbf{w}$:

$\mathbf{v}_{\text{ray}} = c\mathbf{n} + \mathbf{w}$ ($\mathbf{w}$ = local mean flow or wind velocity).

Units m s^{-1}.

Re – (1) operator symbol for the REAL PART of a complex number.

Re – (2) symbol for REYNOLDS NUMBER. The two-letter notation is conventional for named dimensionless groups, and can also be written in roman type, as in the present dictionary: Re. *Units* none.

reactance – the imaginary part of an IMPEDANCE.

reactive intensity – a measure of the oscillatory INSTANTANEOUS ACOUSTIC INTENSITY associated with particle velocity components in quadrature with the acoustic pressure. The term is generally restricted to harmonic sound fields, where it is contrasted with ACTIVE INTENSITY. *Units* W m^{-2}.

Note (1): For a single-frequency sound field whose acoustic pressure and particle velocity at position **x** are represented by

$$p = \text{Re}[P(\mathbf{x})\, e^{j\omega t}], \qquad \mathbf{u} = \text{Re}[\mathbf{U}(\mathbf{x})\, e^{j\omega t}],$$

the *reactive intensity vector* **J** at point **x** may be written in terms of the complex numbers P and **U**, or the mean square pressure p_{rms}^2 and its gradient, as

$$\mathbf{J} = -\frac{1}{2}\text{Im}(P^*\mathbf{U}), \qquad (* = \text{complex conjugate})$$

$$= -\frac{1}{2\omega\rho}\nabla(p_{\text{rms}})^2.$$

Here ω is the angular frequency, and ρ is the density.

Note (2): The instantaneous product of the particle velocity, **u**, with the pressure phase-delayed by $\pi/2$, i.e.

$$p^Q = \text{Re}[P(\mathbf{x})\, e^{j(\omega t - \pi/2)}],$$

is an oscillatory quantity with zero mean value: $\langle p^Q \mathbf{u}\rangle = 0$. Its amplitude of oscillation is **J** and its frequency is twice the acoustic frequency.

Note (3): In a quasi-monochromatic sound field, the reactive intensity vector is defined by the time-average product

$$\mathbf{J} = \langle p_H \mathbf{u}\rangle.$$

Here p_H is the HILBERT TRANSFORM of the local acoustic pressure, and **u** is the acoustic particle velocity vector at the same point. One can also define an *instantaneous reactive intensity* as

$$\mathbf{J}(t) = \tfrac{1}{2}(p_H \mathbf{u} - \mathbf{u}_H p);$$

the time average of this quantity is equal to the reactive intensity as defined above.

real axis – any complex number $z = x + jy$ may be represented graphically by a point in the x–y plane, with x plotted horizontally and y vertically. The horizontal axis is then called the real axis.

real part – any complex number z may be represented as $z = x + jy$, where x and y are both real. The quantity x is called the real part of z, denoted by $\mathrm{Re}\, z$ or $\mathcal{R}\, z$. It equals $(z + z^*)/2$.

real-part sufficiency – *for a linear time-invariant system* refers to the fact that, provided the impulse response function of a system is real and causal, the corresponding FREQUENCY RESPONSE FUNCTION $H(\omega)$ can be deduced from its real part $H_R(\omega)$ alone. An example in acoustics is the fact that the DISPERSION COEFFICIENT $\beta(\omega)$ for a time-invariant medium can be deduced from the attenuation coefficient $\alpha(\omega)$, provided the latter is known at all frequencies. See KRAMERS–KRONIG RELATIONS.

receiver – a transducer (microphone or hydrophone) designed to aid detection and measurement of acoustic pressure signals by converting them into electrical signals. A receiver may also take the form of a transducer array.

receptance – *of a point-excited mechanical system* a generic term for the complex ratio of local displacement to input force, at a given frequency. The terms *cross* ~, *direct* ~, and *driving-point* ~ are defined in a similar way to the corresponding terms for impedance. The ratio of the displacement at a remote point to the force at the driving point is called the ***transfer receptance***. Units m N^{-1}.

reciprocity principle – (1) *for linear acoustics* the property of linear sound fields (with or without boundaries) whereby the complex pressure at $\mathcal{A}$, due to a harmonic point monopole at $\mathcal{B}$, equals the complex pressure at $\mathcal{B}$ when the same source is placed at $\mathcal{A}$. Similar reciprocal relations exist for dipole sources, for linear mechanical systems, and for electroacoustic systems. The principle is limited to systems whose response matrix is symmetric.

reciprocity principle – (2) *for linear elastodynamics of continuous media* the statement that any two single-frequency elastodynamic fields, separately excited in the same medium at the same frequency, are related by the following integral over an arbitrary closed surface S:

$$\int_S [u_i^{(1)} T_{ij}^{(2)} - u_i^{(2)} T_{ij}^{(1)}] n_j dS = 0.$$

Here the two separate single-frequency fields are denoted by superscripts (1) and (2); u_i are cartesian displacement components, T_{ij} are stress components, and the unit normal on S has components n_j ($i, j = 1, 2, 3$). The integral relation above is also known as the ***reciprocity theorem***.

record – a SIGNAL, or segment of a signal, that has a finite duration called the ***record length***.

recruitment – *in psychoacoustics* an unusually rapid growth in the perceived loudness of a tone as the sound pressure level at the ear is increased. The phenomenon occurs in people with cochlear hearing loss, where the sensitivity of the affected ear can be almost normal at high levels even when the absolute threshold is well above normal (e.g. raised as much as 60 dB). A similar effect (though not termed recruitment) occurs in normal listeners at the extremes of the audio frequency range, where the dynamic range of the auditory system is compressed.

rectangular coordinates – a set of coordinates (x, y, z) based on distances x, y, z measured from the origin along three mutually perpendicular axes. The (x, y, z) axes usually form a **right-handed system**, which means that a rotation from the $+x$ direction towards the $+y$ direction appears as clockwise to an observer looking in the $+z$ direction. An equivalent name is **cartesian coordinates**.

rectangular window – *in signal processing* a WEIGHTING FUNCTION that is constant over a specified interval (of time or frequency, for example), and is zero outside the interval.

rectified diffusion – the migration of dissolved gas into or out of a bubble under the action of a sound field.

reduced frequency – *in unsteady airfoil theory* the dimensionless quantity $\omega l / 2 u_\infty$, where ω is the angular frequency associated with an incident gust or with unsteady motion of an airfoil, l is the airfoil CHORD, and u_∞ is the undisturbed value of the free-stream relative velocity. *Units* none.

reduced path length – *in nonlinear ray acoustics* an equivalent plane-wave propagation distance. The reduced path length produces the same amount of waveform distortion in a plane progressive wave, for the same initial amplitude, as occurs between two given points on a ray path. *Units* m.

Note: The reduced path length is a function of the starting position chosen, as well as the distance propagated along the ray. For example, cylindrical and spherical outgoing waves in a uniform medium, starting at radius r_+ and propagating to radius r, have reduced path lengths

$$\tilde{x} = 2r_+ \left[\left(\frac{r}{r_+} \right)^{1/2} - 1 \right] \quad \text{(cylindrical)},$$

$$\tilde{x} = r_+ \ln \left(\frac{r}{r_+} \right) \quad \text{(spherical)}.$$

reduced pressure – *in nonlinear ray acoustics* a modified pressure variable that remains constant following a propagating wavelet, as a consequence of energy conservation. In a stationary medium, the reduced pressure $\tilde{p}$ is related to the acoustic pressure p by

$$\tilde{p} = \left[\frac{A/\rho c}{(A/\rho c)_+}\right]^{1/2}$$

where A is the ray-tube area, ρ and c are the density and sound speed of the medium, and the $+$ subscript denotes a reference point on the ray. *Units* Pa.

reference distance – *in underwater acoustics* a standard distance at which either the radiated sound pressure level from a source, or the scattered sound pressure level from a target, is measured (or to which measurements are scaled). It is usually taken as 1 m, but 1 yd ($= 0.9144$ m) was often used in the past. A reference distance appears in the definitions of SOURCE LEVEL, ENERGY SOURCE LEVEL, and TRANSMISSION LOSS (3). *Units* m.

reference equivalent threshold levels – *for pure-tone air-conduction audiometry* a set of standardized levels, measured in a particular EAR SIMULATOR or similar device at different frequencies across the audio frequency range, to which "zero" on an AUDIOMETER should be calibrated. *Units* dB.

Note: A reference equivalent threshold level at any given frequency is meant to represent the sound pressure level produced by the earphone when measured in a specified ear simulator or acoustic coupler, such that the same earphone driven by the same input and applied to the ears of young otologically normal adults would be expected to produce a barely detectable auditory sensation.

reference value – the normalizing value used to form a LEVEL in decibels. Examples are the *reference sound power*, W_{ref}, that appears in the equation

$L_W = 10 \log_{10} W/W_{\text{ref}}$ (sound power level);

the *reference acoustic intensity*, I_{ref}, in the equation

$L_I = 10 \log_{10} (I/I_{\text{ref}})$ (intensity level);

and the *reference rms sound pressure*, p_{ref}, in the equation

$L_p = 10 \log_{10} (p_{\text{rms}}/p_{\text{ref}})^2$ (sound pressure level).

A quantity equal to the reference value has a level of 0 dB. Standard reference values are listed in the table below, for sound in gases (including air), for sound in other media (including water), and for vibroacoustics.

Note: In underwater acoustics, I_{ref} is sometimes defined as the intensity in a plane progressive sound wave with rms pressure equal to the reference pressure. This gives

$I_{\text{ref}} = (p_{\text{ref}})^2/\rho c$ (with $p_{\text{ref}} = 1$ μPa), rather than 1 pW m^{-2} as in the table below.

Quantity	Sound in gases	Other media
Power W_{ref}	1 pW = 10^{-12} W	1 pW
Intensity I_{ref}	1 pW m^{-2} = 10^{-12} W m^{-2}	1 pW m^{-2}
Energy density w_{ref}	1 pJ m^{-3} = 10^{-12} J m^{-3}	1 pJ m^{-3}
Pressure p_{ref}	20 μPa = 2 × 10^{-5} Pa	1 μPa
	Vibroacoustics	
Force F_{ref}	1 μN	
Particle velocity u_{ref}	1 nm s^{-1}	
Acceleration a_{ref}	1 μm s^{-2}	

reflectance – *for sound waves incident on a boundary* an equivalent term for SOUND POWER REFLECTION COEFFICIENT. *Units* none.

reflection – *of waves by a boundary* the process whereby waves arriving at the boundary (***incident*** waves) interact with it to produce waves travelling away from the boundary (***reflected*** or ***scattered*** waves). See DIFFUSE REFLECTION, SPECULAR REFLECTION.

Note: For the special case of specular reflection in which plane incident waves are reflected by a uniform plane boundary, a simple rule applies: the normal wavenumber component of the incident field is reversed on reflection, and the wavenumber component parallel to the boundary is unaltered.

reflection coefficient – *for sound pressure or particle velocity* the complex ratios

$R_p = p_{refl}/p_{inc}$ (pressure),

$R_u = u_{refl}/u_{inc}$ (velocity),

defined for single-frequency plane-wave reflection at a plane boundary or two-fluid interface. In the first equation p_{refl} and p_{inc} represent the complex amplitudes of the reflected and incident acoustic pressure at the boundary; the second equation is similar, but p (sound pressure) is replaced by u (normal particle velocity). *Units* none.

Note: More precise names for these coefficients are ***sound pressure reflection coefficient*** and ***normal velocity reflection coefficient***. The term reflection coefficient, when used in acoustics without qualification, generally refers to the PRESSURE REFLECTION COEFFICIENT.

reflection coefficient matrix – *for a termination or discontinuity in a multimode waveguide* a square matrix of complex coefficients that relates a column vector of modal reflected pressures **p**⁻ to a column vector of modal incident pressures **p**⁺, as in the relations

$$\mathbf{p}^- = \mathbf{R}\mathbf{p}^+ \quad \text{or} \quad p_m^- = R_{ml}p_l^+.$$

Compare MULTIMODE SCATTERING MATRIX. *Units* none.

reflection factor – (1) a term used by some authors for a complex REFLECTION COEFFICIENT, i.e. the complex ratio of an acoustic variable (e.g. sound pressure) in a reflected wave to that in the incident wave, evaluated at a single frequency. Authors who use this terminology sometimes redefine the term reflection coefficient to mean the modulus of the reflection factor. *Units* none.

reflection factor – (2) an alternative term for SOUND POWER REFLECTION COEFFICIENT. *Units* none.

reflection loss – *for structural or acoustic waves in a one-dimensional system* the reciprocal of the POWER TRANSMISSION FACTOR, expressed in decibels. In symbols,

$$\text{RL} = 10 \log_{10} W_{\text{trans}}/W_{\text{inc}},$$

where W_{trans} is the net transmitted power and W_{inc} is the forward-wave or incident power. *Units* dB.

reflection phase – the PHASE of a reflected wave relative to the incident wave, when both waves are measured at the same point (e.g. on a reflecting boundary). *Units* rad.

refraction – *in geometrical acoustics* refers to the change in wavenormal direction that occurs when a wavefront passes from one region into another region of different sound speed (or different mean flow velocity). See SNELL'S LAW, RAY PARAMETER.

refractive index – *in acoustics* the ratio c_{ref}/c, where c is the local sound speed and c_{ref} is a reference sound speed. *Units* none.

relative error – if a quantity x is approximated by $\hat{x}$, the error is $\delta = \hat{x} - x$, and the relative error is δ/x.

relative molar mass, relative molecular mass – the ratio

$$M = \bar{m}/m_0,$$

where $\bar{m}$ is the MOLAR MASS and $m_0 = 10^{-3}$ kg. On this basis, neutral ^{12}C atoms have a relative molar mass of exactly 12. An equivalent older term is *molecular weight*. *Units* none.

Note: In a gas, M is the average mass of individual molecules, normalized as above. Molecular nitrogen-14 has $M = 28.013\,4$, and dry atmospheric air at sea level has $M = 28.965$.

relative phase – *of two sinusoidal signals at the same frequency* the amount by which one signal leads (or lags) the other. See PHASE LEAD, PHASE LAG. *Units* rad.

Note: Relative phase values that differ by an integer multiple of 2π are equivalent. For this reason, relative phase is commonly quoted modulo 2π, with values lying in the range $[-\pi, \pi]$ or $[0, 2\pi]$. For the reverse of this process, see PHASE UNWRAPPING.

relative total sound level – *in an auditorium* an equivalent term for SOUND STRENGTH. *Units* dB.

relative transmissibility – *of a vibration isolator excited by a displacement input* the ratio of the relative displacement amplitude, between the isolator output and input, to the amplitude of the input displacement. Compare TRANSMISSIBILITY. *Units* none.

relaxation – *following a disturbance* an exponential approach towards equilibrium, characterized by a time constant called the ***relaxation time***. See also MOLECULAR RELAXATION.

relaxation absorption – ABSORPTION (1) of sound in a medium due to relaxation processes. Examples are: (in air) absorption of audio-frequency sound due to vibrational relaxation of oxygen molecules in the presence of water vapour; (in seawater) absorption of sound below about 100 kHz due to chemical relaxation in dissolved magnesium sulphate.

relaxation dispersion – DISPERSION of sound in a medium due to relaxation processes. See also RELAXATION ABSORPTION.

relaxation frequency – the quantity $(2\pi\tau)^{-1}$, where τ is the RELAXATION TIME. *Units* Hz.

relaxation function – *for a linear viscoelastic material* abbreviation for STRESS-RELAXATION FUNCTION. *Units* Pa.

relaxation model – *of a linear viscoelastic material with time-independent properties* refers to a stress-strain relationship in which DEVIATORIC STRESS components are given by an integral of the form

$$\sigma(t) = \int_{-\infty}^{t} \Phi(t-\tau)\,\dot{\varepsilon}(\tau)\,d\tau,$$

where $\varepsilon(t)$ is the corresponding component of the DEVIATORIC STRAIN tensor. The STRESS-RELAXATION FUNCTION $\Phi(t)$ in the relaxation model takes the form of a sum of exponential decay terms,

$$\Phi(t) = 2[G_\infty e^{-\beta_\infty t} + (\Delta G)_1 e^{-\beta_1 t} + (\Delta G)_2 e^{-\beta_2 t} + \ldots],$$

in which G_∞ represents the **long-term shear modulus** and is usually associated with a vanishingly small decay constant ($\beta_\infty \to 0$). The sum $G_\infty + \Sigma(\Delta G)_i$, often denoted by G_0, represents the **short-term shear modulus** and determines the deviatoric strain response immediately after a load is applied.

relaxation spectrum – *for a linear viscoelastic material modelled by a continuous distribution of relaxation processes* the function $F(\lambda)$ in the following model equation for the STRESS-RELAXATION FUNCTION $\Phi(t)$ of the material:

$$\Phi(t) = \Phi_\infty + \int_0^\infty F(\lambda)\, e^{-t/\lambda}\, d\lambda.$$

Here t is time, and λ is a variable RELAXATION TIME. *Units* Pa s^{-1}.

relaxation time – the TIME CONSTANT associated with a RELAXATION process. *Units* s.

repetency – *for a quantity that varies sinusoidally with position in a specified direction* the spatial frequency, equal to $1/\lambda$ where λ is the WAVELENGTH in the direction concerned. Compare WAVENUMBER (1), (2). *Units* m^{-1}.

residual intensity index – an alternative term for PRESSURE–RESIDUAL INTENSITY INDEX. *Units* dB re $p_{\text{ref}}^2/I_{\text{ref}}$.

residual noise – *for purposes of environmental noise assessment* the AMBIENT NOISE (2) that is measured at a given location, when any specific noise sources being investigated have either been turned off, or suppressed to the point where their contribution is insignificant.

Note: Residual noise is commonly measured as an A-weighted L_{eq}.

residual noise level – the level of RESIDUAL NOISE, commonly expressed as an equivalent continuous sound pressure level, L_{eq}, with A-weighting applied. *Units* dB re $(20\ \mu\text{Pa})^2$.

resistance – the real part of an IMPEDANCE.

resistivity – *of a porous material* abbreviation for FLOW RESISTIVITY. *Units* Pa s m^{-2}.

resonance – a peak in the magnitude of the FREQUENCY RESPONSE FUNCTION that describes the FORCED RESPONSE of a system, as the frequency is varied.

For example, a single-degree-of-freedom system with viscous damping has a forced response that is described mathematically by the standard equation

$$m\ddot{x} + R\dot{x} + kx = F(t);$$

this equation gives the displacement $x(t)$ due to a force $F(t)$, where t is time and dots denote time derivatives. If we choose $F(t)$ proportional to $e^{j\omega t}$, then the complex FREQUENCY RESPONSE FUNCTION $\dot{x}/F$ (ratio velocity/force = mechanical admittance) has a maximum absolute value when $\omega = \sqrt{k/m} = \omega_0$, i.e. when the system is driven at its UNDAMPED NATURAL FREQUENCY. The ratio of the DISSIPATION RATE $\langle \dot{x}^2 \rangle R$ to the mean square force $\langle F^2 \rangle$ is also a maximum at this frequency.

Note: The related frequency response functions x/F (displacement/force = receptance) and $\ddot{x}/F$ (acceleration/force = accelerance) have maximum absolute values at $\omega = \omega_0(1 - 2\zeta^2)^{1/2}$ and $\omega = \omega_0(1 - 2\zeta^2)^{-1/2}$ respectively, where ζ is the DAMPING RATIO of the system.

resonance frequency – any frequency at which RESONANCE occurs, for a specified measure of the system response. *Units* Hz.

resonant frequency – alternative term for RESONANCE FREQUENCY; deprecated by some authors as ungrammatical. *Units* Hz.

response – *of a system* the departure from equilibrium that results from a given input. An alternative term is the *output* of the system.

response curve – a graph in which the square of a frequency response function magnitude is plotted against frequency. Compare BODE PLOT.

retarded time – *for signals received from an acoustic source* the time at which the signal was emitted, expressed as a function of the reception time and the source–receiver distance. In the case where both source and observer are stationary with respect to the medium, the retarded time $[t]$ is given by

$$[t] = t - r/c \quad (t = \text{reception time of the signal}).$$

Here r is the source–receiver distance and c is the speed of sound, assumed uniform. See also DOPPLER FACTOR, DOPPLER FREQUENCY RELATION. *Units* s.

reticulated foam – open-cell foam of plastic or other solid material, whose skeleton has the form of a three-dimensional net.

reverberance – the sense of reverberation in a space, perceived either directly or in a recording. It can be conveyed purely as a consequence of the temporal smearing that results from reverberation; alternatively it can arise from spatial effects that are perceived binaurally.

reverberant field – *in an enclosed space* the part of the sound field contributed by reflections from the boundaries. Provided the acoustic space is lightly damped, i.e. the total absorbing area is small in comparison with the actual surface area of the enclosure, the reverberant field may usefully be regarded as representing the contribution of resonant modes to the total response.

reverberant level – *in room acoustics* a sound pressure level obtained by spatially averaging the mean square pressure in a REVERBERANT FIELD, and expressing the result as a LEVEL. See also AVERAGE SOUND PRESSURE LEVEL. *Units* dB re $(20 \, \mu\text{Pa})^2$.

reverberation – (1) multiple reflection of sound waves in a live room, particularly where the arriving echoes are too closely-spaced to be distinguished apart by ear. Also, multiple reflection of vibrational waves in a lightly-damped structure.

reverberation – (2) the build-up of energy within a lightly-damped system due to multiple internal reflection, when the system is continuously excited. The system reaches a steady state when the rate of dissipation equals the power supplied; the steady-state energy is inversely proportional to the system damping.

reverberation – (3) abbreviation for REVERBERATION DECAY or TERMINAL REVERBERATION.

reverberation – (4) *in underwater acoustics* a continuous unwanted echo caused by scattering of a transmitted pulse from rough boundaries or volume inhomogeneities.

reverberation chamber – an equivalent term for REVERBERATION ROOM.

reverberation decay – the slow decay of reverberant-field energy, within a lightly-damped system, that occurs after the excitation is removed. Related terms special to room acoustics are TERMINAL REVERBERATION and RUNNING REVERBERATION.

reverberation level (RL) – (1) *in active sonar* a measure of the background noise from REVERBERATION (4) that arrives in the same time interval and frequency band as the desired signal, after signal processing. *Units* dB re $1 \, \mu\text{Pa}^2 \, \text{s}$.

reverberation level (RL) – (2) *in underwater acoustics* the SOUND PRESSURE LEVEL of background noise due to REVERBERATION (4). *Units* dB re $1 \, \mu\text{Pa}^2$.

reverberation radius – *for a source in a room* the distance, measured from the centre of the sound source, at which the direct and reverberant levels are equal. *Units* m.

reverberation room – a room with low absorption (i.e. the total ROOM ABSORPTION is small in comparison with the actual surface area) and large volume (i.e. the smallest dimension is large in comparison with the sound wavelength), designed to promote the production of a nearly DIFFUSE reverberant sound field.

reverberation time – the time taken for the energy in an initially-steady reverberant sound field to decay by 60 dB. If the slope of the decay curve, in dB per second, is Δ, the reverberation time is $T = 60/\Delta$. *Units* s.

Note (1): In room acoustics, the slope is normally obtained by fitting a straight line between the -5 dB and -35 dB points relative to the initial level; the result is denoted by T or T_{30}. An alternative is to use the -5 dB and -25 dB points, in which case the result is denoted by T_{20}.

Note (2): The reverberation time of a room is the average of duplicated measurements with several source-receiver combinations. For measurement techniques, see INTERRUPTED NOISE METHOD, INTEGRATED IMPULSE RESPONSE METHOD.

reverberation time equations – a class of semi-empirical equations that relate the reverberation time of a live room to the room volume and the properties of the room boundaries, under conditions where a large number of resonant room modes are excited.

Common to all such equations is the concept of ROOM ABSORPTION, regarded as a property of a particular room. The reverberation time, T, and the room absorption, A, are related by

$$T = \frac{55.26 V}{Ac}, \qquad (55.26 = 24 \ln 10)$$

where c is the speed of sound in the room. Two common approximations for A, due to Sabine and Eyring, are described below.

(1) In the *Sabine equation* (**W C SABINE 1898, 1900**),

$$A = S\bar{\alpha}.$$

Here S is the total surface area of the room boundaries, and $\bar{\alpha}$ is the average value of the statistical absorption coefficient:

$$S\bar{\alpha} = \sum_i \alpha_i S_i$$

(summation over elements $i = 1, 2, 3, \ldots$, etc. of the boundary).

(2) In the *Eyring equation* (**C F EYRING 1930**),

$$A = S \ln\left(\frac{1}{1-\bar{\alpha}}\right),$$

where $\bar{\alpha}$ is defined above. This is also known as the ***Norris–Eyring equation*** (**R F NORRIS 1930–32**).

reversible process – *for a given system* a process in which no part of the system departs significantly from thermodynamic equilibrium at any stage. Such a process can be reversed completely by reversing the motion of the system boundary, and also reversing all energy flows across the boundary; the system then returns to its original state.

Reynolds number – *in fluid flow* a measure of the relative contributions made by REYNOLDS STRESSES and viscous stresses to the rate of momentum transfer across fixed surfaces. In a flow characterized by length scale L and velocity scale U, the Reynolds number is defined as

$$\mathrm{Re} = \frac{\rho U L}{\mu},$$

where μ is the viscosity of the fluid and ρ is the density (**O REYNOLDS 1883**). The name was proposed by Sommerfeld in 1908. *Units* none.

Note: In any particular flow the onset of instability, and the process of transition from laminar to turbulent flow, both require the Reynolds number to exceed a minimum value; e.g. steady flow in a pipe of diameter d will not become turbulent unless $\mu^{-1}\rho u_m d$ is greater than 2000 (where u_m = average velocity over the pipe cross-section). See also AEOLIAN TONES.

Reynolds stress – *in unsteady fluid flow* the product $\rho u_i u_j$; here ρ is the fluid density, and u_i, u_j are fluid velocity components in any of the three orthogonal coordinate directions labelled by $i, j = (1, 2, 3)$. In physical terms, $\rho u_i u_j$ represents the transfer rate (advective flux) of j-component fluid momentum, across an imaginary surface x_i = constant, per unit surface area; the mechanism of momentum transfer is convection. *Units* Pa.

Note: The double divergence of $\rho u_i u_j$ represents a source term in Lighthill's ACOUSTIC ANALOGY for aerodynamic sound generation. Another key application is to turbulent flows: the time-average Reynolds stress $\overline{\rho u'_i u'_j}$, based on the fluctuating local velocity components $u_i' = u_i - \bar{u}_i$ where $\bar{u}_i$ is the time-average velocity, appears in the time-averaged momentum equation as the negative of an effective stress. In contrast to the full Reynolds stress, $\overline{\rho u'_i u'_j}$ represents the mean momentum flux due to turbulent eddies.

Riccati equation – a type of first-order nonlinear differential equation, encountered in acoustics as the governing equation for the impedance (or admittance) in a standing-wave system. It can be derived from two coupled linear differential equations of the form

$$\left. \begin{array}{l} P'(x) = AQ(x) + CP(x) \\ Q'(x) = BP(x) + DQ(x) \end{array} \right\} \quad (A, B, C, D \text{ known functions of } x).$$

Rearranging these equations into a single equation for the complex ratio $P/Q = z(x)$ gives

$$\frac{dz}{dx} = A - Bz^2 + (C - D)z,$$

which is the general Riccati equation (**J Riccati** 1723). In acoustics, $P(x)$ and $Q(x)$ typically represent the complex pressure and volume velocity in a one-dimensional system, and $z(x)$ represents the acoustic impedance.

Riemann invariants – *in one-dimensional unsteady compressible flow* two quantities, analogous to $u \pm p/\rho_0 c_0$ in linear acoustics, that are invariant along the forward and backward characteristics C_+, C_- respectively (**G F B Riemann** 1860). They are defined by

$$J_+ = u + \int_{P_0}^{P_0+p} dP/\rho c, \qquad J_- = u - \int_{P_0}^{P_0+p} dP/\rho c.$$

Here P denotes the absolute pressure and u the fluid velocity; p is the pressure perturbation; ρ and c are the local density and sound speed, both evaluated along an isentropic compression curve through P_0, and subscript 0 denotes the undisturbed state. *Units* m s^{-1}.

Note (1): C_+, C_- are defined by $dx/dt = u \pm c$.

Note (2): The definitions of J_+, J_- are not standardized. Often a factor $\tfrac{1}{2}$ is introduced, and some authors reverse the sign of J_-.

right-handed system – see RECTANGULAR COORDINATES.

rigid body – an idealized material object that can be translated and rotated, but does not deform under applied loads. Its shape and size therefore remain fixed.

rigid boundary – a boundary that imposes zero normal velocity on an adjacent acoustic medium. Also known as a **hard boundary**. Compare NEUMANN BOUNDARY CONDITION.

Rijke tube – a vertical cylindrical tube, open at both ends, that contains a heat source towards its lower end. Thermoacoustic oscillations are set up in the tube in accordance with RAYLEIGH'S CRITERION: the acoustic standing-wave field modulates the upward flow through the tube (which acts like a chimney), causing the heat output of the source to vary in phase with the local acoustic pressure (**P L Rijke** 1874).

ring frequency – *of a thin-walled pipe or circular cylindrical shell* the frequency defined by

$$f_R = \frac{c_P}{\pi d},$$

where d is the pipe diameter (mean of inner and outer values) and c_P is the PLATE LONGITUDINAL-WAVE SPEED of the wall material. *Units* Hz.

ringing – *of a system after the excitation is turned off* see TRANSIENT (3).

Ritz method – see RAYLEIGH–RITZ METHOD.

rms – abbreviation for ***root mean square***. The rms value of a fluctuating quantity q is the square root of $\langle q^2 \rangle$, where the angle brackets denote an AVERAGE (temporal, spatial or both).

rms deviation – the square root of the MEAN SQUARE DEVIATION.

Robin boundary condition – an alternative term for IMPEDANCE BOUNDARY CONDITION.

rod longitudinal-wave speed – the propagation speed at which free longitudinal waves, with wavelength long compared with the rod's transverse dimensions, travel along a uniform thin rod when the sides of the rod are stress-free. It is a property of the rod material, given by $c_L = \sqrt{E/\rho}$ where E is the YOUNG'S MODULUS and ρ is the density. Compare PLATE LONGITUDINAL-WAVE SPEED. *Units* m s^{-1}.

room absorption – a measure of the total EQUIVALENT ABSORPTION AREA of a room. In mathematical terms it is the factor A in the equation $W_{abs} = AcE/4V$, which relates the rate at which acoustic energy is absorbed from a diffuse sound field in the room to the total reverberant energy in the field, E, for a room of volume V filled with air (or any fluid) of sound speed c. *Units* m^2.

Note: The room absorption may be divided into the VOLUME ABSORPTION, A_{vol}, which accounts for energy absorbed within the body of the room by the air, and the BOUNDARY ABSORPTION, A_b, which accounts for the remainder of the absorbed energy. Typically A_{vol} becomes significant towards the upper end of the audio frequency range, in reverberation rooms and in very large auditoria.

room constant – the ratio between the SOURCE POWER in a room and the diffuse-field normal intensity, or ACOUSTIC IRRADIANCE, at the boundaries. In mathematical form, the room constant $\bar{A}$ is defined by

$$W_s = \bar{A}cE/4V,$$

where W_s is the steady-state source power required in a room of volume V, in order to maintain a diffuse sound field of reverberant energy E; the sound speed in the room is denoted by c. *Units* m^2.

Note: The source power exceeds the rate of energy loss from the reverberant field, because some power is absorbed at the first reflection by the boundaries; accordingly the room constant is greater than the ROOM ABSORPTION. In live rooms with many resonant modes excited, the field is approximately diffuse and $\bar{A} \approx Ae^{A/S}$, where A is the room absorption and S is the surface area of the room boundaries.

root mean square – *of a fluctuating quantity that varies with time, or position, or both* the square root of the mean square value; abbreviated ***rms***.

roots – the roots of the equation $f(x) = 0$ are the values of the variable x that satisfy the equation. Equivalently, the roots are the ZEROS of $f(x)$.

rotating diffuser – a sound-reflecting structure that is mounted on a turntable in a reverberation chamber, and slowly rotated. Time averaged measurements with a rotating diffuser present are equivalent to measurements averaged over an ensemble of room geometries.

rotational impedance – *of a mechanical system* the complex ratio of a moment to an angular velocity, for single-frequency rotational excitation at a point. Also known as ***torsional impedance***. *Units* N m s rad^{-1}.

rotational receptance – *of a mechanical system* the complex ratio of an angular displacement to an applied moment or torque, for single-frequency rotational excitation at a point. *Units* rad m^{-1} N^{-1}.

rotational stiffness – *of an elastic lumped element* the rate of change of torque with relative angular displacement. *Units* N m rad^{-1}.

Note: Compare TORSIONAL STIFFNESS, which has different dimensions.

rotational velocity field – *in fluid dynamics* a velocity field whose curl is non-zero. See VORTICITY.

rotational waves – an equivalent term for SHEAR WAVES. Rotational waves involve local rotation of the medium, with no compression or expansion; this is expressed mathematically by

$\nabla \times \mathbf{u} \neq 0, \quad \nabla \cdot \mathbf{u} = 0 \quad (\mathbf{u} =$ particle displacement or velocity).

rotor-alone tone – *from an axial-flow fan or propeller* sound radiated at the blade-passing frequency and its higher harmonics, as a result of the thickness and steady loading of the rotor blades. Sometimes called ***Gutin***

noise (L GUTIN 1936). The radiation process is highly inefficient at low tip Mach numbers; compare ROTOR–STATOR INTERACTION NOISE.

rotor–stator interaction noise – *from an axial-flow fan* sound radiated at the blade-passing frequency and its higher harmonics, as a result of interaction between the flow fields of neighbouring rotor and stator blade rows.

Note: Rotor–stator interaction is particularly important for subsonic fans (tip relative Mach numbers less than 1), since the ROTOR-ALONE TONE is then cut off. The interaction process breaks the symmetry of the rotating pressure field attached to the rotor, and generates modes of lower azimuthal order that have higher circumferential phase speeds than the rotor. As a result, these modes can radiate efficiently even when the rotor is subsonic.

round-off error – see QUANTIZATION ERROR.

run – a sequence of consecutive observations in a SAMPLE that either have the same numerical value, or have some other property in common. For example, 00000111 is a run of 5 zeros followed by a run of 3 ones.

running reverberation – acoustic REVERBERATION heard during continuous speech or music. Running reverberation is responsible for sounds being heard against the decaying background of sounds that have preceded them.

S

sabin – see METRIC SABIN.

Sabine absorption – *of a room* an equivalent term for the ROOM ABSORPTION. *Units* m^2.

Sabine absorption coefficient – *of a sample of sound-absorbing material* the STATISTICAL ABSORPTION COEFFICIENT deduced from reverberation-time measurements via the Sabine equation (see REVERBERATION TIME EQUATIONS). Specifically, the Sabine absorption coefficient of a test sample of surface area S_{test} is

$$\alpha_s = (A_1 - A_0)/S_{\text{test}} + A_0/S,$$

where A_0 is the ROOM ABSORPTION of the test room without the sample, A_1 is the room absorption measured with the sample installed, and S is the total area of the room boundaries. *Units* none.

Note: In the equation above, the sample is assumed to cover up a fraction (S_{test}/S) of the absorption originally present in the test room.

Sabine acoustics – an alternative term for STATISTICAL ROOM ACOUSTICS, as applied to the multimode response of reverberant acoustic spaces. In such spaces, under steady-state conditions, an approximately DIFFUSE sound field prevails (**W C SABINE 1898**).

Sabine average absorption coefficient – *for a room* the absorption coefficient averaged over the room surfaces, $\bar{\alpha}$, deduced from the Sabine equation (see REVERBERATION TIME EQUATIONS). Thus $\bar{\alpha} = A/S$, where A is the total contribution of the boundaries to the ROOM ABSORPTION and S is the room surface area. *Units* none.

Sabine equation for reverberation time – see REVERBERATION TIME EQUATIONS.

salinity – a measure of the dissolved salts content of seawater. The modern definition relates salinity to the electrical conductivity of seawater, as measured at 15°C and 1 atm pressure. The numerical salinity value thus obtained is very close to the mass fraction of total dissolved salts in parts per thousand, which is the original (but less practical) definition. *Units* none.

Note: Salinity has a significant effect on the speed of sound in the ocean; for details see SOUND SPEED IN SEAWATER.

sample – a data set of finite size produced by SAMPLING.

sample correlation coefficient – *for a sample of N paired observations* a normalized measure of the extent to which the two observed variables are correlated. The standard definition, for discrete variables $x[i] = x_i$ and $y[i] = y_i$, is

$$r_{xy} = \frac{1}{(N-1)s_x s_y} \sum_{i=1}^{N}(x_i - m_x)(y_i - m_y).$$

Here s_x, m_x are respectively the sample standard deviation (i.e. the square root of the SAMPLE VARIANCE) and the MEAN of $x[i]$, and likewise for $y[i]$.

Note: The sample correlation coefficient, and the related CORRELATION COEFFICIENT for two random variables, are to be distinguished from the CROSS-CORRELATION COEFFICIENT; the latter is defined as a function of time delay, for a pair of continuous time-stationary random functions.

sampled signal – a DIGITAL SIGNAL produced by sampling a CONTINUOUS SIGNAL.

sample function – a waveform or SIGNAL that is one possible realization of a STOCHASTIC PROCESS. An ENSEMBLE is the totality of all possible sample functions for a particular random process.

sample mean – *in statistics* the order-1 SAMPLE MOMENT about the origin.

sample moment – *in statistics* for a set of N data values $(x_1, x_2, \ldots, x_N)$, the sample moment of order k about $x = 0$ is defined as the normalized sum

$$M_k = \frac{1}{N}\sum_i x_i^k;$$

compare MOMENT.

An equivalent name for M_k is the **kth sample moment** about $x = 0$. Sample moments about any other value, e.g. $x = c$, are defined in the same way except that $(x_i - c)$ replaces x_i in the equation above.

Note: For the special case $k = 2$, see SAMPLE VARIANCE. Also compare POPULATION MOMENTS; as the sample size N tends to infinity, each sample moment converges to the corresponding population moment.

sample points – the outcome of discretely SAMPLING a time-waveform or sample function $x(t)$; the term refers to the individual pairs of (x_i, t_i) values produced by this process.

sample statistics – STATISTICS (2) based on a particular data SAMPLE.

sample values – *for a random variable* the individual data values obtained by sampling the random variable.

sample variance – the second MOMENT of a set of SAMPLE VALUES x_i ($i = 1$ to N), taken about the MEAN and normalized by $N - 1$. The defining equation is

$$s^2 = \frac{\sum(x_i - m_x)^2}{N - 1},$$

where m_x is the sample mean $(1/N)\sum x_i$. The sample variance s^2 defined in this way is an unbiased estimator of the population variance.

Note: The **sample covariance** is defined in a similar way; compare COVARIANCE.

sampling – the process of extracting a finite number of data points from a RANDOM SIGNAL, or measuring a finite number of values from a RANDOM VARIABLE. The intention in either case is that the sample shall be statistically representative of the underlying signal or random variable. Usually a CONTINUOUS SIGNAL is sampled at equal intervals of time: see SAMPLING FREQUENCY, SAMPLING THEOREM.

sampling distribution – a PROBABILITY DENSITY FUNCTION for which the underlying random variable is a STATISTIC, formed by SAMPLING another random variable.

sampling frequency – *for a continuous time-dependent signal* the number of SAMPLE POINTS acquired per unit time; i.e. the reciprocal of the time-interval used for SAMPLING the signal. *Units* Hz.

sampling interval – the interval of time between successive SAMPLE POINTS, when a CONTINUOUS SIGNAL is discretely sampled. *Units* s.

sampling rate – an equivalent term for SAMPLING FREQUENCY. *Units* Hz.

sampling theorem – *for a continuous time-dependent signal* states that if the original CONTINUOUS SIGNAL is to be sampled to produce a DIGITAL SIGNAL with equivalent spectral statistics, the SAMPLING FREQUENCY must be at least twice the highest frequency present in the original signal, in order to avoid ALIASING. Also known as the **Nyquist theorem** or **Shannon sampling theorem**.

saturation – *of a finite-amplitude sound field* describes a situation where the pressure waveform received at a finite distance from a sound source is unaltered by increasing the source amplitude. Any additional energy put in at the source is entirely dissipated between the source and the receiver.

Periodic waveforms always exhibit saturation if the source amplitude is sufficiently large; see SAWTOOTH WAVE.

Note: Saturation can also occur under transient conditions. If the initial time-waveform of a transient pressure pulse contains at least one downward zero-crossing such that $\int p \, dt$ is *positive* between the start of the pulse and the zero-crossing point, and is *negative* from that point to the end of the waveform, the pulse will evolve at sufficiently large amplitudes into an N-WAVE by the time it reaches the observation point.

sawtooth wave – (1) a waveform characterized by a sequence of adjacent triangular pulses or "teeth". In a *regular sawtooth wave*, the teeth are identical and form an infinite sequence.

Note (1): A limiting case of particular interest in acoustics has each sawtooth starting with a vertical jump, rather than a ramp. See SAWTOOTH WAVE (2).

Note (2): Such a sawtooth wave in the time domain has downward-sloping ramp-function elements of the form

$$x_i(t) = -a(t - t_i), \qquad (t_i - d_i < t < t_{i+1} - d_{i+1}).$$

Here t_i is the zero-crossing time for the ith ramp segment, a is the downward slope of the ramp function (usually the same for all i), and d_i is the time by which the start of segment i leads the zero crossing.

sawtooth wave – (2) *in nonlinear acoustics* a progressive wave of finite amplitude whose pressure waveform consists of a series of shocks, separated by downward-sloping ramps; compare SAWTOOTH WAVE (1).

Note (1): A periodic plane-progressive sound wave, driven with sufficient amplitude that the GOL'DBERG NUMBER is much larger than 1, distorts and becomes a regular sawtooth wave at propagation distances $x \gg \bar{x}$, where $\bar{x}$ is the SHOCK-FORMATION DISTANCE.

Note (2): A transient pulse whose pressure waveform at $x = 0$ is one cycle of a negative sine wave, with $p = -p_0 \sin(2\pi t/T)$ for $0 < t < T$, distorts into a *single sawtooth* that begins and ends with a zero crossing. (But note that if the sign is changed from $-$ to $+$, the otherwise identical transient distorts into an N WAVE.)

In both of these examples, the sawtooth profile of the acoustic waveform indicates SATURATION: the peak pressure amplitude under these conditions is independent of source amplitude, and is predicted by weak-shock theory as

$$p_{\max} = \frac{\rho c^3}{2\beta} \frac{T}{x}.$$

Here β is the coefficient of nonlinearity, ρ is the fluid density, and c is the sound speed.

SBS – *in underwater acoustics* abbreviation for the SURFACE BACKSCATTERING STRENGTH of a rough ocean surface. *Units* dB.

scalar – a quantity that has magnitude but no direction, and can therefore be represented by a single number. The components of a vector, or the elements of a matrix, are scalars.

scalar potential – see POTENTIAL.

scalar wave equation – see WAVE EQUATION.

scatter diagram – see CORRELATION.

scattered field – *in acoustics* the additional sound field produced by a scattering object; it is the difference between the *total field* that exists when the object is present, and the *incident field* which is what remains when the object is removed. The field variable may be any linear acoustic variable: e.g. *scattered pressure*, *scattered particle velocity*.

scatterer – *in acoustics* abbreviation for **acoustic scatterer**, meaning a passive obstacle (or an inhomogeneity in the medium) that produces a SCATTERED FIELD when irradiated by sound waves.

scattering – reradiation or reflection of an incident wave field, for example by boundaries or inhomogeneities in the medium. Where surfaces larger than a wavelength are involved, scattering often refers specifically to reradiation in directions other than the one corresponding to SPECULAR REFLECTION. Compare DIFFRACTION.

scattering coefficient – see BULK $\sim$, SURFACE $\sim$, VOLUME $\sim$.

scattering cross-section – *of an object in a plane-wave incident field* the ratio of the scattered power, W_{scat}, to the incident acoustic intensity, I_{inc}, when the object is irradiated by plane progressive waves at a given frequency and direction of incidence. In symbols, the scattering cross-section is $\sigma_{\text{scat}} = W_{\text{scat}}/I_{\text{inc}}$. *Units* m^2.

scattering differential – *in underwater acoustics* an equivalent term for SCATTERING STRENGTH. If no scattering direction is specified, the scattering differential is defined as an average over all directions. Thus an object of SCATTERING CROSS-SECTION σ has a scattering differential (in the average sense) numerically equal to

$$\Delta_s = 10\log_{10}\frac{\sigma}{4\pi r_{\text{ref}}^2},$$

where r_{ref} is the reference distance ($r_{\text{ref}} = 1$ m). *Units* dB @ 1 m, or dB re 1 m^2.

scattering matrix – (1) *for a junction connecting two or more single-mode waveguides* an abbreviation for MULTIPORT SINGLE-MODE SCATTERING MATRIX. *Units* none.

scattering matrix – (2) *for a one-port acoustical system, e.g. a termination in a multimode acoustic waveguide* see REFLECTION COEFFICIENT MATRIX. *Units* none.

scattering matrix – (3) ◆ *for a two-port acoustical system or junction connecting two waveguides* an abbreviation for MULTIMODE SCATTERING MATRIX. *Units* none.

scattering strength – *of a scattering object, in underwater acoustics* a generic term for the SOURCE LEVEL of the scattered field radiated in a specified direction from the object, minus the sound pressure level of the incident plane wave in dB re 1 μPa^2. The scattering strength may refer specifically to the backscattering direction (in this case the scattering strength of an object is called the TARGET STRENGTH); otherwise, the prefix BISTATIC is used to indicate scattering in an arbitrary direction.

Scattering strengths may also be normalized per unit scattering area or volume: thus scattering by the sea bed is described by a BISTATIC BOTTOM SCATTERING STRENGTH, or a BOTTOM BACKSCATTERING STRENGTH. Likewise, scattering by the ocean surface or by a shoal of fish is described by similar terms in which *bottom* is replaced by *surface* or *volume*. All these scattering strengths are conventionally expressed in logarithmic form.

Scholte wave – *at a plane solid–fluid interface* a nondispersive interface wave that propagates along a plane boundary between two semi-infinite regions, one region being a fluid and the other a solid. Scholte waves propagate with a speed lower than either the fluid compressional wave speed or the solid shear wave speed; they are evanescent away from the interface in both media (**J G SCHOLTE** 1949).

Note: In the limit of low fluid density, Scholte waves do not exist; they are essentially acoustic surface waves at an elastic boundary. From a mathematical viewpoint, they correspond to a pole in the acoustic pressure reflection coefficient. Compare LEAKY RAYLEIGH WAVE.

Schroeder diffuser – a generic name for a class of acoustic reflector in which a plane boundary is modified by a set of parallel recesses, whose depth (or width) varies according to a mathematical sequence (**M R SCHROEDER 1979, 1984**). Schroeder diffusers can be tailored to provide a variety of non-specular reflection characteristics over a specified range of frequencies. Specific examples are the MAXIMUM LENGTH SEQUENCE DIFFUSER and the QUADRATIC RESIDUE DIFFUSER.

Note: If the recesses have sharp edges, significant absorption may result from the high particle velocities in the edge regions. The diffuser absorption can be enhanced by adding a porous facing layer (e.g. gauze), with a specific flow resistance as low as 0.1 times the characteristic impedance of air.

Schroeder frequency – *for pure-tone sound fields in live rooms* the frequency above which significant overlap occurs, on average, between adjacent modal resonances; the frequency response of the room above this frequency is no longer dominated by separate resonant peaks (**M R SCHROEDER 1954; M R SCHROEDER, K H KUTTRUFF 1962**). The criterion for modal overlap may be stated either as $B_h n(f) \geq 3$ or (with equivalent accuracy) as $B_e n(f) \geq 4$; here $n(f)$ is the MODAL DENSITY, and B_h, B_e are the bandwidths (half-power and effective) of a typical modal response curve whose resonance is close to f. Using the second criterion gives the Schroeder frequency as follows:

$$f_{\min}/c \approx 4/\sqrt{\pi A} - S/(16V)$$

(A = room absorption, S = surface area of boundaries, V = volume). Also known as the **Schroeder large-room frequency**. *Units* Hz.

Schroeder plot – *for a room* a plot of the reverberation decay curve obtained by the INTEGRATED IMPULSE RESPONSE METHOD, with filtering applied either to the input or to the output signal. The final result, after backward integration of the squared output, is a band-limited decay curve. Compare ENERGY-TIME CURVE, which provides similar information for a room but reveals the presence of distinct reflections.

scientific notation – *for real decimal numbers* the representation of numbers in the form $a \times 10^n$, where a is a number with magnitude between 1 and 10, and n is an integer. For example, 1200 in scientific notation is 1.2×10^3.

screen – a type of sound BARRIER used indoors, typically in open-plan offices, to provide acoustic privacy.

seat dip effect – *in an auditorium* the attenuation of sound, especially at frequencies in the range 100–400 Hz, that occurs when sound passes over audience seating, with or without audience present. The effect occurs both for the direct sound and also for reflections that pass over the audience. An equivalent term is **attenuation due to grazing incidence**.

second sound – a longitudinal thermal wave that propagates in bulk liquid ^{4}He at temperatures between about 0.5 K and 2.2 K, and also in ^{3}He–^{4}He mixtures at similar temperatures. The wave is characterized by variations in the proportions of normal and superfluid components at a fixed point; the total mass density and momentum density remain constant, while the temperature and the entropy per unit volume vary with time. Second sound displays many wave-type phenomena, including diffraction, nonlinear distortion, and shock formation. See also SUPERFLUIDITY.

seepage velocity – *through a layer of porous material* the volume flowrate per unit total cross-sectional area. At the face of the layer, the seepage velocity equals the FACE VELOCITY. See also DARCY'S LAW. *Units* m s^{-1}.

seismic exploration – the use of artificially generated stress waves in the Earth's crust to prospect for oil, gas, or other minerals. Detection of deposits makes use of reflection, refraction, or channelling along seams of low wave speed (compare SOFAR propagation in the ocean). The typical frequency range used in seismic exploration is 1–10 Hz.

seismic mass – an inertial element that is used to convert an acceleration into a force, e.g. in an accelerometer.

seismic waves – stress waves in the Earth, produced by explosions, earthquakes, or seismic exploration equipment.

seismogram – a continuous record of vibrations in the Earth's crust, particularly at frequencies low enough to propagate over several kilometres (typically below 5 Hz).

seismology – (1) the science of stress wave propagation in the Earth; (2) the study of earthquakes in particular.

seismometer – a generic term for any instrument that measures seismic waves, either in the Earth's crust (see also GEOPHONE) or underwater (see HYDROPHONE).

self-adjoint differential equation – any equation for the unknown function u that has the general form

$$L(u) \equiv \operatorname{div}(a \operatorname{grad} u) + b\, u = 0.$$

The differential operator L defined above is said to be ***self-adjoint***. Here the independent variable on which u depends is a VECTOR (1), e.g. position $\mathbf{x}$ in three-dimensional space; the coefficients a and b are given, but are allowed to vary with $\mathbf{x}$. The HELMHOLTZ EQUATION is a special case of the equation above; it follows by taking $a = 1$ and $b = \kappa^2$ (a constant).

self-adjoint differential operator – see the previous entry.

self-adjoint matrix – an equivalent term for a HERMITIAN MATRIX.

self-excited oscillation/vibration – alternative terms for SELF-INDUCED OSCILLATION/VIBRATION.

self-exciting system – an unstable nonlinear system whose free oscillations approach a LIMIT CYCLE. Also known as a *self-sustaining system*.

self-impedance – (1) *of a particular mode of vibration of a fluid-loaded surface* one of the diagonal terms in the MODAL RADIATION IMPEDANCE matrix; (2) *at a given cross-section in a multimode waveguide* one of the diagonal terms in a MODAL IMPEDANCE MATRIX or TERMINATION IMPEDANCE MATRIX. Units Pa s m^{-1}.

self-induced oscillation/vibration – oscillations that occur spontaneously, in a system that exhibits instability. See STABILITY, LIMIT CYCLE.

Note: The instability may be either linear or nonlinear; in either case, the growth in amplitude is eventually limited by nonlinearity.

self-noise – (1) *in underwater acoustics* see SONAR SELF-NOISE.

self-noise – (2) *in aeroacoustics* the broadband noise radiated from a fan rotor or stator in the absence of inflow disturbances, or from a single airfoil in a smooth steady incident flow. Compare AIRFRAME NOISE, ROTOR-ALONE TONE.

self-sustaining system – an equivalent term for SELF-EXCITING SYSTEM.

semi-anechoic room – *for measurement of sound power output* a room in which reflections from the boundaries are partially suppressed by absorbing treatment. When the room is excited by a steady-state source, e.g. for sound power measurement, the sound field contains a residual reflected component that has to be corrected for in estimating the source power. Compare HEMI-ANECHOIC ROOM.

semicircular canals – three fluid-filled semicircular passages in the INNER EAR, that lie in approximately orthogonal planes and are used to sense angular acceleration of the head. See VESTIBULAR SYSTEM.

sensitivity – *of a linear transducer* the ratio of the output magnitude to the input magnitude, at a specified frequency.

sensitivity analysis – a numerical investigation in which parameters are varied systematically, in order to establish their relative influence on a particular output variable.

sensorineural hearing loss – HEARING LOSS due to a lesion or disorder of the inner ear or of the auditory nervous system.

separable solutions – *of a partial differential equation* solutions whose mathematical form is a product of two or more factors, with each factor containing only one of the independent variables.

separation – *in fluid dynamics* fluid flow along a solid surface is said to separate when the direction of the flow ceases to follow the surface. The onset of separation is indicated by the wall shear stress falling to zero.

separation impedance – *of an acoustically-driven plate, partition, or thin-shell structure* an equivalent term for the DIFFERENTIAL IMPEDANCE of the partition. It is defined as the complex ratio

$$z_t = \frac{p_1 - p_2}{\dot{w}};$$

here p_1 and p_2 are the pressures on the two sides of the partition, and $\dot{w}$ is the normal velocity of the partition directed from side 1 to side 2. Also known as ***transmission impedance***. *Units* Pa s m^{-1}.

sequence – a set of terms arranged in a specified order. A sequence may be finite or infinite.

series – a set of terms arranged in a specified order and summed. A series may be finite or infinite. If the first N terms are used as an approximation to the sum of the series, the series is said to be ***truncated*** after N terms.

sgn – symbol for signum function. The function sgn x, where x is real, is defined by

$$\operatorname{sgn} x = \begin{cases} -1 & (x < 0), \\ 1 & (x > 0). \end{cases}$$

shadow zone – *for sound radiated from a specified source position* a region into which RAYS emitted by the source cannot penetrate. The ray paths may be blocked by an opaque object or boundary, or be deflected away from the shadow zone by refraction. See also CREEPING WAVES.

shaft order – (1) a frequency that is an integer multiple of the shaft rotational speed in an engine; (2) sound or vibration occurring at such a frequency.

shaker – a vibration generator designed to apply an oscillatory force to a test object.

Shannon frequency – equivalent term for the NYQUIST FREQUENCY or FOLDING FREQUENCY of a discretely sampled signal.

Shannon sampling theorem – see SAMPLING THEOREM.

shape factor – *for a porous material* the ratio Λ_{v0}/Λ_v, where Λ_v is the CHARACTERISTIC LENGTH FOR VISCOUS EFFECTS, and

$$\Lambda_{v0} = \left(\frac{8\alpha_\infty \mu}{\sigma \phi}\right)^{1/2} \quad (\phi = \text{porosity})$$

is a reference value. In the expression above, σ is the FLOW RESISTIVITY of the material, μ is the fluid viscosity, and α_∞ is the frame TORTUOSITY. *Units* none.

Note: For a material with cylindrical pores of circular cross-section, the shape factor is 1.

shear modulus – *of a solid* the ratio of the DEVIATORIC STRESS to the DEVIATORIC STRAIN. Symbol G. *Units* Pa.

shear number – *for oscillatory viscous flow in a pipe* a dimensionless parameter given by

$$\text{Sh} = \sqrt{\frac{\nu}{\omega a^2}} = \frac{1}{\sqrt{2}} \frac{\delta}{a}.$$

Here ω is the angular frequency of the flow oscillation, a is the radius of the pipe internal cross-section, and ν is the kinematic viscosity of the fluid; $\delta = \sqrt{2\nu/\omega}$ is the viscous penetration depth. *Units* none.

Note: The shear number as defined above is the square root of the WOMERSLEY PARAMETER. Some authors invert the definition to read $\text{Sh} = \sqrt{2}a/\delta$, which then makes the shear number proportional to the square root of the STOKES NUMBER.

shear-thickening fluid – *in viscoelasticity* a fluid whose apparent viscosity increases as the STRAIN RATE increases.

shear-thinning fluid – *in viscoelasticity* a fluid whose apparent viscosity decreases as the STRAIN RATE increases.

shear waves – *in a fluid or solid* waves that involve local rotation of the medium, without compression or expansion; also known as ROTATIONAL WAVES. Plane shear waves are TRANSVERSE WAVES, meaning that the particle displacement is transverse to the propagation direction. See also POLARIZATION.

Note: Shear waves in lossless elastic solids propagate without attenuation. Inviscid fluids do not support shear waves, but viscous fluids do (although they are heavily damped). Superfluid helium-3 (^{3}He–B) is unique among fluids in supporting shear waves with relatively low damping, but the phenomenon occurs only over a narrow temperature range.

shear wave speed – *in a uniform isotropic elastic medium* in a bulk medium, plane shear waves of small amplitude travel at speed b given by $b = \sqrt{G/\rho}$, where G is the SHEAR MODULUS and ρ is the density of the material. *Units* m s^{-1}.

shielding – partial sound isolation achieved by placing a barrier between a sound source and receiver, so that the receiver is in a SHADOW ZONE.

shipping noise – *in underwater acoustics* the contribution to AMBIENT NOISE (2) from distant shipping; its frequencies typically lie in the range 50–500 Hz.

shock – (1) transient loading applied to a system in the form of an impulsive force, particularly where the impulse duration is short compared with the fundamental period of the system.

shock – (2) abbreviation for SHOCK WAVE.

shock-cell noise – broadband noise radiated from an IMPROPERLY EXPANDED JET, as a result of interaction between turbulence in the jet mixing layer and stationary shock waves in the jet. Also known as ***broadband shock-associated noise***. Compare JET SCREECH.

shock-formation distance – *for an initially-sinusoidal plane progressive wave of finite amplitude* the propagation distance at which acoustic plane waves in a fluid (or longitudinal waves in a solid) would first become discontinuous, if the medium were lossless:

$$\bar{x} = \frac{c^2}{\beta \omega u_{\max}} = \frac{1}{\beta \varepsilon k}, \qquad (k = \omega/c, \; \varepsilon = u_{\max}/c).$$

Here β is the COEFFICIENT OF NONLINEARITY for the medium (fluid or solid), c is the speed of small-amplitude waves, ω is the angular frequency of the sinusoidal waveform, and $u_{\max}$ is the particle velocity amplitude. The shock-formation distance is also known as the ***discontinuity distance***. *Units* m.

Note (1): In the second of the two expressions for $\bar{x}$, ε is the acoustic Mach number of the unshocked wave. The expressions are valid for weakly nonlinear waves, i.e. $\varepsilon \ll 1$.

Note (2): In the general case where the initial particle velocity waveform $u(t)$ is not sinsuoidal, it is the maximum time derivative $\dot{u}_{max}$ that determines the shock-formation distance:

$$\bar{x} = \frac{c^2}{\dot{u}_{max}}.$$

shock wave – a wavefront that propagates through a fluid and is characterized by sudden finite changes in fluid velocity, pressure, density, and temperature from one side of the front to the other. The thickness of a stable shock front depends on the size of the pressure jump, and on dissipative processes within the shock.

short-term shear modulus – see RELAXATION MODEL. *Units* Pa.

SH waves – *in a horizontally stratified medium* horizontally-polarized S WAVES; i.e. shear waves in which the particle motion lies entirely in the horizontal plane.

SI – abbreviation signifying the International System of Units (*Système International d'Unités*), based on the metre, kilogram, and second as units for length, mass, and time respectively. It was adopted in its original form in 1960, when it replaced the mks system.

SI acoustic ohm – 1 N s m^{-5}; the unit of ACOUSTIC IMPEDANCE based on SI fundamental quantities. Not a recognized SI derived unit, despite its name.

sifting relation – see DIRAC DELTA FUNCTION.

signal – a function of time, defined either at discrete points along the time axis (discrete or DIGITAL SIGNAL) or over an interval (CONTINUOUS SIGNAL). The signal may consist of quantized data, or take a continuous range of values. See also SAMPLING, SIGNAL PROCESSING.

Note: Signals that encode desired information, or provide a desired input to a system, are clearly different from NOISE (1) which does neither. From this viewpoint the terms *signal* and *noise* have opposite meanings, as in SIGNAL-TO-NOISE RATIO. The more general definition given above, however, allows noise to be included as a type of signal.

signal excess – *in active sonar* the number of decibels by which the signal-to-noise ratio at the input to a sonar detector exceeds the DETECTION THRESHOLD. *Units* dB.

signal processing – the treatment or manipulation of SIGNALS in order to extract information. A typical sequence of operations is: ANALOGUE SIGNAL → amplification and pre-processing (e.g. filtering) → ANALOGUE-TO-DIGITAL CONVERSION → digital processing of data segments → averaging over segments to yield STATISTICS. According to context, the term *signal processing* can refer either to the entire sequence, or to the **digital signal processing** part alone (i.e. the last two steps). In the latter case, the complete process is referred to as **data acquisition and analysis**.

signal-to-noise ratio – *in a signal consisting of a desired component and an uncorrelated noise component* the ratio of the desired-component POWER (3) to the noise power. Suppose the total signal $x(t)$ is written as

$$x(t) = x_s(t) + n(t),$$

where $n(t)$ represents NOISE (1) and $x_s(t)$ is the desired signal. Then the definition above gives the signal-to-noise ratio as $R = \langle x_s^2 \rangle / \langle n^2 \rangle$, where angle brackets indicate a time average. The ratio R is conventionally expressed in logarithmic form, as $\text{SNR} = 10 \log_{10} R$. *Units* (for R) none; (for SNR) dB.

SIL – abbreviation for SPEECH INTERFERENCE LEVEL. *Units* dB re $(20 \ \mu\text{Pa})^2$.

silencer – any passive device used to limit noise emission from air handling equipment, process plant, internal combustion engines, or fluid machinery in general.

SI mechanical ohm – $1 \ \text{N s m}^{-1}$; the unit of MECHANICAL IMPEDANCE based on SI fundamental quantities. Not a recognized SI derived unit, despite its name.

simple harmonic motion – oscillatory motion in which the displacement of a system varies sinusoidally with time.

simple material – *in viscoelasticity* a material for which the stress depends only on the history of the displacement gradients, and not on higher spatial derivatives of the displacement with respect to material coordinates.

simple source – equivalent term for MONOPOLE SOURCE.

simple wave – *in one-dimensional gas dynamics* a progressive wave of finite amplitude. See also RIEMANN INVARIANTS.

simulation – numerical modelling of a physical system.

simultaneous equations – see NON-HOMOGENEOUS SET OF LINEAR EQUATIONS.

sinc – a sinc function is any function of the form $(ax)^{-1}\sin ax$. The notation sinc x commonly refers either to the function defined by

$$\operatorname{sinc} x = \frac{\sin x}{x},$$

or to the function $(\pi x)^{-1}\sin \pi x$, whose first zero crossings are at $x = \pm 1$.

single-degree-of-freedom system – a dynamical system that requires only one variable, and its time derivatives, to define the state of the system at any instant. See DEGREES OF FREEDOM, GENERALIZED COORDINATES.

single-event emission sound pressure level – *of a transient single-event noise source under controlled conditions* the SOUND EXPOSURE LEVEL produced by the source at a defined measurement location, during a single sound event of specified duration (or over a specified measurement time). *Units* dB re E_{ref}; $E_{\text{ref}} = 4 \times 10^{-10}$ Pa^2 s $= (20\ \mu\text{Pa})^2$ s.

single-sided – *in relation to spectra* limited to positive frequencies. The ***single-sided autospectral density*** of a time-stationary signal $x(t)$, denoted by $G_{xx}(f)$, has the property that

$$\int_0^\infty G_{xx}(f)\, df = \langle x^2(t)\rangle;$$

here f is the frequency and $\langle x^2(t)\rangle$ is the total signal power.

Note: Single-sided spectra are often used in engineering applications, but are not recommended for signals containing dc components.

single-sided spectral density – an autospectral density, co-spectral density, or quad-spectral density that is defined for positive frequencies only. See SINGLE-SIDED.

singular matrix – a SQUARE MATRIX whose determinant is zero.

SI rayl – 1 N s m^{-3} $\equiv$ 1 Pa s m^{-1}; the unit of SPECIFIC ACOUSTIC IMPEDANCE based on SI fundamental quantities. Not a recognized SI derived unit, despite its name.

SI sabin – ◆ 1 m^2; the unit of ROOM ABSORPTION based on the metre, commonly known as the ***metric sabin***. Not a recognized SI derived unit, despite its name.

skeleton – *of a porous material* an equivalent term for FRAME.

skewness coefficient – *of a univariate probability density function* the dimensionless quantity $M_3/(M_2)^{3/2}$, where M_2 and M_3 are the second and third moments of the PROBABILITY DENSITY FUNCTION about the MEAN. *Units* none.

skin depth – *for oscillatory flow near a boundary* see THERMAL ~, VISCOUS ~. *Units* m.

skip distance – *in underwater or atmospheric acoustics* the horizontal distance travelled by a ray in a SURFACE DUCT, between successive encounters with the surface (or with the ground). Compare CYCLE DISTANCE. *Units* m.

SL – *in underwater acoustics* abbreviation for SOURCE LEVEL. *Units* dB re 1 μPa^2 @ 1 m, or dB re 1 μPa^2 m^2.

slowness – *for single-frequency free wave propagation in a given medium* the reciprocal of the PHASE SPEED (1). For plane waves of angular frequency ω, the slowness is given by

$$s = \omega^{-1}\mathrm{Re}[k(\omega)],$$

where $k(\omega)$ is the CHARACTERISTIC WAVENUMBER. *Units* s m^{-1}.

slowness vector – *of single-frequency plane progressive waves* the vector $\mathbf{s} = \mathbf{k}/\omega$, where $\mathbf{k}$ is the WAVENUMBER VECTOR of the plane wave system (assumed to be real), and ω is the angular frequency. *Units* s m^{-1}.

slow weighting – one of the standard settings for TIME WEIGHTING in a sound level meter.

smear – *in signal analysis* the distortion, or loss of resolution, in either the time or frequency domain that is associated with WINDOWING the data in the conjugate domain (i.e. frequency or time, respectively). Also referred to as LEAKAGE.

Snell's law – *for wave refraction at a plane interface* the relation

$$\frac{\sin\theta_1}{c_1} = \frac{\sin\theta_2}{c_2}$$

between the angles of incidence and transmission (SNEL VAN ROYEN 1620). Here θ_1, θ_2 are the angles that the wavenormal on either side of the interface makes with the normal to the interface, and c_1, c_2 are the sound speeds on either side.

SNR – abbreviation for a SIGNAL-TO-NOISE RATIO expressed in decibels. *Units* dB.

SOFAR channel – a SOUND CHANNEL in the ocean that allows long-range transmission entirely by refracted paths, without involving reflection at the surface or bottom. The ***channel axis***, where the sound speed is a minimum, varies in depth from a maximum of around 1000 m at mid-latitudes to zero in polar regions; the effectiveness of the SOFAR channel for sound transmission is directly related to the channel axis depth.

soft boundary – a boundary on which the acoustic pressure equals zero at all times. Also known as a *pressure-release boundary*. Compare DIRICHLET BOUNDARY CONDITION.

solenoidal – a term used to describe vector fields with zero divergence.

solid-angle distribution of intensity – *in acoustics* see ANGULAR INTENSITY DISTRIBUTION. *Units* W m^{-2} sr^{-1}.

solid-angle distribution of sound power – see ANGULAR POWER DISTRIBUTION. *Units* W sr^{-1}.

solid horn – a solid rod of tapering cross-section, used to match a load of low mechanical impedance to a transducer of high impedance. The transducer drives the wide end of the horn, usually at frequencies well below the first transverse-mode cutoff so that the motion is approximately one-dimensional. The horn then converts the high-force low-displacement input into a low-force high-displacement output. Solid horns are mainly used at ultrasonic frequencies; see also HORN, ACOUSTIC HORN.

soliton – a nonlinear one-dimensional wave pulse in a dispersive medium, with the property that the effects of nonlinearity and dispersion counteract one another and the waveform preserves its shape during propagation. Two solitons approaching one another from opposite directions emerge from the collision unaltered.

Sommerfeld integral – an integral representation of spherically symmetric outgoing waves in terms of cylindrical waves (**A SOMMERFELD 1909; H WEYL 1919**), also known as the *Sommerfeld–Weyl integral*. The Sommerfeld integral for outgoing waves from a point monopole is

$$\frac{e^{-jkR}}{R} = \frac{1}{j}\int_0^\infty J_0(\kappa r) e^{-j\gamma|z|} \frac{\kappa d\kappa}{\gamma},$$

where $R = \sqrt{r^2 + z^2}$ and the monopole is located at $r = 0$, $z = 0$ in cylindrical coordinates. Symbol J_0 denotes the Bessel function of order zero; see CYLINDER FUNCTIONS. The complex numbers k (acoustic wavenumber) and γ (wavenumber in the z direction) are related by $j\gamma = \sqrt{\kappa^2 - k^2}$, and their arguments are subject to the constraints

$\mathrm{Re}(jk) \geq 0$ (k corresponds to non-amplifying waves),

$$-\frac{\pi}{2} < \arg\sqrt{\kappa^2 - k^2} \leq \frac{\pi}{2}$$

(γ corresponds to waves that decay in the $\pm z$ directions).

Note: A related result, equivalent to the integral given above, is

$$\frac{e^{-jkR}}{R} = \frac{1}{2j}\int_{-\infty}^{\infty} H_0^{(2)}(\kappa r) e^{-j\gamma|z|} \frac{\kappa d\kappa}{\gamma}.$$

Here $H_0^{(2)}$ is the Hankel function of type 2, of order zero. This result is sometimes referred to as the **Weyrich integral**. Compare the WEYL INTEGRAL, which represents the same left-hand side in terms of plane waves.

Sommerfeld radiation condition – *for free-field acoustic radiation in three dimensions* a relation between acoustic pressure and radial particle velocity that holds at large distances from the source region in a lossless fluid:

$$\lim_{r\to\infty}\left[r\left(\frac{p}{\rho c} - u_r\right)\right] = 0.$$

Here p denotes sound pressure; r is the distance from the centre of the source region; u_r is the radial particle velocity; and ρc is the CHARACTERISTIC IMPEDANCE of the medium. An alternative version, restricted to *single-frequency* sound fields but applicable to lossy media, is

$$\lim_{r\to\infty}\left[r\left(\frac{\partial p}{\partial r} + jkp\right)\right] = 0 \quad \text{(time factor } e^{j\omega t}\text{)},$$

where $\partial p/\partial r$ is the pressure gradient in the radially outward direction and k is the CHARACTERISTIC WAVENUMBER for plane waves in the medium (A SOMMERFELD 1912).

sonar – the use of underwater sound to detect objects in the sea, either by listening for the sound they make, or by sending out an acoustic signal and listening for reflections. The term *sonar* is also used as an abbreviation for *sonar equipment*. See PASSIVE SONAR, ACTIVE SONAR.

sonar equations – standardized equations used to estimate the performance of SONAR equipment in underwater detection tasks. They also provide a framework for sonar equipment design. The equations combine separate multiplicative parameters (which typically relate to the equipment, to the target and transmitting medium, and to signal detection requirements) into an overall prediction scheme. By convention, parameters in the sonar equations are always expressed in decibel units.

sonar projector – an underwater TRANSDUCER, or array of transducers, used to generate acoustic signals for ACTIVE SONAR.

sonar range – (1) *in underwater acoustics* an equivalent term for the DETECTION RANGE of a sonar system. *Units* m.

sonar range – (2) a measurement range used for testing sonar equipment, as in RANGE (2).

sonar self-noise – the BACKGROUND NOISE that occurs in SONAR equipment as a result of processes within the equipment, the ship carrying the equipment, or the flow around the ship.

sonar source level – *in active sonar* an equivalent term for SOURCE LEVEL. *Units* dB re 1 μPa^2 @ 1 m, or dB re 1 μPa^2 m^2.

sone – the unit of LOUDNESS. A reference sound of frequency 1 kHz and sound pressure level 40 dB re (20 μPa)^2, presented to the observer as a plane wave arriving from directly in front, has by definition a LOUDNESS LEVEL of 40 phons. The loudness of this reference sound is assigned the value 1 sone, and the loudness of any other sound that is judged by the listener to be *n* times that of the reference sound is *n* sones. An increment of 10 phons corresponds roughly to a two-fold increase of loudness in sones. *Unit symbol* sone (not an SI unit).

sonic – (1) related to SOUND, as in *sonic frequency*, or operated by sound, as in *sonic drilling*. Compare ULTRASONIC.

sonic – (2) associated with flow at the speed of sound, as in *sonic boom* or *sonic throat*.

sonicate – to disrupt by irradiation with ultrasound; usually applied to cells, bacteria, etc. Compare SONOLYSOGENESIS.

sonic boom – the transient pressure signal heard by a stationary observer, due to the passage of the shock wave attached to a supersonic aircraft. Similar shock waves are produced by any object travelling through the atmosphere faster than the speed of sound, e.g. a supersonic land vehicle.

sonics – the use of sound, often at ultrasonic frequencies, to perform some function (such as cleaning or drilling) that is not inherently acoustical; also, the technology of such applications. Compare MACROSONICS.

sonic throat – a constriction in a nozzle or flow passage, through which steady flow passes at sonic speed.

sonic wind – the ACOUSTIC STREAMING that occurs when a beam of high-intensity ultrasound is absorbed in a body of fluid, transferring its momentum to the fluid to produce a dc flow in the beam direction.

sonification – the use of varying sounds, tones, and pitches to present information (by analogy with visualization). Compare AURALIZATION.

Note: Despite their similar appearance, sonification and INSONIFICATION have entirely different meanings.

sonobuoy – a small SONAR set, suspended from a floating buoy, that is deployed by dropping it from an aircraft; signals are relayed back to the aircraft by a radio transmitter in the buoy.

sonoluminescence – emission of light pulses from a collapsing bubble during acoustically-excited INERTIAL CAVITATION.

sonolysis – alternative term for SONOLYSOGENESIS.

sonolysogenesis – breakup and dissolution of cells, or of other biological tissue, as a result of acoustically-induced CAVITATION.

SOSUS – *in underwater acoustics* abbreviation for ***sound surveillance system***. The term refers to a network of arrays deployed on the sea bed by the United States Navy, originally used for tracking submarines and now routinely used for scientific research (e.g. whale migration studies, ATOC).

sound – (1) a disturbance in pressure that propagates through a compressible medium. More generally, sound can refer to any type of mechanical wave motion, in a solid or fluid medium, that propagates via the action of elastic stresses and that involves local compression and expansion of the medium. See also STRUCTURE-BORNE SOUND, SUPERFLUIDITY.

sound – (2) the auditory sensation produced by transient or oscillatory pressures acting on the ear, or by mechanical vibration of the cranial bones at audio frequencies.

sound ~ – see also ACOUSTIC ~.

sound absorber – a sound-absorbing structure that relies on either fluid or material damping to dissipate acoustic energy. See FUNCTIONAL ABSORBER, HELMHOLTZ RESONANCE ABSORBER, MEMBRANE ABSORBER, MULTILAYER ABSORBER, PANEL ABSORBER, PERFORATED-PLATE ABSORBER.

sound absorption – the process of energy DISSIPATION that occurs when sound is transmitted in a lossy material. Often abbreviated, in an acoustical context, as ABSORPTION.

sound channel – *in a horizontally stratified medium* a layer within which sound is trapped; for example near-horizontal rays can be trapped in a surface layer bounded above by the ocean surface, and below by a region of higher sound speed (see SURFACE DUCT). A sound channel is also formed whenever the sound speed passes through a minimum as a function of depth; an example is the SOFAR CHANNEL. Compare ABNORMAL AUDIBILITY ZONE.

sound energy – an equivalent term for ACOUSTIC ENERGY. *Units* J.

sound energy density – an equivalent term for ACOUSTIC ENERGY DENSITY. *Units* J m^{-3}.

sound energy flux – an equivalent term for SOUND POWER, in the time-average sense. *Units* W.

sound exposure – a measure of the ear's exposure to audio-frequency sound over a specified interval of time. It is defined as the integral, with respect to duration, of the mean square FREQUENCY-WEIGHTED sound pressure. A sound of level 85 dB re (20 μPa)^2 lasting for 8 h produces approximately 1 Pa^2 h of sound exposure. Symbol E. *Units* Pa^2 s, Pa^2 h (the pascal second is the SI unit; the pascal hour is the practical unit for sound exposure meters).

Note: A-WEIGHTING is normally used for sound exposure; any other frequency weighting needs to be stated explicitly. Compare PRESSURE-SQUARED INTEGRAL.

sound exposure level – a measure of the SOUND EXPOSURE in decibels. On this scale, 0 dB corresponds to a steady sound whose rms frequency-weighted sound pressure equals the reference pressure of 20 μPa, persisting for a reference time of 1 s. Note that sound exposure level and NOISE EXPOSURE LEVEL are similar quantities, differing only by a constant (namely 44.6 dB). Sound exposure level can be applied to single events, as well as to noise of a continuing character; it was first defined for the purpose of describing the overflight noise exposure from individual aircraft, and given the symbol L_{AX}. *Units* dB re E_{ref}; $E_{ref} = 4 \times 10^{-10}$ Pa^2 s = (20 μPa)^2 s.

Note: The current symbol for sound exposure level is L_E. Normally A-WEIGHTING is implied; it may be made explicit by writing L_{AE}. If a different frequency weighting is used, it is indicated by the appropriate subscript, e.g. L_{CE}, L_{DE}.

sound field – a FIELD (2) of acoustic pressure, particle velocity, and any other sound-related quantities; usually defined over a specific region, e.g. in a room.

sound field audiometry – the determination of a person's hearing threshold levels when listening in an ambient sound field (as opposed to conventional audiometry using earphones).

sound insulation – (1) *in buildings* a generic term for the ability of partitions, floors, or ceilings to prevent sound energy from leaking across them. Good sound insulation means that sound generated in one part of a building does not escape into other parts where it is not wanted.

sound insulation – (2) *of a connecting partition between two rooms* an approximation to the diffuse-field TRANSMISSION LOSS of the partition in a specified frequency band; it is deduced from the reverberant LEVEL DIFFERENCE measured between a source room and a receiving room, when the rooms are separated by the partition in question. Also known as the *laboratory transmission loss*. For the usual situation where the sound speed and air density are the same in both rooms, the sound insulation is related to the level difference, D, by the equation

$$R_D = D + 10 \log_{10} S_w/A_r;$$

here S_w is the partition area and A_r is the total ROOM ABSORPTION of the receiving room. *Units* dB.

Note: A more accurate way to obtain the true diffuse-field transmission loss of the partition would be to replace D in the equation above with $10 \log_{10}(10^{D/10} - 1)$, and to define A_r as the receiving-room absorption measured with the connecting partition rigidly blocked.

sound intensity – an equivalent term for ACOUSTIC INTENSITY. *Units* W m^{-2}.

sound isolation between two rooms – an equivalent term for LEVEL DIFFERENCE (2). *Units* dB.

sound level – an abbreviation for *weighted sound pressure level*, as read by a SOUND LEVEL METER with standard FREQUENCY WEIGHTING and (except for L_{eq} measurements) with exponentially-weighted time averaging. When exponential time weighting is used, the type (i.e. fast or slow – see the note below) needs to be specified. *Units* dB re (20 μPa)^2.

Note: Typically, sound levels are obtained from a sound level meter set to A-WEIGHTING, with either fast or slow time averaging. Exponential weighting with a time constant τ is applied to the square of the "instantaneous" A-weighted pressure $p_A(t)$, to yield a time-varying *A-weighted running-average mean square pressure*. Its value $q_A(t)$ at time t is given by

$$q_A(t) = \frac{1}{\tau} \int_0^\infty p_A^2(t-u) e^{-u/\tau} du.$$

The corresponding sound level at time t is then given by

$$L_{A\tau}(t) = 10 \log_{10} \frac{q_A(t)}{p_{ref}^2} \quad [\text{dB re } p_{ref}^2].$$

For fast (F) weighting, τ equals 125 ms; for slow (S) weighting, τ equals 1 s.

Symbol L_A implies F weighting, but this may be made explicit by writing L_{AF}. If a standard time weighting other than F is used, it is indicated by the appropriate subscript; e.g. L_{AS}, L_{Aeq}.

sound level meter – an instrument designed to measure a frequency- and time-weighted value of sound pressure level. It consists of a microphone, amplifier, square-law rectifier, averaging circuits, and an indicating meter or visual display. Its performance is standardized in respect of directivity, frequency response, and the rectification and weighting characteristics. See also FREQUENCY WEIGHTING, TIME WEIGHTING.

sound particle displacement, sound particle velocity – equivalent terms for the acoustic quantities PARTICLE DISPLACEMENT, PARTICLE VELOCITY respectively.

sound power – the rate of acoustic energy flow across a specified surface, or emitted by a specified sound source. The latter quantity is also called the *source power*. The *sound power in a frequency band* is the energy flow rate associated with sound frequencies lying within the band. *Units* W.

sound power absorption coefficient – *at a boundary* see ABSORPTION COEFFICIENT, which has the same meaning. Compare SABINE ABSORPTION COEFFICIENT. *Units* none.

sound power level – the level of SOUND POWER, expressed in decibels relative to a stated reference value. See REFERENCE VALUE, LEVEL. Abbreviated as *PWL*. *Units* dB re 1 pW.

sound power reduction index – ◆ *for a partition* the reciprocal of the SOUND POWER TRANSMISSION COEFFICIENT (2), expressed in decibels. An equivalent term is TRANSMISSION LOSS (2). *Units* dB.

sound power reflection coefficient – *of a boundary* the ratio of reflected to incident-wave power (or energy, or intensity) at the boundary; i.e. the fraction of incident power that is reflected. The sound field is either incident randomly on the boundary, or else is restricted to plane waves arriving from a specified direction. Equivalent terms are *reflectance* and *reflection factor*. *Units* none.

sound power spectral density – the spectral distribution of SOUND POWER, expressed as a function of frequency. *Units* W Hz^{-1}.

sound power transmission coefficient – (1) *in a one-dimensional system or transmission line* the ratio of net to incident-wave power flow, at a specified point in the system. Also known as the *energy transmission coefficient*. *Units* none.

350 sound power transmission coefficient

Note: The equivalent terms *transmittance* and *transmission factor* may also be used for the sound power transmission coefficient, in any of the senses (1)–(3) defined here. Compare HORN TRANSMISSION FACTOR.

sound power transmission coefficient – (2) *of a partition excited by plane incident sound waves* the ratio of the sound power transmitted by the partition to the sound power incident on it, when the transmitting side of the partition radiates into a free field. With the arrival direction of incident waves defined by spherical polar angles (θ, ϕ), the transmission coefficient from side (1) to side (2) may be written as

$$\tau_{12}(\theta, \phi, f) = \frac{W_{trans}}{W_{inc}} \quad \text{(at frequency } f\text{)}.$$

Here W_{trans}, is the transmitted sound power on side (2), and W_{inc} is the sound power incident on the partition from side (1). *Units* none.

Note (1): The sound power transmission coefficient is a function of frequency and incident wave direction, as indicated above. For practical purposes it is often quoted in one-third octave bands. Its value for RANDOM INCIDENCE is commonly called the *sound transmission coefficient* in architectural acoustics.

Note (2): For a uniform flat partition of infinite extent (not necessarily isotropic), with lossless media on both sides, reciprocity connects the sound-power transmission coefficients in the two directions. The reciprocal relationship is

$$\tau_{12}(\mathbf{v}_t, f) = \tau_{21}(\mathbf{v}_t, f);$$

it applies to plane waves incident from either side of the partition with the same TRACE VELOCITY along the partition, $\mathbf{v}_t$, in each case. The media on the two sides need not have the same acoustic properties.

Note (3): For a partition irradiated by plane waves at random incidence, a different reciprocal relationship applies:

$$\frac{\tau_{12,av}}{\tau_{21,av}} = \left(\frac{c_1}{c_2}\right)^2.$$

Here $\tau_{12,av}$ is the STATISTICAL SOUND POWER TRANSMISSION COEFFICIENT from side 1 to side 2, and $\tau_{21,av}$ is the same in the opposite direction.

sound power transmission coefficient – (3) *at a plane interface between two media* the ratio of the power transmitted across the interface to the incident power, for transmission into a half-space (i.e. the transmitting medium extends to infinity). It is the quantity τ in the equation

$$\tau = I_{trans}/I_{inc},$$

where I_{trans} and I_{inc} denote the normal intensity, or sound power per unit interface area, associated with the transmitted and incident waves. *Units* none.

sound pressure – the difference between the instantaneous pressure at a fixed point in a sound field, and the pressure at the same point with the sound absent. *Units* Pa.

Note: The term *instantaneous sound pressure* is sometimes used to distinguish this quantity from the *effective sound pressure*, i.e. the rms sound pressure averaged over a given time interval.

sound pressure level (SPL) – *at a given point* the quantity L_p defined by

$$L_p = 10 \log_{10}(p_{\text{rms}}/p_{\text{ref}})^2 = 20 \log_{10}(p_{\text{rms}}/p_{\text{ref}}).$$

Here p_{rms} is the ROOT MEAN SQUARE sound pressure, and p_{ref} is the reference rms sound pressure. An equivalent (less common but more precise) term is *sound pressure-squared level*. For **peak** $\sim$, see note (2) under LEVEL. *Units* dB re $(20\ \mu\text{Pa})^2$ (air or other gases); dB re $1\ \mu\text{Pa}^2$ (water or other media).

Note: The term *average sound pressure level* refers to a spatial average over a finite region. It is defined by averaging the mean square sound pressure over the region concerned, before calculating the level as above. See also REVERBERANT LEVEL, MAXIMUM SOUND PRESSURE LEVEL, PEAK SOUND PRESSURE LEVEL.

sound pressure reflection coefficient – *for single-frequency incident waves* the complex ratio of reflected to incident sound pressure, given by

$$R_p = p_{\text{refl}}/p_{\text{inc}}.$$

It is defined at a specified position in a standing-wave field. For a fuller discussion, see COMPLEX PRESSURE REFLECTION COEFFICIENT. *Units* none.

sound pressure spectrum level – see note (2) under SPECTRUM LEVEL. *Units* dB re $(20\ \mu\text{Pa})^2\ \text{Hz}^{-1}$ (air and other gases); dB re $1\ \mu\text{Pa}^2\ \text{Hz}^{-1}$ (water and other media).

sound pressure transmission coefficient – *for single-frequency waves incident on a discontinuity* the complex ratio of the transmitted pressure on the output side to the incident pressure on the input side, when the transmitted field consists of outgoing waves. In equation form,

$$T_p = p_{\text{trans}}/p_{\text{inc}}.$$

It is defined either for plane waves incident at a given angle on a uniform plane boundary, or for plane waves transmitted between specified points in a one-dimensional system (e.g. a pipe or waveguide). For a fuller discussion, see COMPLEX PRESSURE TRANSMISSION COEFFICIENT. *Units* none.

sound propagation coefficient – *for a single-frequency progressive sound wave* the complex PROPAGATION COEFFICIENT of the sound wave. Its real part is the attenuation coefficient and its imaginary part is the acoustic phase coefficient. *Units* (real part) Np m^{-1}; (imaginary part) rad m^{-1}.

sound quality – (1) a generic term used to describe all those features of a sound that contribute to the subjective impression made on the listener.

sound quality – (2) *of an engineering product* the suitability of the sound associated with the product, in the context of a particular set of design goals. Sound quality in this sense is judged by users of the product according to whether they find the sound acceptable, or adequate, for the intended purpose; it is intended to reflect their expectations of the product, recognizing that these will depend on the user's cognitive and emotional state at the time.

sound reduction index – *of a partition separating two diffuse sound fields* an equivalent term for TRANSMISSION LOSS (2), as measured by techniques standard in building acoustics. Symbol *R*. *Units* dB.

Note (1): Use of the term sound reduction index implies that it is measured in a laboratory under controlled conditions, by measuring the level difference between two rooms separated by the partition under test. In this sense the term equates to LABORATORY TRANSMISSION LOSS.

Note (2): Compare APPARENT SOUND REDUCTION INDEX, WEIGHTED SOUND REDUCTION INDEX.

sound reinforcement – amplification of sound for the benefit of an audience. The sound may be either live or pre-recorded. Compare PUBLIC ADDRESS SYSTEM.

sound source – an equivalent term for ACOUSTIC SOURCE, in senses (1) or (2).

sound speed – (1) *for a fluid* the thermodynamic property $\sqrt{B/\rho}$, evaluated under equilibrium conditions (B = isentropic bulk modulus, ρ = density). An equivalent expression is $\sqrt{(\partial P/\partial \rho)_s}$, where P and ρ are the fluid pressure and density, and subscript s indicates that the derivative is evaluated holding the specific entropy s constant. See also SOUND SPEED IN AIR. *Units* m s^{-1}.

Note: The terms sound speed and ***speed of sound*** are interchangeable.

sound speed – (2) *for any medium, at a specified frequency* the PROPAGATION SPEED of free plane compressional waves. If the medium is dispersive, it is the PHASE SPEED that is understood. *Units* m s^{-1}.

sound speed in air – the thermodynamic property $\sqrt{B/\rho}$, evaluated for air under equilibrium conditions (B = isentropic bulk modulus, ρ = density). Air at atmospheric pressures and temperatures behaves approximately as an IDEAL GAS, and the speed of sound deviates only slightly from the ideal-gas value (i.e. the value approached in the limit of zero pressure, for a given temperature). *Units* m s^{-1}.

Note: The numerical values listed below illustrate how small the deviation is; here γ is the specific-heat ratio, κ is the ISENTROPIC EXPONENT, and c is the speed of sound. For an ideal gas, κ equals γ, and c is a function of temperature (T) with no dependence on pressure (P).

T [K]	P [bar]	γ	κ	c [m s^{-1}]
273.15	0	1.4005	1.4005	331.38
273.15	1.01325	1.4027	1.4010	331.44

sound speed in seawater – the thermodynamic property $\sqrt{B/\rho}$, evaluated for seawater under equilibrium conditions (B = isentropic bulk modulus, ρ = density). The following approximate expressions relate sound speed, c, to SALINITY, S, Celsius temperature, T, and depth in metres, z:

Medwin equation (**H MEDWIN 1975**):

$$c = 1449.2 + 4.6T - 0.055T^2 + 0.00029T^3 \\ + (1.34 - 0.01T)(S - 35) + 0.016z;$$

$$22 < S < 45, 0 < T < 32, 0 < z < 1000.$$

The estimated rms error over the range indicated is about 0.2 m s^{-1}.

Mackenzie equation (**K V MACKENZIE 1981**):

$$c = 1448.96 + 4.591T - 0.05304T^2 + 0.0002374T^3 \\ + (1.34 - 0.01025T)(S - 35) + 0.0163z \\ + 1.675 \times 10^{-7}z^2 - 7.139 \times 10^{-13}Tz^3;$$

$$25 < S < 40, -2 < T < 30, 0 < z < 8000.$$

The estimated rms error over the range indicated is about 0.07 m s^{-1}. *Units* m s^{-1}.

sound speed profile – *in the ocean or atmosphere* the profile of sound speed as a function of vertical coordinate (depth or height). The short-time average is always implied. Note that in underwater acoustics, **velocity profile** is sometimes used as an equivalent term (meaning *velocity of sound* profile).

sound strength – *at a specified listening position in an auditorium* a measure of the amplification provided by a room; it is defined as the sound pressure level produced by an omnidirectional sound source located at a representative position in the auditorium, relative to the level of direct sound at 10 m from the same source. Sound strength is a frequency-dependent quantity, typically measured in octave bands from 125 Hz to 4 kHz. An alternative term is *relative total sound level*. Symbol G. *Units* dB.

sound transmission coefficient – *of a partition excited by plane incident sound waves* the fraction of the power incident on a partition that is transmitted. Use of this term generally implies that the incident field is diffuse (i.e. randomly incident). Compare STATISTICAL SOUND POWER TRANSMISSION COEFFICIENT. *Units* none.

sound transmission loss – *of a partition excited by plane incident sound waves* an alternative term for TRANSMISSION LOSS (2); use of this term generally implies that the incident field is diffuse (i.e. randomly incident). *Units* dB.

sound velocimeter – see VELOCIMETER (1).

source – (1) *as input to a linear system* refers to one of various types of idealized input to a system: e.g. *velocity* (or *displacement*) ∼, *force* ∼, *pressure* ∼, *volume-velocity* ∼. In each of these cases the amplitude of the exciting velocity, force, etc. at each frequency is fixed in advance, i.e. it does not depend on the parameters of the system being driven. An alternative term is *generator*.

source – (2) abbreviation for ACOUSTIC SOURCE (1,2).

source – (3) *in an acoustic medium* a localized ACOUSTIC SOURCE (3) of monopole type; e.g. a volume velocity input, or some other local excitation that is acoustically equivalent, such as a heat input. See also ACOUSTIC SOURCE STRENGTH, POINT SOURCE, DISTRIBUTED SOURCE.

source broadening – the sense that a sound source, especially the orchestra or players in a concert hall, occupies a larger region than its physical extent. It is one component of SPATIAL IMPRESSION. Source broadening is generally thought to depend on the listener receiving early reflections from the side, and to increase with sound level at the listener.

source density – *in acoustics* abbreviation for ACOUSTIC SOURCE DENSITY.

source distribution – *in acoustics* an equivalent term for ACOUSTIC SOURCE DENSITY.

source energy level – *in underwater acoustics, for a directional transient sound source* the level of received signal ENERGY (3) (i.e. the pressure-squared integral with respect to time), scaled to a reference distance of $r_{ref} = 1$ m.

For additional comments, see the entry for SOURCE LEVEL. *Units* dB re 1 μPa^2 m^2 s.

Note: Source energy level (SL$_E$) can also be viewed as a decibel measure of the ANGULAR RADIATED-ENERGY DISTRIBUTION E_Ω, although conversion between the two requires a knowledge of the characteristic impedance ρc of the medium:

$$\mathrm{SL}_E = 10\log_{10}\frac{\rho c E_\Omega}{p_{\mathrm{ref}}^2 r_{\mathrm{ref}}^2 t_{\mathrm{ref}}}.$$

source level (SL) – *in underwater acoustics, for a directional sound source* the sound pressure level measured in a given radiation direction, corrected for absorption and scaled to a reference distance of $r_{\mathrm{ref}} = 1$ m. In equation form, the source level is given by

$$\mathrm{SL} = 10\log_{10}\left[\frac{\bar{r}p_{\mathrm{rms}}(\bar{r})}{r_{\mathrm{ref}}p_{\mathrm{ref}}}\right]^2,$$

where $p_{\mathrm{rms}}(\bar{r})$ is the rms pressure radiated to the far field (in the relevant direction) at distance $\bar{r}$ from the source. Related terms are **projector** ~, **target** ~; note that these usually refer to the source level measured in the peak radiation direction. The explicit term *axial source level* is also used for a radiating transducer, particularly where ambiguity might arise as to direction. *Units* dB re 1 μPa^2 @ 1 m, equivalent to expressing the level of $\bar{r}^2 p_{\mathrm{rms}}^2(\bar{r})$ in dB re 1 μPa^2 m^2.

Note (1): In a similar manner, source levels can also be defined for either the peak sound pressure or the PEAK-TO-PEAK pressure radiated by a transient source. Compare ENERGY SOURCE LEVEL, which is a measure of signal energy applied to transient sources.

Note (2): When it is required to express the source level in spectral form, the appropriate units are dB re 1 μPa^2 Hz^{-1} @ 1 m, equivalent to dB re 1 μPa^2 m^2 Hz^{-1}. No standard nomenclature appears to have been established for this quantity; a suggestion is *source spectrum level*.

Note (3): To convert between SL and the radiated acoustic power per unit solid angle in SI units (denoted by W_Ω and measured in W sr^{-1}, or equivalently W m^{-2} @ 1 m), the relevant equation is

$$\mathrm{SL} = 10\log_{10}\frac{\rho c W_\Omega}{r_{\mathrm{ref}}^2 p_{\mathrm{ref}}^2};$$

here ρ and c are the density and sound speed at the source location.

source moments – *in theoretical acoustics* the MOMENTS of a three-dimensional ACOUSTIC SOURCE DISTRIBUTION. For example, the monopole, dipole, and quadrupole moments of a source distribution $\mathcal{F}(\mathbf{y}, t)$ are given by the following volume integrals

$$S = \int \mathcal{F} \, d^3\mathbf{y} \quad \text{(monopole moment)},$$

$$D_i = \int y_i \mathcal{F} \, d^3\mathbf{y} \quad \text{(dipole moments)},$$

$$Q_{ij} = \tfrac{1}{2} \int y_i y_j \mathcal{F} \, d^3\mathbf{y} \quad \text{(quadrupole moments)}.$$

Here y_i, y_j are cartesian components (i, j = 1, 2, 3) of the position vector $\mathbf{y}$, and $d^3\mathbf{y}$ denotes a volume element. The vector with components D_i is called the ***dipole-moment vector***, and the tensor with components Q_{ij} is called the ***quadrupole-moment tensor***.

source power – the SOUND POWER emitted by an acoustic source. *Units* W.

source spectrum level – *in underwater acoustics, for a directional steady-state sound source* the sound pressure spectrum level radiated from the source, scaled to a reference distance of $r_{\text{ref}} = 1$ m. See also SOURCE LEVEL. *Units* dB re 1 $\mu\text{Pa}^2 \, \text{m}^2 \, \text{Hz}^{-1}$.

source strength – *in theoretical acoustics* abbreviation for ACOUSTIC SOURCE STRENGTH (1). *Units* $\text{m}^3 \, \text{s}^{-1}$.

spaciousness – an equivalent term for SPATIAL IMPRESSION.

spatial impression – the sense, due to sound heard by the ears, of being in a three-dimensional space. Spatial impression in concert halls is currently thought to consist of two separate components: SOURCE BROADENING and LISTENER ENVELOPMENT. See also APPARENT SOURCE WIDTH.

spatial phase – *of a field that varies sinusoidally with time* the part of the PHASE that varies according to position in the field. For example, a plane progressive wave field has spatial phase $\mathbf{k} \cdot \mathbf{x}$ at vector position $\mathbf{x}$, where $\mathbf{k}$ is the WAVENUMBER VECTOR. *Units* rad.

speaker – abbreviation for LOUDSPEAKER.

specific – (1) *in acoustics* usually denotes a quantity normalized per unit surface area, as in SPECIFIC ACOUSTIC IMPEDANCE.

specific – (2) *in thermodynamics* denotes an extensive quantity normalized per unit mass, as in SPECIFIC VOLUME.

specific acoustic admittance – *of a given surface or boundary* the complex ratio u_n/p, where u_n is the particle velocity normal to the boundary (positive into the boundary), and p is the acoustic pressure at the same point. The real part of the specific acoustic admittance is the ***specific acoustic conductance***, and the imaginary part is the ***specific acoustic susceptance***. *Units* $\text{m s}^{-1} \, \text{Pa}^{-1}$.

specific acoustic impedance – *in a given direction in a single-frequency sound field* the complex ratio $z = p/u$, where p is the local acoustic pressure and u is the particle velocity in the specified direction at the same point. The real part of the specific acoustic impedance is the ***specific acoustic resistance***, and the imaginary part is the ***specific acoustic reactance***. *Units* Pa s m^{-1}.

Note: At a boundary, the fluid velocity u is generally understood to refer to the component normal to the boundary; see SPECIFIC BOUNDARY IMPEDANCE. In other situations, the direction of the velocity component u has to be specified; for example, p/u_x is the specific acoustic impedance in the x direction.

specific boundary admittance – *at a point on a boundary* the reciprocal of the SPECIFIC BOUNDARY IMPEDANCE. *Units* Pa s m^{-1}.

specific boundary impedance – *at a point on a boundary* the SPECIFIC ACOUSTIC IMPEDANCE that the boundary surface presents to an adjacent sound field; it equals the complex ratio of the acoustic surface pressure to the normal velocity of the fluid at the surface (positive into the boundary). *Units* Pa s m^{-1}.

Note: In general, the specific boundary impedance at a given frequency depends on the spatial pressure distribution over the boundary. For a homogeneous isotropic plane boundary, it is a function of wavenumber magnitude parallel to the surface (or equivalently, a function of the angle of incidence).

specific flow resistance – *of a uniform porous layer* the ratio

$$r = \frac{\Delta p}{\bar{u}},$$

where Δp is the pressure drop through the layer under conditions of slow steady flow, and $\bar{u}$ is the FACE VELOCITY of the flow relative to the layer (equivalent to the volume flowrate per unit surface area). Compare FLOW RESISTIVITY. *Units* Pa s m^{-1}.

Note (1): The quantity r is the FLOW RESISTANCE of a sample of the layer that has unit cross-sectional area measured transversely to the flow; hence the label "specific".

Note (2): For related quantities under dynamic conditions, see COMPLEX SPECIFIC FLOW IMPEDANCE, DYNAMIC SPECIFIC FLOW RESISTANCE.

specific gas constant – *for a gas of given composition* the GAS CONSTANT for unit mass of gas. *Units* J K^{-1} kg^{-1}.

specific-heat ratio – *for a fluid* the ratio of the specific heat at constant pressure to that at constant volume. Symbol γ. *Units* none.

Note: The specific-heat ratio of a stable fluid cannot be less than 1. It is related to the sound speed, c, and the VOLUME THERMAL EXPANSIVITY, α, by

$$\gamma - 1 = \frac{\alpha^2 T c^2}{C_p};$$

and since T (the absolute temperature of the fluid) and C_p (the constant-pressure specific heat) are both positive, the right-hand side is positive or zero.

specific impedance – abbreviation for SPECIFIC ACOUSTIC IMPEDANCE. *Units* Pa s m^{-1}.

specific impedance ratio – *on a given surface or boundary* the ratio of the local SPECIFIC ACOUSTIC IMPEDANCE to the CHARACTERISTIC IMPEDANCE of the medium. *Units* none.

specific radiation impedance – *for a fluid-loaded boundary that vibrates at a single frequency* the complex ratio of the pressure at the boundary to the local normal velocity of the boundary. In symbols,

$$z_{\text{rad}} = \frac{p}{\dot{w}}.$$

Here p is the acoustic pressure in the fluid next to the boundary, and $\dot{w}$ is the normal velocity measured at the same point. *Units* Pa s m^{-1}.

Note: The specific radiation impedance is a local measure of fluid loading, and varies with position over the surface (except in special cases – see WAVE IMPEDANCE). It is not a LOCALLY-REACTING impedance.

specific volume – *of a fluid or solid* the volume per unit mass; the reciprocal of the local DENSITY (1). *Units* m^3 kg^{-1}.

spectral coefficients – *of a periodic discrete-time signal* the complex coefficients in the DISCRETE FOURIER SERIES representation of the signal. The magnitudes of the spectral coefficients define the signal's MAGNITUDE SPECTRUM. Compare COMPLEX FOURIER AMPLITUDES.

spectral density – *of a steady-state signal whose frequency spectrum is continuous* the limiting value of the signal POWER (3) in a frequency band normalized by the width of the band, as the bandwidth is reduced towards zero. The resulting function of frequency is generally called the AUTOSPECTRAL DENSITY or POWER SPECTRAL DENSITY of the signal; its integral over the whole frequency band is the total signal power. Compare SOUND POWER SPECTRAL DENSITY.

spectrogram – *of a transient signal* a time–frequency representation of the signal. A sequence of spectra from successive samples of the signal shows how the frequency content of the signal varies over time. The information is presented as a grey-scale (or colour-scale) contour map of intensity in the frequency–time plane, also known as a *sonogram*. An alternative is to display the spectra of successive signal samples on a waterfall plot.

spectrum level – *of a time-stationary random signal, at a specified frequency* the level of the single-sided AUTOSPECTRAL DENSITY, expressed in decibels relative to a reference value. Thus if a signal $x(t)$ has single-sided autospectral density $G_{xx}(f)$, the spectrum level at frequency f is

$$L_{x,s}(f) = 10 \log_{10} [G_{xx}(f)/G_{ref}] \quad [\text{dB re } G_{ref}].$$

Note (1): For any signal whose reference rms value is x_{ref}, the quantity G_{ref} is taken as $(x_{ref}^2)/B_{ref}$ where B_{ref} is the standard bandwidth of 1 Hz. This leads to a convenient relation between *band levels*, defined in the usual way as

$$L_{x,\text{band}} = 10 \log_{10} [\langle x^2 \rangle_{\text{band}}/x_{ref}^2] \quad [\text{dB re } x_{ref}^2],$$

and spectrum levels. At the band centre frequency, $G_{xx}(f)$ may be estimated as

$$G_{xx}(f) \approx \langle x^2 \rangle_{\text{band}}/B \quad (B = \text{filter energy bandwidth}).$$

Then the spectrum level and the corresponding *band level* $L_{x,\text{band}}$ are related by

$$L_{x,s}(f) \approx L_{x,\text{band}} - 10 \log_{10}(B/B_{ref}).$$

Note (2): If the signal in question is the sound pressure at a particular point, the equation above defines the *sound pressure spectrum level*. For sound in air, the reference value G_{ref} would be taken as $(20 \ \mu\text{Pa})^2 \ \text{Hz}^{-1}$; in water, it would be $1 \ \mu\text{Pa}^2 \ \text{Hz}^{-1}$.

specular reflection – a reflection process in which an incident sound field is reflected from a locally-plane boundary like light from a mirror, so that the angle of reflection of an incident plane wave is equal to the ANGLE OF INCIDENCE.

Note: Specular reflection implies that the reflection of sound incident from a source onto a plane surface can be described by adding an IMAGE SOURCE (of appropriately modified strength compared to the original), with both sources radiating into a free field.

speech audiogram – a chart or graph depicting a person's speech recognition scores as a function of speech level.

speech audiometry – the presentation of speech material, usually in the form of word lists, to determine the percentage of material correctly identified (*speech recognition score*) over a range of levels. In the simplest form, recorded material is presented monaurally by earphone in quiet listening conditions. Variations include live-voice presentation, sound-field binaural listening, added noise, etc. Results are displayed on a SPEECH AUDIOGRAM.

speech interference level – the arithmetic average of the sound pressure levels in the four octave bands whose nominal midband frequencies are 500 Hz, 1 kHz, 2 kHz, and 4 kHz. Abbreviated as ***SIL***. *Units* dB re $(20\ \mu\text{Pa})^2$.

speech transmission index (STI) – an objective measure used for predicting the output INTELLIGIBILITY of a speech transmission channel. The STI, on a scale of 0 to 1, measures the faithfulness with which the transmission path preserves the envelope of typical speech signals; it is designed to take account of the room response, as well as background noise. It is determined from the single-channel MODULATION TRANSFER FUNCTION (MTF), measured using sinusoidal power modulation applied to separate octave bands of noise covering the frequency range 125 Hz to 8 kHz. Compare ARTICULATION INDEX. *Units* none.

speed of sound – an equivalent term for SOUND SPEED. *Units* m s^{-1}.

spherical Bessel functions – a set of functions, analogous to the CYLINDER FUNCTIONS of the first, second, or third kind, that are solutions of the ***spherical Bessel equation***

$$R'' + \frac{2}{r}R' + \left[\kappa^2 - \frac{\nu(\nu+1)}{r^2}\right]R = 0.$$

Here $R(r)$ is a function of the spherical radius r, and primes denote derivatives of the function; κ and ν are constants, restricted only by $\kappa \neq 0$. The general solution of this equation is written in terms of spherical Bessel functions of ***order*** ν as

$$R(r) = A j_\nu(\kappa r) + B y_\nu(\kappa r),$$

or alternatively as

$$R(r) = C_1 h_\nu^{(1)}(\kappa r) + C_2 h_\nu^{(2)}(\kappa r).$$

The ***spherical Hankel functions***, $h_\nu^{(1)}$ and $h_\nu^{(2)}$, are related to the spherical Bessel and Neumann functions by

$$h_\nu^{(1)}(z) = j_\nu(z) + j y_\nu(z); \quad h_\nu^{(2)}(z) = j_\nu(z) - j y_\nu(z).$$

The various kinds of spherical Bessel function, namely j_ν, y_ν, $h_\nu^{(1)}$, and $h_\nu^{(2)}$, are listed below; note that the argument z and the order ν may be real or complex.

Function	Name	Order	Symbol
1st kind	Spherical Bessel function	ν	$j_\nu(z)$
2nd kind	Spherical Neumann function	ν	$y_\nu(z)$ or $n_\nu(z)$
3rd kind	Spherical Hankel functions	ν	$h_\nu^{(1)}(z), h_\nu^{(2)}(z)$

spherical harmonics – *in spherical polar coordinates* a set of directional factors $D(\theta, \phi)$ that appear in separable solutions of either the wave equation or the Helmholtz equation. They have the property that for appropriate choices of the functions F and R,

(1) $D(\theta, \phi) F(r, t)$ is a solution of the wave equation;

(2) $D(\theta, \phi) R(r)$ is a solution of the Helmholtz equation.

Note that in both cases the directional dependence of the field is the same at every radius r.

The specific directional factors that have this property are called spherical surface harmonics, or spherical harmonics. They form a family, labelled by two integers m and n:

$$D(\theta, \phi) = Y_n^m(\theta, \phi) = P_n^m(\cos \theta) \, e^{-jm\phi}.$$

Here n is sometimes called the **order** of the spherical harmonic, and can take values $0, 1, 2, 3, \ldots$; the upper index m is limited to values $0, \pm 1, \pm 2, \ldots, \pm n$. The functions P_n^m are the ASSOCIATED LEGENDRE FUNCTIONS. The set of $Y_n^m(\theta, \phi)$ functions is complete, in the sense that any reasonable directivity function $D(\theta, \phi)$ on the surface of a sphere can be expressed as a weighted sum of the spherical surface harmonics Y_n^m:

$$D(\theta, \phi) = \sum_{m,n} a_{mn} Y_n^m(\theta, \phi).$$

Note (1): The process of finding the coefficients a_{mn} is made easier by the orthogonality property of the Y_n^m functions. When the product Y^*Y is integrated over a sphere centred on the origin, the result is zero *unless* the two Y functions are the same. Expressed mathematically, orthogonality means that

$$\int_{\text{unit sphere}} (Y_n^m)^* Y_j^i \, dS = C_{nm} \quad (\text{for } j = n \text{ and } i = m);$$

$$= 0 \quad (\text{otherwise}).$$

With Y_n^m defined as in the first equation, and P_n^m defined in the standard manner, the constant C_{nm} is

$$C_{nm} = \frac{4\pi}{2n+1}\frac{(n+m)!}{(n-m)!} \quad (m \text{ positive}).$$

Note (2): There is a connection between the Y_n^m angular functions in spherical coordinates, and the nth-order derivatives of $1/r$ in rectangular coordinates. For example, in the axisymmetric case

$$Y_n^0(\theta, \phi) = P_n(\cos\theta)$$
$$= \frac{(-)^n}{n!} r^{n+1} \left(\frac{\partial}{\partial z}\right)^n \frac{1}{r},$$

provided n is not negative. In the equation above, $z = r\cos\theta$ is the coordinate along the polar axis, and $\partial r/\partial z$ equals z/r. More general x, y, z derivatives, like

$$\left(\frac{\partial}{\partial x}\right)^a \left(\frac{\partial}{\partial y}\right)^b \left(\frac{\partial}{\partial z}\right)^c \frac{1}{r},$$

can also be expressed in terms of the $Y_n^m(\theta, \phi)$ functions; the harmonic order n is always given by the total order of the spatial derivatives, i.e.

$$n = a + b + c.$$

Note (3): An alternative definition of the spherical harmonic functions uses $\cos m\phi$ and $\sin m\phi$ as separate azimuthal factors, and restricts m to positive values or zero:

$$Y_{mn}^e(\theta, \phi) = P_n^m \cos m\phi, \quad Y_{mn}^o(\theta, \phi) = P_n^m \sin m\phi.$$

spherically-symmetric outgoing wave – a sound field whose wavefronts are concentric spheres, propagating outwards from the origin. In a non-dispersive medium, the pressure in this type of outgoing wave has the mathematical form

$$p = \frac{1}{r} f(t - r/c)$$

where r is the distance from the centre of the wave system (the apparent source position), c is the sound speed, t is time, and f is any function.

Single-frequency waves of this type are described by a complex pressure field

$$p = \frac{A}{r} e^{j(\omega t - kx)} \quad (A = \text{const.}),$$

where ω is the angular frequency and k is the PROPAGATION WAVENUMBER.

spherical polar coordinates – a set of coordinates (r, θ, ϕ) based on radial distance r from the origin, angular rotation ϕ about a specified axis, and polar angle θ measured from the axis direction.

spherical progressive wave – *of arbitrary multipole order* a wave field whose space-time dependence consists of a directional factor in the form of an order-*n* SPHERICAL HARMONIC, multiplied by a function $F_n(r, t)$ that combines the spherical radius (r) and time (t) in a particular manner (see the note below for details). At large distances, $F_n(r, t)$ has the following asymptotic form for all orders n:

$$F_n(r, t) \approx \frac{1}{r} f(t - r/c) \quad \text{(for outgoing waves), or}$$

$$F_n(r, t) \approx \frac{1}{r} f(t + r/c) \quad \text{(for incoming waves).}$$

Note: For outgoing waves, $F_n(r, t)$ has the general form

$$F_n(r, t) = L_n \left\{ \frac{1}{r} g(t - r/c) \right\}, \quad (n = 1, 2, 3 \ldots)$$

where L_n is a polynomial operator of degree n in the separate operators (r^{-1}, $c^{-1} \partial/\partial t$), and c is the sound speed. For incoming waves, the sign of c is reversed. The L_n operators for $n = 0, 1, 2$ are

$$L_0 = 1,$$

$$L_1 = \frac{1}{r} + \frac{1}{c} \frac{\partial}{\partial t},$$

$$L_2 = \frac{1}{r^2} + \frac{1}{cr} \frac{\partial}{\partial t} + \frac{1}{3c^2} \frac{\partial^2}{\partial t^2}.$$

spherical radius – *in spherical polar coordinates* the distance of a point from the origin. *Units* m.

spherical spreading – *in an outgoing wave field* a $1/r^2$ dependence of intensity on radius (r), as in the far field of a SPHERICAL PROGRESSIVE WAVE. It may be regarded as a consequence of energy conservation applied to wavefronts whose area increases in proportion to radius squared. Equivalently, in spherical spreading the level falls off at 6 dB per doubling of distance from the source.

spherical wave field – a combination of outgoing and incoming spherical progressive waves.

SPL – abbreviation for SOUND PRESSURE LEVEL. *Units* dB re $(20 \, \mu\text{Pa})^2$ (air or other gases); dB re 1 μPa^2 (water or other media).

s-plane – the COMPLEX PLANE whose coordinates are the real and imaginary parts of the LAPLACE VARIABLE *s*.

splitter – *in a duct silencer* a slab of sound-absorbing material placed edge-on to the flow direction, in such a way as to divide the duct cross-section into multiple passages. In order to reduce flow losses, the leading and trailing edges are commonly faired. Erosion of the sound-absorbing material is prevented by facing the material with a protective perforated (or porous) sheet.

splitter silencer – a DUCT SILENCER in which the flow is subdivided into parallel passages by sound-absorbing splitters.

spreading loss – (1) an equivalent term for DIVERGENCE LOSS (1). *Units* dB.

spreading loss – (2) *in underwater acoustics* an equivalent term for DIVERGENCE LOSS (2), i.e. the part of the TRANSMISSION LOSS (3) due to wavefront spreading. *Units* dB re 1 m^2.

SPR function – abbreviation for STRICTLY POSITIVE-REAL FUNCTION.

spring – an elastic element, capable of elongation or deflection under an applied load (*longitudinal* ~, *rectilinear* ~) or angular deformation under an applied torque (*torsional* ~). Unless otherwise qualified, the term usually refers to a spring of the first type.

spring constant – *for a spring that responds linearly to an applied load* the constant of proportionality, k, in an equation of the form

$$F = k\,\Delta x \quad \text{or} \quad G = k\,\Delta\theta.$$

The first equation relates the transmitted force, F, to relative displacement, Δx, in a longitudinal or rectilinear spring, and the second relates the transmitted torque, G, to relative angular displacement, $\Delta\theta$, in a torsional spring. *Units* N m^{-1} (longitudinal), N m rad^{-1} (torsional).

square-integrable – a function $f(t)$ is said to be square-integrable if

$$\int_{-\infty}^{\infty} |f(t)|^2\,dt < \infty.$$

Here t is a real variable that can take any value between $-\infty$ and ∞.

square matrix – a MATRIX with the same number of rows as columns. Important scalar properties of a square matrix **A** are its *determinant*, det **A**, and its TRACE, tr **A**. For the 2×2 case, these are defined by

$$\mathbf{A} = \begin{bmatrix} a_{11} & a_{12} \\ a_{21} & a_{22} \end{bmatrix}, \ \det \mathbf{A} = a_{11}a_{22} - a_{12}a_{21};\ \operatorname{tr} \mathbf{A} = a_{11} + a_{22}.$$

square root – the square root of a non-zero number (real or complex) has two possible values, one the negative of the other. The following two cases can usefully be distinguished.

(1) When x is real and positive, the symbol $\sqrt{x}$ is interpreted as the positive square root.

(2) More generally, the usual convention for complex numbers z (adopted in this dictionary) is that

$$-\pi/2 < \arg\sqrt{z} \leq \pi/2;$$

this implies that $\mathrm{Re}\sqrt{z}$ is always non-negative. For describing wave propagation in terms of complex wavenumbers with $e^{j(\omega t - kx)}$ as the propagation factor, however, the alternative convention

$$-\pi < \arg\sqrt{z} \leq 0$$

has advantages; it implies that $\mathrm{Im}\sqrt{z}$ is never positive, and therefore a propagation wavenumber k obtained from an expression like $k = \sqrt{k_0^2 - \kappa^2}$ always corresponds to non-amplifying waves.

Sr – symbol for STROUHAL NUMBER; also written Sr. *Units* none.

SSP – *in underwater acoustics* abbreviation for SOUND SPEED PROFILE.

stability – *of a system* a tendency for the system to return to its original position following a disturbance.

stable system – a LINEAR SYSTEM is said to be stable if any bounded input produces a bounded output. A nonlinear system that is ***linearly stable*** is able to respond to small disturbances without the response amplitude increasing as time progresses: a bounded input produces a bounded output. Some nonlinear systems are linearly stable, but exhibit ***nonlinear instability***, meaning that finite disturbances can precipitate large changes in configuration from which the system does not recover.

Note: A stable linear time-invariant system has a transfer function $\tilde{H}(s)$ that is an ANALYTIC FUNCTION of s (i.e. differentiable, and therefore free of poles) in the right-hand half of the complex S-PLANE. The corresponding FREQUENCY RESPONSE FUNCTION $H(\omega)$ is then defined by

$$H(\omega) = \tilde{H}(j\omega) \quad (\omega \text{ real}).$$

Compare MINIMUM PHASE SYSTEM.

stagnation pressure – (1) *in a compressible fluid flow* the pressure produced by compressing the fluid isentropically from its local value of specific ENTHALPY, h, to the stagnation value $h + \frac{1}{2}u^2$. Here u is the fluid velocity magnitude. Equivalent names are ***total pressure*** and ***reservoir pressure***. *Units* Pa.

Note (1): The stagnation pressure in a steady subsonic flow can be measured using a Pitot tube aligned with the flow.

Note (2): If the Mach number of the local flow is much less than 1, the difference $P_t - P$ between the stagnation and static pressures is close to the dynamic pressure $\frac{1}{2}\rho u^2$, where ρ is the fluid density. A better approximation for finite Mach numbers M is

$$P_t - P \approx \frac{1}{2}\rho u^2 \left(1 + \frac{1}{4}M^2\right).$$

stagnation pressure – (2) *in an incompressible fluid flow* the sum $P + \frac{1}{2}\rho u^2$, where P is the local pressure, ρ is the fluid density, and u is the fluid velocity magnitude. This is a limiting special case of the more general definition above. *Units* Pa.

standard ambient conditions – *in air at sea level* a pressure of 1 atmosphere, a temperature of 20°C, and a relative humidity of 65%.

standard deviation – *of a random variable* the square root of the VARIANCE. For a quantity with zero mean, the standard deviation and rms value are the same.

standard linear solid – an idealized model of a linear viscoelastic material with time-independent properties, that contains the Maxwell and Voigt models as special cases. The stress-strain relationship has the general form

$$a\sigma + b\dot{\sigma} = a'\varepsilon + b'\dot{\varepsilon} \quad (a, b, a', b' \text{ constants}).$$

Also known as a ***Maxwell–Kelvin model*** or ***Maxwell–Voigt model***. Compare the more general RELAXATION MODEL.

standing wave – (1) *in one dimension* a combination of two progressive waves at the same frequency, propagating in opposite directions. Interference produces a standing (i.e. fixed in space) pattern of peaks and troughs in the wave amplitude, with a spatial period equal to half the wavelength of the progressive waves.

standing wave – (2) *in two or three dimensions* interference between waves travelling in different directions, leading to a spatial variation in the amplitude of a single-frequency wave field (for example the sound pressure in a room). An equivalent term is ***interference pattern***.

standing wave ratio (SWR) – *at a given frequency in a one-dimensional standing wave* the ratio of maximum to minimum pressure amplitude. *Units* none.

Note: The standing wave ratio is an important parameter in STANDING WAVE TUBE measurements.

standing wave tube – an alternative term for IMPEDANCE TUBE.

stapedial reflex – reflex contraction of the STAPEDIUS MUSCLE in response to any stimulus. Compare ACOUSTIC REFLEX.

stapedius muscle – a muscle attached to the stapes, responsible for the ACOUSTIC REFLEX.

starting transient – see TRANSIENT (3).

static pressure – *in fluid dynamics* an alternative term for PRESSURE; it is used to emphasize the distinction between pressure (in its ordinary sense) and STAGNATION PRESSURE. *Units* Pa.

Note: Static pressure is not to be confused with AMBIENT, MEAN, or steady pressure.

stationarity – *of a random process* invariance of statistical properties with respect to the time origin. More specifically, if $X(t)$ is a time-dependent RANDOM PROCESS with individual realizations $x(t)$, stationarity implies that the JOINT PROBABILITY DENSITY FUNCTION of $x(t_1)$ and $x(t_2)$ depends only on the time difference $t_2 - t_1$. It follows that the mean and variance of $x(t)$, regarded as ENSEMBLE AVERAGES, are independent of t; also that the covariance of $x(t_1)$ and $x(t_2)$ depends only on the time difference. See also ERGODICITY.

Note: A distinction can be drawn between *wide-sense stationary* (or *weakly stationary*) processes, for which the BIVARIATE joint pdf is invariant with respect to the time origin as described above, and *strict-sense stationary* (or *strongly stationary*) processes, for which the multivariate or general N-fold pdf is similarly invariant. The former does not imply the latter, except for Gaussian random processes.

stationary phase – a method for the asymptotic estimation of integrals of the form $\int e^{jKh(x)} g(x)\, dx$, in the limit $|K| \to \infty$. The idea of the method is that for large $|K|$, the value of the integral is dominated by regions of x for which $h'(x) \approx 0$, i.e. where the phase is a slowly-varying function of x. Elsewhere, the rapid oscillations of the integrand lead to cancellation.

Note: The stationary phase method is particularly useful in wave problems, where the limit $|K| \to \infty$ typically corresponds to the far field or to the high-frequency limit.

stationary signal, stationary stochastic process – a signal or stochastic process that exhibits STATIONARITY.

stationary wave – an alternative term for PURE STANDING WAVE.

statistic – a quantity, such as the MEAN or STANDARD DEVIATION, that provides an overall or aggregate measure of some kind. Statistics may be based on an entire POPULATION or on a SAMPLE.

statistical absorption coefficient – *of a boundary in an acoustic field* the directionally-averaged value defined by averaging the plane-wave sound power absorption coefficient over the entire hemisphere of incident directions, with equal weighting for all directions. In mathematical form, under the assumption that α depends on the angle of incidence, θ, but not on azimuthal direction ϕ, the statistical absorption coefficient is given by

$$\alpha_{\text{stat}} = 2 \int_0^{\pi/2} \alpha(\theta) \sin\theta \cos\theta \, d\theta$$

where $\alpha(\theta)$ is the plane-wave absorption coefficient for sound waves incident at angle θ. Also known as the **random-incidence absorption coefficient**. *Units* none.

statistical energy analysis – a technique for estimating time-average and ensemble-average vibrational energy levels within a population of similar subsystems making up a complex structure or assembly, based on energy exchange relationships between the subsystems. The technique works reliably only if the subsystems are weakly coupled and each subsystem has several resonances within the excitation bandwidth.

Note: In practice, frequency averaging is often substituted for ensemble averaging, although the two are equivalent only for weakly-coupled systems.

statistically homogeneous – having statistical properties that are independent of position.

statistically independent – (1) *referring to two random variables* implies that the value taken by one variable has no influence on the value taken by the other. In terms of CONDITIONAL PROBABILITIES,

$P(X \leq x \mid Y \leq y) = P(X \leq x)$

where X and Y denote the two random variables. Equivalently, the joint cdf of X and Y is the product of the separate CUMULATIVE DISTRIBUTION FUNCTIONS:

$F_{XY}(x, y) = F_X(x) F_Y(y)$.

statistically independent – (2) *referring to two events* implies that the outcome of one event has no influence on the outcome of the other. In terms of CONDITIONAL PROBABILITIES,

$P(A \mid B) = P(A)$

where *A* and *B* are the two events. Equivalently, the JOINT PROBABILITY of *A* and *B* is the product of the separate probabilities:

$P(A, B) = P(A) P(B)$.

statistically isotropic – having statistical properties that are independent of direction (in a plane, or in three dimensions).

statistical model – an idealized description, based on a limited number of parameters, of the STATISTICS (2) of a random variable or process.

statistical room acoustics – the study of reverberation and sound transmission processes in rooms, on the basis of an assumed steady-state distribution of energy among the room modes. To a first approximation, this approach requires a knowledge of only the volume, surface area, and EQUIVALENT ABSORPTION AREA of the room. Also known as SABINE ACOUSTICS.

statistical sound power transmission coefficient – *of a partition* the directionally-averaged value defined by averaging the plane-wave SOUND POWER TRANSMISSION COEFFICIENT over the entire hemisphere of incident directions, with equal weighting for all directions. In mathematical form, under the assumption that τ depends on the angle of incidence, θ, but not on azimuthal direction ϕ, the statistical absorption coefficient is given by

$$\tau_{\text{stat}} = 2 \int_0^{\pi/2} \tau(\theta) \sin\theta \cos\theta \, d\theta$$

where $\tau(\theta)$ is the plane-wave sound transmission coefficient for sound waves incident at angle θ. Also known as the ***random-incidence sound power transmission coefficient***, or simply the ***sound transmission coefficient***. *Units* none.

statistical weight – *applied to an observation in statistics* an equivalent term for a WEIGHTING FACTOR. *Units* none.

statistics – (1) *used as a collective singular noun* the branch of mathematics concerned with data analysis and interpretation, where the data arise from multiple observations of an event or variable.

statistics – (2) *used in the plural* a collection of derived parameters obtained from statistical analysis of a data set or SAMPLE; the plural of STATISTIC.

steady-state response – *of a system to a steady sinusoidal input* the response that persists after any starting transients have decayed to zero.

stepped-plate transducer – a circular plate of stepped profile in the radial direction, driven at its centre by a shaker. At its design frequency, the plate resonates in an axisymmetric flexural mode, with high radiation efficiency owing to the phase shift introduced by the stepped profile. By suitable design of the profile, the radiation can be formed into a parallel beam or focused to a point.

STI – abbreviation for SPEECH TRANSMISSION INDEX. *Units* none.

stiffness – *of an elastic lumped element* the rate of change of force with linear extension (as in a rectilinear or longitudinal spring), or the rate of change of torque with relative angular displacement. If the elastic element behaves linearly, an equivalent term is SPRING CONSTANT. *Units* N m^{-1} (rectilinear), N m rad^{-1} (rotational).

stiffness matrix – *of a linear viscoelastic material* the frequency-dependent complex matrix **D** in the following stress-strain relationship for a general anisotropic material under sinusoidal excitation:

$$\mathbf{s} = \mathbf{D}\mathbf{e}, \quad \text{or equivalently} \quad s_i = D_{ij}e_j.$$

Here $\mathbf{s} = \{s_i\}$ is a column vector of the 6 stress components, represented as complex amplitudes, and $\mathbf{e} = \{e_j\}$ is the column vector of corresponding strain components. *Units* Pa.

Note: In an ISOTROPIC material the elements of **D** are determined by the complex shear modulus, G, and the complex bulk modulus, B. See also COMPLEX MODULUS.

stochastic process – a family of time-dependent SIGNALS for which the value $X(t_0)$ at a specified time t_0 may be regarded as a RANDOM VARIABLE, since infinitely many different realizations of the complete signal $X(t)$ are possible. Each realization of $X(t)$, called a **sample function**, is a random waveform whose future evolution cannot be predicted with certainty from a knowledge of its earlier behaviour. An equivalent term is **random process**.

Note that the statistics of a stochastic process will in general evolve over time; see ENSEMBLE AVERAGE, STATIONARITY. Examples of stochastic processes are speech signals, radar signals, television signals, and noise.

Stokes drag – *on a solid spherical particle* the steady viscous force exerted on the surrounding fluid by a rigid sphere, when it moves at low Reynolds number through a uniform fluid. The Stokes drag for a particle of diameter d, moving through a fluid of viscosity μ, is

$$\mathbf{F} = 3\pi\mu d(\mathbf{u}_p - \mathbf{u}_\infty),$$

where $\mathbf{u}_\infty$ is the velocity of the fluid far from the particle and $\mathbf{u}_p$ is the velocity of the particle (**G G STOKES 1845**). *Units* N.

Stokes number – *for oscillatory viscous flow past a solid body* the dimensionless ratio

$$S = \frac{\omega d^2}{v}$$

where ω is angular frequency, d is a typical linear dimension of the body, and v is the kinematic viscosity of the fluid. It equals $2(d/\delta)^2$, where δ is the viscous penetration depth $(2v/\omega)^{1/2}$. An alternative symbol is Sk. *Units* none.

Note (1): The square root of S is sometimes called the **shear number**.

Note (2): Two other physical interpretations of the Stokes number are:

(a) *for viscous oscillatory flows* the ratio of inertia forces to viscous forces;

(b) *for fluid flows with suspended particles* the ratio τ_d/τ_f, where τ_d is the equilibration time, or relaxation time, for a particle of different density as it moves relative to the fluid, and τ_f is a typical time scale for the unsteady flow.

Stoneley wave – *at a plane interface between two solid half-spaces in welded contact* a type of non-dispersive INTERFACE WAVE that propagates unattenuated (provided both media are lossless), at a speed less than the shear wave speed in either medium (**R STONELEY 1924**).

Note (1): If the shear wave speeds are (b_1, b_2), and the lower shear wave speed (say b_2) belongs to the medium of lower density, the existence of Stoneley waves requires b_2 to lie in the narrow range

$0.8742 b_1 \leq b_2 \leq b_1$ (for Poisson's ratio $v \geq 0$);

the Stoneley wave velocity v_{ST} then lies in the range

$0.8742 b_1 \leq v_{ST} \leq b_2$.

On the other hand, in the less likely event that b_2 belongs to the medium of higher density, Stoneley waves can exist for all b_2/b_1 less than 1. Their velocity then lies in the range

$0.8742 b_2 \leq v_{ST} \leq b_2$.

Note (2): The existence of Stoneley waves always requires the shear wave speed in the lighter medium to exceed the RAYLEIGH WAVE speed v_R in the denser medium (compare PSEUDO-STONELEY WAVES). The Stoneley wave speed v_{ST} then lies between v_R and b_2 (the lower shear wave speed).

Note (3): Compare SCHOLTE WAVE. Scholte waves propagate along a fluid–solid interface at an entirely different speed (slower than the sound speed in the fluid, as well as slower than the shear wave speed in the solid). However, in geophysics and geoacoustics any kind of interface wave at a fluid–solid or solid–solid boundary tends to be referred to as a Stoneley wave.

stop band – (1) *of a periodic filter with lossless elements* a frequency band over which the transmission loss is infinite. An acoustical example is provided by a periodic waveguide with purely reactive elements, which transmits no sound power in a stop band; see also BLOCH WAVE.

stop band – (2) *of a filter designed to reject a range of frequencies* the frequency range over which the TRANSMISSION LOSS is large (analogue filter), or the frequency response function of the filter has a magnitude close to zero (digital filter).

storage modulus – the real part of a COMPLEX MODULUS. *Units* Pa.

strain – a measure of deformation in a material. For extensions in one direction only, the strain is the increase in length divided by the original length. For more general three-dimensional deformations one can define a ***strain tensor***, whose components ε_{ij} are related to gradients of the displacement field as follows (under conditions of small deformation):

$$\varepsilon_{ij} = \frac{1}{2}\left(\frac{\partial u_i}{\partial x_j} + \frac{\partial u_j}{\partial x_i}\right).$$

Here x_i are cartesian coordinates fixed in space ($i = 1, 2, 3$), and u_i are components of the vector $\mathbf{u}(\mathbf{x}, t)$ that represents the DISPLACEMENT of the material at position $\mathbf{x}$ and time t from its undeformed position. *Units* none.

Note (1): The strain tensor ε_{ij} can be represented as the sum of a deviatoric component (i.e. with zero TRACE), and an isotropic component proportional to the unit tensor δ_{ij}:

$$\varepsilon_{ij} = \varepsilon'_{ij} + \frac{1}{3}\varepsilon_{kk}\delta_{ij}.$$

The ε'_{ij} term is called the ***deviatoric strain***.

Note (2): Whether the motion is described in Eulerian coordinates – as in the expression above – or in Lagrangian coordinates makes no difference to the expression for strain in linear acoustics. However, for finite deformations (e.g. nonlinear wave propagation in elastic solids), Lagrangian coordinates are generally used, in combination with a general nonlinear strain expression.

strain rate – an equivalent term for DEFORMATION RATE. In linear viscoelasticity, the term is also used for the partial time derivative of the STRAIN, since the two quantities are indistinguishable for small strains. *Units* s^{-1}.

stratified medium – a medium whose properties vary spatially in one coordinate direction, but are uniform in the other two directions. To a first approximation, the atmosphere and oceans are ***horizontally stratified***, meaning that properties such as mean density and sound speed are uniform in horizontal planes.

stratospheric channel, stratospheric duct – a SOUND CHANNEL in the Earth's atmosphere, associated with a sound speed minimum in the tropopause (occurring typically between 11 km and 20 km altitude). Infrasound at frequencies of up to a few Hz can propagate in the channel for hundreds of km.

strength – *in auditorium acoustics* see SOUND STRENGTH. *Units* dB.

stress – *in a continuous material* the force per unit area exerted across any surface. If the force is perpendicular to the surface, the stress is called ***direct*** or ***normal***; if the force is parallel to the surface, the stress is called ***shear*** or ***tangential***. *Units* Pa.

Note (1): In a cartesian system of coordinates x_i (with $i = 1, 2, 3$), the stress in the j direction on a surface whose normal points in the i direction is denoted by σ_{ij}.

Note (2): By convention, the symbol σ_{ij} is interpreted as the force per unit area exerted on the surface by the material on the $+x_i$ side (so that a positive value of σ_{11} represents a tensile stress); but some authors use a compressive stress notation, with $p_{ij} = -\sigma_{ij}$.

stress-relaxation function – *for a linear viscoelastic material with time-independent properties* the function $\Phi(t)$ that appears in the integral

$$\sigma(t) = \int_{-\infty}^{t} \Phi(t - \tau)\dot{\varepsilon}(\tau)d\tau,$$

where $\sigma(t)$ and $\varepsilon(t)$ are the time-dependent stress and strain at a point in the material and the dot denotes a time derivative. This expression gives the stress at time t that is produced by a strain history $\varepsilon(\tau)$; compare CREEP FUNCTION. See also RELAXATION MODEL. *Units* Pa.

stretch ratio – *in a given direction* the factor by which the distance between two neighbouring points in a material increases under deformation. Also called ***stretch***. Symbol λ. *Units* none.

strictly coherent – see COHERENT.

strictly positive-real function – a complex function whose real part is never negative. Abbreviated as ***SPR function***.

stridulation – *in animal bioacoustics* production of a shrill creaking sound by rubbing together special parts of the body; grasshoppers, for example, rub a hind leg against a wing, and crickets rub their wings together.

string transverse-wave speed – *in a uniform string under tension* the propagation speed of free transverse waves of small amplitude, given by $\sqrt{T/m}$ where T is the tension in the string and m is the mass per unit length. *Units* m s^{-1}.

Strouhal number – *in aeroacoustics and unsteady fluid dynamics* the dimensionless ratio

$$S = \frac{fd}{U},$$

where f denotes frequency, d is a typical linear dimension, and U is a typical flow velocity (or velocity of an object relative to the surrounding fluid). It functions as a dimensionless frequency, in which the actual frequency f is scaled on the inviscid flow parameters U and d (**V STROUHAL** 1878). See also AEOLIAN TONES. Alternative symbol Sr. *Units* none.

structural acoustic coupling – the coupling of waves in an adjacent fluid to the vibration of a structure. The term is also used for the coupling of *in vacuo* structural modes that results from fluid loading. See VIBROACOUSTIC.

structural damping – refers to any damping mechanism in which the amplitude of the damping force, over a range of frequencies, is modelled as being proportional to the displacement amplitude (rather than the velocity amplitude). An equivalent term is ***hysteretic damping***.

structure-borne path – a sound transmission path that depends at some point on vibrational energy propagation through a solid structure, rather than being entirely fluid-borne.

structure-borne sound – vibrational energy that propagates in the form of elastic waves through a solid structure. The term generally implies eventual coupling of the elastic wave energy to sound waves in an acoustic medium, as in STRUCTURE-BORNE PATH.

Note: Structure-borne sound usually refers to energy in the audio frequency range.

structure form factor – an alternative term for TORTUOSITY.

subharmonic – *of a given fundamental frequency* an integer submultiple of the fundamental; e.g. $f/2$, $f/3$, etc., where f stands for the fundamental frequency.

subharmonic response – *of a nonlinear system* frequency components of the system response appearing at any of the integer submultiples $1/n$ of the driving frequency, where $n = 2, 3, 4$ etc.

subjective – relating to human responses to a stimulus. For example, a *subjective experiment* involves listeners making judgments on what they hear.

subjective diffuseness – *in room acoustics* a qualitative term related to the sense that sound is arriving at a listener from all directions. Compare SPATIAL IMPRESSION, LISTENER ENVELOPMENT.

submatrix – an element of a PARTITIONED MATRIX.

substantial derivative, substantive derivative – *in continuum mechanics* alternative terms for the MATERIAL DERIVATIVE.

summation convention – the convention that repeated subscripts, for example in cartesian subscript or matrix subscript notation, imply summation of the term in which they appear. Thus if a_{ij} are the elements of a 3×3 matrix (with $i, j = 1, 2, 3$), the notation a_{kk} stands for $a_{11} + a_{22} + a_{33}$, i.e. the TRACE of the matrix. Also known as the *Einstein summation convention*.

superfluidity – the phenomenon of frictionless flow that is observed in liquid ^{4}He at temperatures below about 2.2 K, and also in ^{3}He–^{4}He mixtures. The liquid helium behaves like a 2-fluid mixture: a normal viscous component coexists with a *superfluid* component, that has zero entropy and zero viscosity. The two types of bulk longitudinal wave motion that can propagate in such a medium are called *first sound* and SECOND SOUND: in first sound the two fluids move in the same direction, but in second sound they move in opposite directions in such a way that the local momentum density remains zero. A further type of superfluid longitudinal wave, called *third sound*, has been observed in surface films of both ^{3}He and ^{4}He; here the superfluid component oscillates parallel to the solid substrate, while the normal fluid component is held stationary by viscosity.

superposition – *for linear systems* the *principle of superposition* states that if $y_1(t)$ and $y_2(t)$ are the responses of a linear system to inputs $x_1(t)$ and $x_2(t)$ respectively, the responses can be combined additively: thus $y_1(t) + y_2(t)$ is the response to $x_1(t) + x_2(t)$.

superposition sum – an equivalent term for CONVOLUTION SUM.

superresolution – a group of techniques used to recover information about an acoustic source from far-field measurements, at scales that are smaller than the DIFFRACTION LIMIT OF RESOLUTION.

Note: Superresolution is theoretically possible for source distributions that are spatially bounded, because the spatial Fourier transform of a bounded function is analytic everywhere. If such a function is known over a finite interval (in this case, for wavenumbers with magnitude less than the acoustic wavenumber), it is in principle known for all arguments. The missing wavenumber information could therefore be reconstructed, given perfect noise-free data in the far field, by analytic continuation. However, the difficulties involved in achieving superresolution with real data are severe: resolution of the source down to a quarter of an acoustic wavelength ($\lambda/4$) probably represents a practical limit based on far-field measurements.

support – *in an auditorium* the sensation for performers that the auditorium is responding to the sound they produce. The degree of support is strongly influenced by the design of the performing platform, and of the reflecting surfaces in its vicinity.

supra-aural earphone – an EARPHONE that fits over the pinna and flattens it against the head.

surface acoustic wave – *at the surface of a solid* a generic term for an elastic wave that propagates along a solid surface, with or without fluid loading. See RAYLEIGH WAVE, PSEUDO-RAYLEIGH WAVE.

surface backscattering strength – *of the ocean surface, in underwater acoustics* the SOURCE LEVEL of the backscattered field radiated from unit area of the surface, minus the sound pressure level of the incident plane wave in dB re 1 μPa^2. In equation form, the surface backscattering strength SBS is given by

$$\text{SBS} = 10\log_{10}\left[\frac{\bar{r}^2 p_{\text{back}}^2(\bar{r})}{S p_{\text{inc}}^2}\right],$$

where $p_{\text{back}}(\bar{r})$ is the rms backscattered pressure measured at distance $\bar{r}$ from the scattering region and corrected for absorption, S is the scattering surface area, and p_{inc} is the rms pressure of the incident wave. Compare BISTATIC SURFACE SCATTERING STRENGTH. *Units* dB.

Note (1): Provided consistent units are used, the surface backscattering strength is numerically related to the DIFFERENTIAL SCATTERING CROSS-SECTION s_d of the scattering region (with dimensions of area per unit solid angle) by the equation

$$\text{SBS} = 10\log_{10}(s_d/S).$$

Note (2): If the scattering is isotropic, the bistatic surface scattering strength (SSS) is the same in any direction over the hemisphere. Its value is then related to the SURFACE SCATTERING COEFFICIENT, m_s, by

$$\text{SSS} = \text{SBS} = 10\log_{10}\frac{m_s}{2\pi}.$$

surface duct – *in underwater acoustics* a SOUND CHANNEL associated with an isothermal ***mixed layer*** next to the surface, that extends downwards between 25 m and 250 m (depending on wind and climatic conditions), and is maintained by wind-driven turbulent mixing. The increase in sound speed with depth (due to hydrostatic pressure) causes the isothermal layer to act as an acoustic waveguide, since the resulting upward ray curvature combines with reflection from the surface to trap shallow-angle rays close to the surface.

Note: Because of the free-surface boundary, there is a cutoff frequency below which the field in a surface duct is evanescent and no energy can propagate. The low-frequency cutoff is a wave phenomenon that normal ray models are unable to describe.

surface gravity wave – *at the free surface of a liquid* a wave that propagates along the free surface under the combined influence of fluid inertia and gravity. The phase speed c of such waves, in a uniform fluid layer of depth h over a rigid bottom, is a function of the wavenumber k given by

$$\frac{v_{ph}^2}{gh} = \frac{1}{kh}\tanh kh \quad (g = \text{gravitational acceleration}).$$

Compare INTERNAL WAVES.

surface layer – *of the ocean* the region extending a few tens of metres down from the surface, within which the temperature (and hence also the sound speed) exhibits significant diurnal and seasonal variability.

surface scattering coefficient – *in underwater acoustics* the coefficient m_s in the equation

$$W_s = m_s I_{inc}.$$

The left-hand side represents the total power scattered from a rough sea surface, per unit surface area, when the surface is irradiated by acoustic plane waves of intensity I_{inc} travelling in a specified direction. *Units* none.

surface scattering strength – *in underwater acoustics* see SURFACE BACK-SCATTERING STRENGTH, BISTATIC SURFACE SCATTERING STRENGTH. *Units* dB.

surface wave – *at the surface of a solid or liquid* a wave that propagates along a free surface, with a disturbance amplitude that decays with distance into the solid (e.g. a RAYLEIGH WAVE) or into the liquid (e.g. a SURFACE GRAVITY WAVE). The particle motion in such a wave follows elliptical orbits. Compare INTERFACE WAVE.

susceptance – the imaginary part of an ADMITTANCE.

SV waves – *in a horizontally stratified medium* vertically-polarized S WAVES; i.e. shear waves in which the particle motion lies entirely in the vertical plane that contains the wavenormal direction (or mode propagation direction, in the case of guided waves in a layer).

S waves – *in a solid medium* an equivalent term for SHEAR WAVES; see also SH WAVES and SV WAVES, which are particular types of S wave in a horizontally stratified medium. Compare P WAVES.

swept sine – a signal, usually of constant amplitude, that locally resembles a sine wave but whose instantaneous frequency changes with time. The dependence of frequency on time may take many forms, e.g. linear or logarithmic, and the rate of change will be much higher in transient testing than in quasi-steady state testing.

swept sinewave testing, swept-sine testing – there are two types of swept sinewave testing:

(1) In *quasi-steady state testing*, the instantaneous frequency of a constant amplitude sine-wave is slowly varied through the frequency range of interest. The technique is used in resonance searches, often as part of a test carried out to a vibration test specification, and can be used in resonance testing. In the latter application, care has to be taken that the frequency sweep rate is low enough that quasi steady state response conditions are achieved in the resonant system under test. If the sweep rate is too high, the principal errors are (a) that the peak response of a resonant system will be less than the true steady state maximum, (b) the frequency at which maximum response apparently occurs is shifted in the direction in which the excitation frequency is changing, and (c) the measured damping coefficient will be greater than the true value.

(2) In *transient testing*, a short duration, rapid frequency sweep signal (sometimes called a *chirp*) is used to excite a system over a wide range of frequencies. The electronically generated signal is usually of constant amplitude, with a linear variation of instantaneous frequency over time. System excitations are typically of about 1 s duration and cover wide frequency ranges, without the high frequency "roll off" or zeros in the Fourier transform modulus spectrum that occur in mechanically generated (impact) pulses of simple shape. A *linear rapid frequency sweep* (linear variation of instantaneous frequency with time) of constant amplitude has an essentially rectangular Fourier transform modulus spectrum, with high cut-off rates at the initial and final frequencies. The latter are readily controlled, making this form of excitation attractive in frequency response measurement, as well as for shock testing.

See also TIME DELAY SPECTROMETRY, which is a related time and frequency domain technique.

SWR – abbreviation for STANDING WAVE RATIO. *Units* none.

symmetric matrix – a SQUARE MATRIX that is symmetric about the main DIAGONAL. The square matrix **A**, with elements a_{ij}, is symmetric if $a_{ji} = a_{ij}$ (for all combinations of i and j). An equivalent statement is $\mathbf{A}^T = \mathbf{A}$, where $\mathbf{A}^T$ is the TRANSPOSE of **A**.

sympathetic response – resonant or near-resonant response of a mechanical or acoustical system, excited by energy from an adjoining system in steady-state vibration. Also known as *sympathetic vibration*.

synaesthesia, synesthesia – *in human sensory perception* the linking of different senses, so that (for example) a particular taste, or sound, is visualized by an individual as a shape or colour. An example is CHROMATIC AUDITION.

synchronous averaging – extraction of a periodic signal from a noisy waveform, by dividing the data record into contiguous segments with each segment starting at the same point in the cycle, and averaging over many segments to cancel out the noise and leave the desired signal. Compare EDUCTION, which operates on the same principle but relies on conditional sampling since the underlying signal to be extracted is not periodic.

system function – *of a linear time-invariant system* an alternative term for the TRANSFER FUNCTION of the system.

T

Tait–Kirkwood isentrope equation – a model equation of state for liquids under isentropic compression. It generalizes the corresponding model equation for gases, $PV^\kappa = $ const. (where the isentropic exponent κ is treated as a constant), by adding a term to the pressure (**P G Tait 1888, J G Kirkwood 1942**). Thus the pressure P and specific volume $V = 1/\rho$ are related by

$$(P + \Pi)V^k = \text{const.} \qquad \text{(along an isentrope)},$$

where Π and k are fluid properties that depend only on the specific entropy. The isentropic BULK MODULUS B is $k(P + \Pi)$, and the speed of sound c varies with pressure or density along the isentrope according to

$$\frac{c^2}{c_0^2} = \left(\frac{P + \Pi}{P_0 + \Pi}\right)^{1 - 1/k} = \left(\frac{\rho}{\rho_0}\right)^{k-1};$$

here subscript 0 denotes a reference state on the isentrope. The COEFFICIENT OF NONLINEARITY for a fluid that obeys the Tait–Kirkwood equation is

$$\beta = \frac{k + 1}{2} \qquad \text{(note that } \beta \text{ is constant along an isentrope)}.$$

Note: An alternative equivalent form of the equation is

$$P - P_0 = \frac{\rho_0 c_0^2}{k}\left[\left(\frac{\rho}{\rho_0}\right)^k - 1\right],$$

from which it follows that the successive nonlinearity parameters of order 2, 3, 4, etc. are $(k-1)$, $(k-1)(k-2)$, $(k-1)(k-2)(k-3)$, etc. Actual values of these parameters for water at $T_0 = 10\,°C$, $P_0 = 1$ atm are $B/A = 4.69$, $C/A = 57$, $D/A = 420$, showing poor agreement with the model beyond second order. See ISENTROPIC EQUATION OF STATE IN SERIES FORM.

tangential mode – *in a hard-walled room of cylindrical or prismatic shape with end-walls normal to the room axis* an acoustic mode or eigenfunction whose only spatial pressure variation is parallel to the end walls (i.e. transverse to the room axis). The mode may therefore be thought of as consisting of wavefronts that propagate parallel to the end walls of the room or enclosure.

Note (1): The term cylindrical is here interpreted in a general sense, meaning a three-dimensional object formed by translating a plane curve (the cross-section) in a direction normal to the plane (the axial direction). The cross-section need not be circular.

Note (2): Rectangular (cuboidal) rooms form a special case in which there are three room axes, and therefore three groups of tangential modes.

tangential mode count – *in a hard-walled room of cylindrical shape with orthogonal end walls* the number of tangential mode EIGENVALUES less than a stated value. A smoothed high-frequency estimate for rectangular rooms of total surface area S is $N_{tan} \approx k^2 S/(8\pi)$; this gives the number of eigenvalues less than k^2. For rooms of non-rectangular cross-section, N_{tan} is estimated by replacing the total area S in the expression above with the area of the end walls alone.

Note: To estimate the number of **tangential-mode natural frequencies** occurring below a given frequency, one replaces k above by ω/c, where ω is the angular frequency and c is the sound speed in the room.

tapering – *in signal processing* an equivalent term for WINDOWING. The term **linear tapering** is used when a window is applied to the data directly, and **quadratic tapering** when a window is applied to the autocorrelation function.

target – *in active sonar* any object from which echoes are returned.

target strength – *in active sonar* the SOURCE LEVEL of the backscattered field from a target, expressed in dB re 1 μPa^2 m^2, minus the sound pressure level of the incident plane wave in dB re 1 μPa^2. In equation form, the target strength TS is given by

$$TS = 10 \log_{10} \left[\frac{\bar{r} p_{back}(\bar{r})}{r_{ref} p_{inc}} \right]^2,$$

where $p_{back}(\bar{r})$ is the rms backscattered pressure measured at distance $\bar{r}$ from the target and corrected for absorption, r_{ref} is the reference distance of 1 m, and p_{inc} is the rms pressure of the incident wave. *Units* dB @ 1 m, or dB re 1 m^2.

Note: The target strength is a function of frequency. It is numerically related to the BACKSCATTERING CROSS-SECTION σ_{back} (expressed in m^2) by $TS = 10 \log_{10}(\sigma_{back}/4\pi r_{ref}^2)$.

Taylor series – an expansion of a continuous function $f(x)$ about an arbitrary point $x = a$, in the form of a power SERIES:

$$f(x) = f(a) + f'(a)\frac{(x-a)}{1!} + f''(a)\frac{(x-a)^2}{2!} + \ldots$$

(B TAYLOR 1715).

Taylor's transformation – *for sound propagation in irrotational flows* a transformation of space–time coordinates that removes the effect of an irrotational mean flow on the acoustic wave equation for the velocity potential, up to second order in the flow Mach number M (K TAYLOR 1978). The transformed variables ($\mathbf{X}$, T) are related to the original variables ($\mathbf{x}$, t) by

$$\mathbf{X} = \mathbf{x}, \quad T = t + \bar{\varphi}/c^2;$$

here $\bar{\varphi}$ is the velocity potential of the mean flow, with $\nabla \bar{\varphi}$ equal to the flow velocity.

Note: The propagation of acoustic disturbances in a homentropic potential flow is described by the following equation in transformed coordinates:

$$\frac{1}{c_0^2} \frac{\partial^2 \Phi}{\partial T^2} - \frac{\partial^2 \Phi}{\partial X_i^2} \approx 0 \quad \text{(relative error of order } M^2\text{)}.$$

Here c_0 is the unperturbed sound speed, and the velocity potential perturbation as a function of the new variables is denoted by $\Phi(\mathbf{X}, T)$. Note that the relationship of Φ to the acoustic pressure, and the conditions to be satisfied by Φ at boundaries, are modified by the transformation.

temperature inversion – *in the Earth's atmosphere* a layer within which the temperature increases with increasing height. A strong temperature inversion is generally associated with atmospheric stability (for more detail, see BUOYANCY FREQUENCY). Compare THERMOCLINE.

temporal phase – *of a field that varies sinusoidally with time* the product ωt, where ω is angular frequency and t is time. *Units* rad.

temporary threshold shift – the component of THRESHOLD SHIFT that shows a recovery with the passage of time after the apparent cause has been removed. Recovery usually occurs within a period ranging from seconds to hours. Abbreviated as **TTS**. *Units* dB.

tensors – quantities with any number of associated directions. In this sense tensors are a generalization of the scalar and vector quantities used in physics: tensors of order zero are scalars, like temperature; tensors of order one are vectors, like acceleration. An example of an order-2 tensor is the stress in a fluid.

An important feature of tensors of order one or more is that components with respect to different sets of coordinate axes are related in a specific way; in the case of vectors, this amounts to the familiar cosine rule for resolving vector components. See also ORDER (10).

tenth-decade – *used as a noun or adjective* abbreviation for ONE-TENTH DECADE.

terminal reverberation – *in an auditorium* acoustic REVERBERATION that is heard either when a continuous sound is stopped, or when an impulsive sound is produced in a room. For typical sounds, the time for which terminal reverberation remains audible is similar to the REVERBERATION TIME.

termination – the end of a waveguide at which a passive boundary condition is applied. If the waveguide is driven by a source at one end, the other end is called the termination.

termination admittance matrix – *for a termination in a multimode waveguide* the inverse of a TERMINATION IMPEDANCE MATRIX. *Units* $\mathrm{m\,s^{-1}\,Pa^{-1}}$.

termination impedance – *for plane-wave transmission* the ACOUSTIC IMPEDANCE at the end of a one-dimensional system or transmission line. *Units* $\mathrm{Pa\,s\,m^{-3}}$.

Note (1): For a multimode waveguide, see TERMINATION IMPEDANCE MATRIX.

Note (2): Similar difficulties of definition arise with the concept of termination impedance as are discussed under ACOUSTIC RADIATION IMPEDANCE. See also END CORRECTION.

termination impedance matrix – *for a termination in a multimode waveguide* a square matrix of complex coefficients that relates a column vector of modal acoustic pressures **p** to a column vector of modal velocities **u**, as in the relations

$$\mathbf{p} = \mathbf{Z}\mathbf{u} \quad \text{or} \quad p_m = Z_{ml} u_l.$$

Compare MODAL IMPEDANCE MATRIX. *Units* $\mathrm{Pa\,s\,m^{-1}}$.

thermal absorption – *in a composite medium with inhomogeneities much smaller than an acoustic wavelength* the absorption of acoustic energy that occurs due to oscillatory heat transfer between the component phases of the mixture, when sound is transmitted through the composite medium. For typical emulsions, thermal absorption is the dominant sound attenuation mechanism at low frequencies, since the contrast in the thermal-expansion parameter ε (see note below) is generally greater than the contrast in density. The latter is what controls the viscous absorption mechanism (where droplets move relative to the surrounding fluid).

Note: The attenuation coefficient due to thermal absorption, in a dilute suspension of spherical droplets or bubbles with disperse phase volume fraction $x \ll 1$, varies asymptotically as

$$\alpha_{\text{thermal}} \sim x f^2 d^2 (\Delta \varepsilon)^2.$$

Here $\varepsilon = \alpha/\rho C_p$ is a thermal-expansion parameter (with α = thermal expansivity, ρ = density, and C_p = constant-pressure specific heat); $\Delta \varepsilon = \varepsilon_2 - \varepsilon_1$; and subscripts 1, 2 refer to the continuous phase and disperse phase, respectively. For this expression to be valid the frequency f must be sufficiently low that the droplet diameter d is much less than the thermal penetration depth in either fluid; in addition the droplet spacing must be small compared with the acoustic wavelength.

thermal conductivity – *of a fluid or solid* the ratio of the heat flux vector to the negative of the temperature gradient, in an isotropic material. See FOURIER'S LAW OF HEAT CONDUCTION. *Units* $W\,m^{-1}\,K^{-1}$.

thermal diffusivity – *of a fluid or solid* the ratio $\kappa/\rho C_p$, where κ is the THERMAL CONDUCTIVITY, ρ is the density, and C_p is the specific heat at constant pressure. *Units* $m^2\,s^{-1}$.

thermal expansivity – abbreviation for VOLUME THERMAL EXPANSIVITY. *Units* K^{-1}.

thermal noise – (1) *in analogue measuring equipment* electronic noise with a characteristic broadband spectrum, that arises from thermal motions of electrons in conductors. It decreases as the temperature is reduced, and disappears at absolute zero (0 K).

thermal noise – (2) *in a fluid* random fluctuations in pressure that represent departures from a state of local thermodynamic equilibrium, and are due to the random motion of molecules.

thermal penetration depth, thermal skin depth – *in a heat-conducting fluid or solid* the quantity $\sqrt{2\nu_{th}/\omega}$, which is a measure of the distance over which single-frequency temperature disturbances propagate in the medium. Here ν_{th} is the THERMAL DIFFUSIVITY and ω is the angular frequency. Compare VISCOUS PENETRATION DEPTH. *Units* m.

thermal unsteady boundary layer – *produced by interaction of a sound field with a boundary* a thin region, either side of an interface between two media of different thermal properties, within which there is unsteady conduction of heat. In the thermal boundary layer, the oscillatory temperature in each medium is the sum of an acoustic component (the value corresponding to zero heat conduction), and a thermal conduction component that allows the temperatures to match across the interface.

Note: The boundary layer qualifies as thin provided $\delta_{th} \ll (R, \lambda)$, where R is the interface radius of curvature, λ is the acoustic wavelength and δ_{th} is the thermal boundary layer thickness. Since δ_{th} is of order $\sqrt{2\nu_{th}/\omega}$, where ν_{th} is the THERMAL DIFFUSIVITY and ω is the angular frequency, this requires $\omega\nu_{th}/c^2$ to be much less than unity.

thermoacoustic – refers to thermal energy exchanges driven by sound (e.g. thermoacoustic refrigeration); also to thermal–acoustic interactions generally (e.g. unsteady combustion coupled to a sound field). For sound produced by optical absorption of a time-modulated laser beam, see OPTOACOUSTIC and PHOTOACOUSTIC.

thermocline – *in the ocean, or any large body of water* a layer within which the temperature decreases with increasing depth. A thermocline is generally associated with stability, although it may be destabilized by an upward gradient of salinity (see BUOYANCY FREQUENCY). Compare TEMPERATURE INVERSION.

thermometric conductivity – *of a fluid or solid* an alternative term for THERMAL DIFFUSIVITY. *Units* $m^2 s^{-1}$.

thermosonication – the combination of ultrasonic irradiation with subsequent heat treatment; usually employed to kill microorganisms.

thermospheric channel, thermospheric duct – a SOUND CHANNEL in the Earth's atmosphere, associated with a sound speed minimum in the thermosphere at around 100 km altitude. It can propagate low-frequency infrasound (typically below 0.5 Hz) for distances of order 1000 km before returning it to ground level.

thermoviscous fluid – a model of a real fluid that takes viscous stresses and heat conduction into account, but no other diffusive or relaxation processes.

third-octave – *used as a noun or adjective* abbreviation for ONE-THIRD OCTAVE.

third sound – a longitudinal surface wave associated with SUPERFLUIDITY.

threshold of hearing – *of a given sound* an equivalent term for ABSOLUTE THRESHOLD. Also known as ***threshold of audibility***. Compare HEARING THRESHOLD LEVEL. *Units* dB re $(20 \, \mu Pa)^2$.

threshold shift – *in psychoacoustics* the amount by which the ABSOLUTE THRESHOLD for a given listener is increased, for example through noise exposure or ototoxic drug administration. The shift may be either temporary, i.e. showing progressive recovery with time, or permanent. *Units* dB.

throat – the narrowest part of a passage or duct. See HORN THROAT, SONIC THROAT.

timbre – (1) the aggregate of attributes that allows a listener to distinguish a sound, in terms of subjective impression, from any other sound having the same loudness, pitch, and duration as well as the same direction of arrival.

timbre – (2) the tone colour or quality of musical sound. The term timbre can be applied to the sound of individual instruments or groups of instruments. It can also refer to the balance between low, middle and high frequencies in the musical sound. More subtle changes of timbre can be caused by the performing environment; see TONE COLOURATION and ACOUSTICAL GLARE.

time-average sound level – *for airborne sounds* an alternative term for EQUIVALENT CONTINUOUS SOUND LEVEL, understood to be based on the A-WEIGHTED sound pressure, with uniform time weighting of the squared pressure over a specified duration. Symbols L_{Aeq}, $L_{Aeq,T}$. *Units* dB re $(20\,\mu Pa)^2$.

time-average sound pressure level – an alternative term for EQUIVALENT CONTINUOUS SOUND PRESSURE LEVEL. Note that it does not imply that the level in decibels is time-averaged. Because the second term conveys the desired meaning more clearly, it is generally preferred. Symbols L_{eq}, $L_{eq,T}$. *Units* dB re p_{ref}^2.

time constant – *of an exponential decay process* the time taken for a specified variable to decay by a factor $1/e$. It is the constant τ by which the time t is normalized, in the exponential decay factor $e^{-t/\tau}$. *Units* s.

time delay response – *for a linear time-invariant system* an alternative term for the ENERGY–TIME CURVE.

time delay spectrometry – *for the characterization of linear time-invariant systems* an implementation of swept sine testing in which a linear frequency sweep is used to drive the system under test, and the output is subjected to combined time–frequency analysis. Delays in the system response (for example, due to echoes) appear as frequency shifts between the input and output signals, and can therefore be recognized in the frequency domain.

time factor – *for single-frequency variables represented as complex quantities* the expression $e^{j\omega t}$ that contains the dependence of the variable on time, t. Here ω is the angular frequency, and j is the SQUARE ROOT of -1. *Units* none.

Note: Some authors, particularly in the physics community, write the time factor as $e^{-i\omega t}$. Symbols i and j have the same meaning, but it is convenient to distinguish the $e^{j\omega t}$ and $e^{-i\omega t}$ conventions by means of a different letter symbol.

time-history – *over a specified time interval* a continuous record of a signal as a function of time. An example would be the time-history of the acceleration measured at a point on a structure, during an impact. The term can also refer to a sequence of discrete sampled values, i.e. a ***sampled time-history***.

time-integrated intensity – ◆ *in a transient sound field* the vector quantity

$$\mathbf{N} = \int_0^T p\mathbf{u}\, dt \qquad (t = \text{time}),$$

which represents the total acoustic energy transfer (in magnitude and direction) per unit area at a fixed point, integrated over the duration T of the transient sound field. Here p is the acoustic pressure and $\mathbf{u}$ is the vector particle velocity at the point in question. Compare ACOUSTIC INTENSITY. *Units* J m^{-2}.

time-integrated sound pressure squared – *for transient sound signals* the integral with respect to time of the instantaneous squared sound pressure, $p^2(t)$; an equivalent term is PRESSURE-SQUARED INTEGRAL. Compare SOUND EXPOSURE, which is a similar quantity but with FREQUENCY WEIGHTING applied to the pressure signal before squaring. *Units* $\text{Pa}^2\,\text{s}$.

time-interval equivalent continuous sound level – *for airborne sounds* an alternative term for EQUIVALENT CONTINUOUS A-WEIGHTED SOUND PRESSURE LEVEL. Also known as *time-average sound level*. Symbols L_{Aeq}, $L_{\text{Aeq},T}$. *Units* $\text{dB re }(20\,\mu\text{Pa})^2$.

time-reversal mirror – *for acoustic waves* an array of sources and receivers that generates a time-reversed replica of any incident wavefront, and sends it back towards the source. In general only a limited-duration sample of the incident field can be reversed, but reversal of a *periodic* field is achievable with no limit on duration. See PHASE CONJUGATION.

time series – a discrete SEQUENCE of values presented as a function of time, usually with equal intervals between successive vales.

time series analysis – the analysis of TIME SERIES data in order to identify trends and periodicities, and possibly to predict future behaviour. Compare SIGNAL PROCESSING.

time weighting – the smoothing or averaging process (usually an exponentially-decaying weighting factor) that is applied to the square-law rectified signal in a SOUND LEVEL METER. The time weightings S and F (*slow* and *fast*) provide exponential smoothing over periods of order 1 s and 125 ms respectively. The I-weighting (*impulse*), characterized by a fast rise time and slow decay time, is sometimes used to measure impulsive sounds.

Note: The S and F weightings both yield the same result for steady sounds, but not for time-varying sounds such as speech or vehicle pass-by noise. For details of the exponential weighting process, see the note under SOUND LEVEL.

tinnitus – a sensation of sound in the ear that exists without any external mechanical or acoustical stimulus. Occasionally, emission of sound from the ear can be detected objectively.

TL – (1) abbreviation for TRANSMISSION LOSS (1, 2). *Units* dB.

TL – (2) *in underwater acoustics, for sound propagation between a source and receiver* abbreviation for TRANSMISSION LOSS (3). *Units* dB re 1 m^2.

tone – a single-frequency sound, usually in the audio frequency range. Sometimes called a *pure tone*.

tone colouration, -oration – *associated with a listening environment* a change in the TIMBRE of music, introduced either by a room or by artificial reverberation. Most commonly, tone colouration refers to the shrill sound caused by strong early reflections, especially those arriving at the listener from above. The effect is particularly noticeable for the sound produced by the strings in an orchestra.

tone-corrected perceived noise level – an adaptation of the PERCEIVED NOISE LEVEL to allow for any prominent 1/3-octave band levels that stand out, in the noise being measured, above a broadband background spectrum. The correction is intended to account for the extra subjective noisiness caused by pronounced audible tones. *Units* dB re $(20\,\mu\text{Pa})^2$.

tone protrusion – *in a measured spectrum* the amount by which the BAND LEVEL containing a particular tone exceeds the adjacent band levels on either side. The amount of tone protrusion depends on the analysis bandwidth. *Units* dB.

Note: A criterion for assessing noise spectra is that a tone in a broadband spectrum is considered subjectively prominent if the band level of the one-third octave containing the tone exceeds the average of the two adjacent band levels by 5 dB or more.

torque – a pure moment, with no net force. *Units* N m.

torsional impedance – an equivalent term for ROTATIONAL IMPEDANCE. *Units* N m s rad^{-1}.

torsional stiffness – *at a given point along a rod or beam* the ratio of the applied twisting moment to the resulting twist per unit length (i.e. rate of change of angle with distance) along the beam axis. The twisting moment is applied about the beam axis. *Units* N m^2 rad^{-1}.

torsional stiffness factor – *for a rod or beam of uniform material* a geometrical factor, related to the shape and dimensions of the rod cross-section; its product with the shear modulus of the material gives the TORSIONAL STIFFNESS of the rod. *Units* m^4.

tortuosity – a geometrical parameter related to the kinetic energy of an ideal incompressible potential flow through the material. For a given mean flow direction, it is the factor by which the fluid kinetic energy is increased, relative to a hypothetical material consisting of parallel uniform channels aligned with the mean flow.

An equivalent mathematical definition of the tortuosity, denoted by α_∞, is written below for mean flow in the x direction:

$$\alpha_\infty = \frac{\langle u^2 \rangle_V}{\langle u_x \rangle_A^2} \quad (\mathbf{u} = \text{incompressible potential velocity field}).$$

Here u, u_x denote the magnitude and x component of $\mathbf{u}$, and angle brackets denote spatial averages: thus $\langle \ldots \rangle_A$ is a 2D average in the plane x = const. (where A is the area of the plane occupied by fluid), and $\langle \ldots \rangle_V$ is a 3D average over the fluid (V represents the fluid-filled pores). Tortuosity is also called **structure form factor**. *Units* none.

Note (1): The tortuosity may be measured acoustically, by transmitting high-frequency sound through a thin layer of the material (of thickness much less than the acoustic wavelength λ in the material); "high-frequency" means that the VISCOUS PENETRATION DEPTH δ should be small compared with the typical pore size L, so that viscous effects require only a minor correction. Alternatively, it may be measured electrically, by using the analogy between potential flow in a fluid and electric current in a uniform conducting material. This method involves embedding a sample of the porous frame in an electrically-conducting matrix, and measuring the bulk electrical resistivity of the resulting composite: e.g. for a slab of total cross-sectional area A and thickness h, whose electrical resistance between opposite faces is R, the bulk resistivity is $\rho_B = RA/h$. The tortuosity then follows as

$$\alpha_\infty = \frac{\rho_B}{\rho_i} \phi \quad \text{(frame assumed non-conducting)},$$

where ϕ is the porosity of the sample and ρ_i is the electrical resistivity of the conducting matrix material.

Note (2): In the limit $\delta \ll L \ll \lambda$, the tortuosity α_∞ and the COMPLEX EFFECTIVE DENSITY $\rho(\omega)$ are asymptotically related:

$$\rho(\omega) \approx \rho_0 \alpha_\infty \quad \text{(ideal-fluid or inviscid limit)}.$$

Here ρ_0 is the equilibrium density of the fluid in the pores. See DYNAMIC

TORTUOSITY for a generalization of this relation to porous materials with significant viscous effects.

total absorbing area – *of a room* an equivalent term for ROOM ABSORPTION. *Units* m^2.

total harmonic distortion – *of a periodic waveform* a measure of how far the waveform departs from a perfect sinusoid. If the total power is P_t and the power in the desired sinusoidal component is P_s, then the **distortion power**, P_d, is defined as the difference $P_t - P_s$. The usual definition of total harmonic distortion, expressed as a percentage, is

$$\text{THD} = 100 \sqrt{P_d/P_s}.$$

The corresponding **distortion level** is calculated as $10\log_{10}(P_d/P_s)$. *Units* (for THD) %; (for distortion level) dB.

Note (1): Total harmonic distortion is commonly used as an indication of nonlinearity in a system or process. The system is driven by a sinusoidal input until a steady state is reached, and the following output quantities are measured:

(a) the power P_s at the input frequency;

(b) the distortion power P_d, defined as previously (i.e. $P_t - P_s$).

Inserting these in the definition gives the total harmonic distortion expressed as a percentage.

Note (2): Alternative terms for total harmonic distortion are **harmonic distortion**, **rms distortion**, and **distortion factor**.

total internal reflection – *at a plane interface between two lossless media* the total reflection of incident energy that occurs when the angle of incidence exceeds the CRITICAL ANGLE FOR TRANSMISSION. The transmitting medium must have a higher wave speed than the incident medium for this to occur.

Note: Total internal reflection is associated with evanescent waves in the transmitting medium. If the region of higher wave speed extends for a finite distance, and on the far side of the layer the wave speed drops again so that the wave is no longer evanescent, **tunnelling** can occur; this means that some of the incident energy penetrates through the evanescent wave region and into the propagating region beyond.

total pressure – (1) *in fluid dynamics* an equivalent term for STAGNATION PRESSURE; compare STATIC PRESSURE. *Units* Pa.

total pressure – (2) *in acoustics* the sum of the AMBIENT pressure and the instantaneous ACOUSTIC PRESSURE. *Units* Pa.

total reflection angle – *at a plane interface between two lossless media* an equivalent term for CRITICAL ANGLE FOR TRANSMISSION. The transmitting medium must have a higher wave speed than the incident medium for this angle to be real. Also known as **critical angle for total reflection.** *Units* rad (but commonly expressed in degrees).

total scattering cross-section – *of an object in a plane-wave incident field* an equivalent term for the SCATTERING CROSS-SECTION of the object, i.e. the total scattered power normalized by the incident intensity. Its value varies with the frequency and direction of the incident sound. Compare DIFFERENTIAL SCATTERING CROSS-SECTION. *Units* m^2.

trace – *of a square matrix* the sum of the DIAGONAL elements.

trace coincidence – an equivalent term for COINCIDENCE.

trace velocity – (1) *of a travelling-wave field over a specified plane* the velocity at which the field travels along the plane. For example, if a plane wave is incident at angle θ on a plane boundary, its trace velocity has magnitude $c/\sin\theta$, where c is the propagation speed of the wave. *Units* m s^{-1}.

Note: By introducing vector notation, an incident plane-wave pressure field with wavenormal **n** can be expressed as $p = f(t - \mathbf{n}\cdot\mathbf{x}/c)$. Its trace on the plane $\mathbf{x}\cdot\mathbf{b} = $ const., with unit normal **b**, has the form $p = f(t - \tau - \mathbf{q}\cdot\mathbf{y}/c)$: the trace velocity has magnitude $c/|\mathbf{q}|$ and direction **q**. Here **q**, equal to $\mathbf{n} - (\mathbf{n}\cdot\mathbf{b})\mathbf{b}$, is the in-plane component of **n**; τ is a constant, and **y** is a position vector in the plane.

trace velocity – (2) *of a wave field on a specified line* the velocity at which the field travels along the line. If a plane wave propagates with its wavenormal at angle θ to a given axis, its trace velocity along that axis is $v_w/\cos\theta$, where v_w is the propagation speed of the wavefront (i.e. the speed of advance normal to the wavefront surface). *Units* m s^{-1}.

Note: Trace velocity and phase speed are interchangeable in this context.

trace-velocity matching principle – (1) *at a plane homogeneous boundary* states that if acoustic waves in the adjacent medium have a trace velocity v_{tr} along the boundary, then the boundary response consists of waves that have the same velocity v_{tr}. The principle holds for any plane boundary provided its acoustic properties are time-invariant and spatially homogeneous.

trace-velocity matching principle – (2) *in the axial direction along a cylindrical boundary* states that if acoustic waves in the exterior (or interior) medium have a trace velocity component v_{tr} along the cylinder parallel to the axis, then the boundary response consists of waves that also have axial velocity

v_{tr}. The principle holds for any cylindrical boundary whose acoustic properties are time-invariant and axially homogeneous.

trace wavelength – *for single-frequency plane progressive waves travelling along a plane boundary* the distance between neighbouring wavefronts, measured along the boundary. A trace wavelength can also be defined along a specified *axis*, for example parallel to the axis of a cylinder on which sound waves are incident. *Units* m.

transducer – *in acoustics* a device for converting an acoustical input signal into an electrical output signal (or vice versa).

transfer admittance – *for a network with two or more ports, or any extended system with inputs and responses at two or more discrete points* the transfer admittances are the off-diagonal terms in a MULTIPORT ADMITTANCE MATRIX.

transfer function – *of a linear time-invariant system* the transfer function, also known as the *system function*, is the function $\tilde{H}(s)$ in the complex S-PLANE that relates the LAPLACE TRANSFORM of the system output, $Y(s)$, to the transformed input $X(s)$:

$$Y(s) = \tilde{H}(s)X(s).$$

The function $\tilde{H}(s)$ is the Laplace transform of the system's IMPULSE RESPONSE FUNCTION $h(t)$ in the time domain:

$$\tilde{H}(s) = \int_0^\infty e^{-st} h(t)\, dt.$$

Note (1): A transfer function gives the ratio of an output to an input quantity.

Note (2): It is common in acoustics to use the term transfer function for the FREQUENCY RESPONSE FUNCTION of a system. However, in control engineering and signal processing the two terms are kept distinct (the frequency response function being a special case of the transfer function, with $s = j\omega$). To emphasize the difference, symbol $\tilde{H}(s)$ is used in the dictionary for transfer functions and symbol $H(\omega)$ for frequency response functions, the equivalence being

$$H(\omega) = \tilde{H}(j\omega) \qquad (\omega \text{ real}).$$

transfer impedance – *for a network with two or more ports, or any extended system with inputs and responses at two or more discrete points* the transfer impedances are the off-diagonal terms in a MULTIPORT IMPEDANCE MATRIX. An example is the *acoustic transfer impedance*

$$Z_{12} = p_2/U_1 \qquad \text{(excitation at 1 and response at 2)},$$

which relates the complex pressure p_2 measured at point 2 to the volume velocity U_1 applied at point 1. See also TRANSFER RECEPTANCE.

Note: Compare the term MUTUAL IMPEDANCE used in the context of radiating arrays.

transfer matrix – *for a two-port mechanical system or network* a square matrix of complex coefficients that relates a vector of quantities measured at the input port ($\mathbf{q}_1$) to a vector of corresponding quantities measured at the output port ($\mathbf{q}_2$), according to the equation

$$\mathbf{q}_1 = \mathbf{T}\mathbf{q}_2.$$

Typically, the elements of $\mathbf{q}_1$ might be the complex pressure and velocity at port 1; $\mathbf{q}_2$ would then represent the same quantities at port 2, and $\mathbf{T}$ in this case is a 2×2 matrix. An older equivalent term is *transmission matrix*. Compare SCATTERING MATRIX.

transfer receptance – *for a point-excited mechanical system* the complex ratio of the displacement at one point to the force applied at another point, at a given frequency. *Units* $\mathrm{m\,N}^{-1}$.

transient – (1) *used as an adjective* of limited or short duration.

transient – (2) *used as a noun* any oscillation or signal of limited duration.

transient – (3) *used as a noun* the damped natural oscillation that forms part of the response of any linear system during the start-up phase of a STEADY-STATE forced oscillation; also called *starting transient*. The transient consists of one or more NATURAL MODES of the system oscillating at their natural frequencies.

Note: A transient in this sense is triggered by a transition between *any* two steady states of forced oscillation, one of which may be of zero amplitude; for example, a transient also occurs when the steady-state forcing is turned off. A more common name in this case is *ringing* or REVERBERATION (3).

transient signal – a DETERMINISTIC signal that is non-periodic and of limited duration.

transient source level – *in underwater acoustics, for a transient sound source* an alternative term for ENERGY SOURCE LEVEL. *Units* dB re $1\,\mu\mathrm{Pa}^2\,\mathrm{s}$ @ 1 m, or dB re $1\,\mu\mathrm{Pa}^2\,\mathrm{m}^2\,\mathrm{s}$.

transition – *in fluid dynamics* the process in which a flow changes from laminar to turbulent. The change in the opposite direction, which is uncommon, is called *reverse transition*.

transmissibility – *of a linear device used for sound or vibration isolation* the ratio of output to input amplitude, under single-frequency excitation. The definition requires both input and output to be of the same type, e.g. both forces or both displacements; the specific terms *force transmissibility* and *motion transmissibility* are used to indicate these two cases. *Units* none.

Note (1): The transmissibility of a device is not an inherent property, but depends on what the device is connected to.

Note (2): The transmissibility and INSERTION LOSS of an isolator are independent performance measures. However, there are two special limiting cases where one can be deduced from the other: these are excitation by a force source, and excitation by a velocity source. The relation under either of these conditions is

IL $= 10 \log_{10}(1/T_a)$;

here IL is the isolator insertion loss and T_a is the force or motion transmissibility (force for a force source, and motion for a velocity source).

transmission coefficient – *for sound pressure or particle velocity* the complex ratios

$T_p = p_\text{trans}/p_\text{inc}$ (pressure),

$T_u = u_\text{trans}/u_\text{inc}$ (velocity),

defined for single-frequency plane-wave transmission across a plane boundary or two-fluid interface. In the first equation p_trans and p_inc represent the complex amplitudes of the transmitted and incident acoustic pressure at the boundary; the second equation is similar, but p (sound pressure) is replaced by u (normal particle velocity). *Units* none.

Note: More precise names for these coefficients are *sound pressure transmission coefficient* and *normal velocity transmission coefficient*.

transmission coefficient matrix – *for a discontinuity in a multimode waveguide* a square matrix of complex coefficients that relates a column vector of modal transmitted pressures on the far side of the discontinuity to a column vector of modal incident pressures. There are two transmission coefficient matrices, one for waves incident from side (1) and one for waves incident from side (2), denoted in the relations below by $\mathbf{T}^{(1)}$ and $\mathbf{T}^{(2)}$:

$$\mathbf{p}_2^+ = \mathbf{T}^{(1)}\mathbf{p}_1^+, \qquad \mathbf{p}_1^- = \mathbf{T}^{(2)}\mathbf{p}_2^-.$$

Here superscript $+$ denotes waves travelling in the direction from (1) to (2), and $-$ denotes waves travelling in the opposite direction. Compare MULTIMODE SCATTERING MATRIX. *Units* none.

transmission factor – *of a horn* see HORN TRANSMISSION FACTOR. *Units* none.

transmission impedance – *of an acoustically-driven plate, partition, or thin-shell structure* an equivalent term for the SEPARATION IMPEDANCE of the partition. *Units* Pa s m^{-1}.

transmission loss – (1) *of a silencer or other element in a one-dimensional system* the reduction in sound power level, at a given frequency, between the incident-wave power arriving at the silencer and the transmitted-wave power leaving the silencer, when the silencer is anechoically terminated. *Units* dB.

transmission loss – (2) *of a partition* the reciprocal of the SOUND POWER TRANSMISSION COEFFICIENT (2), expressed in decibels; symbol TL. In mathematical form,

$$TL = 10 \log_{10}\left(\frac{1}{\tau}\right),$$

where τ is the transmission coefficient. Also known as **sound transmission loss**. *Units* dB.

Note: Compare SOUND REDUCTION INDEX.

transmission loss – (3) *in underwater acoustics, between specified source and receiver locations* the amount by which the sound pressure level at the receiver lies below the SOURCE LEVEL (i.e. the free-field sound pressure level in the radiation direction, scaled to the REFERENCE DISTANCE from the source, normally taken as 1 m). In equation form, the transmission loss TL is given by

$$TL(\bar{r}) = SL - SPL(\bar{r})$$

for a receiver whose vector position relative to the source is $\bar{r}$. Here SL is the source level, and SPL($\bar{r}$) is the sound pressure level at the receiver. Also known as **propagation loss**. *Units* dB re 1 m^2.

Note (1): Equivalently, if $p_{meas}(\bar{r})$ is the rms acoustic pressure measured at the receiver, the transmission loss is given by

$$TL(\bar{r}) = 10 \log_{10}\left[\frac{r_0 p_0}{r_{ref} p_{meas}(\bar{r})}\right]^2.$$

Here p_0 is the rms pressure radiated by the source in a lossless medium, at any far-field distance r_0 in the relevant direction, and r_{ref} is the reference distance. The product $r_0 p_0$ is a measure of the rms source strength.

Note (2): For purposes of defining p_0, the source is assumed to operate in a uniform medium whose properties (density ρ, sound speed c) are those of the actual medium at the source location.

Note (3): The normalized pressure $\tilde{p}_{meas} = p(\bar{r})/r_0 p_0$, called the **transmission loss pressure**, is commonly used as a field variable in analytical and computational studies of sound propagation, in place of the actual acoustic pressure.

transmission loss pressure – a normalized acoustic pressure used in underwater acoustics; see TRANSMISSION LOSS (3). *Units* m^{-1}.

transmission matrix – *for a two-port mechanical system or network* an older term (now becoming obsolete) for the TRANSFER MATRIX that relates the system variables at the input and output ports.

transmittance – *for sound waves incident on a boundary* an equivalent term for SOUND POWER TRANSMISSION COEFFICIENT. *Units* none.

transmitter – *in underwater acoustics* a transducer or transducer array designed to convert an electrical signal into an acoustic signal.

transpose – the transpose of a matrix $\mathbf{A}$, of size $M \times N$, is the $N \times M$ matrix formed by rearranging the elements so that element a_{ij} is relabelled a_{ji}; this process is equivalent to interchanging rows and columns. The resulting matrix is denoted by $\mathbf{A}^T$. Repeating the operation recovers the original matrix: $(\mathbf{A}^T)^T = \mathbf{A}$.

transversely isotropic – describes a material whose elastic properties are independent of direction in a given plane (e.g. the $x-y$ plane), but are different in the direction normal to that plane (z in this case).

transverse mode – *in a fluid-filled waveguide with rigid walls* any mode other than the plane wave.

transverse waves – waves in which displacements of the medium are everywhere parallel to the wavefronts. See also POLARIZATION. Compare LONGITUDINAL WAVES.

transverse wave speed – the propagation speed of free transverse waves. In bulk media, the term is equivalent to SHEAR WAVE SPEED. See also STRING TRANSVERSE-WAVE SPEED, MEMBRANE TRANSVERSE-WAVE SPEED. *Units* $m\,s^{-1}$.

travelling wave – alternative term for PROGRESSIVE WAVE.

trend removal – a SIGNAL PROCESSING operation whose purpose is the removal of any features in the signal whose time scale exceeds the RECORD length. For example, removal of a slowly-changing mean component from an oscillatory signal that suffers from DRIFT might reveal a stationary underlying signal. An alternative term for trend removal is ***detrending***.

truncation – cutting off a series after a certain number of terms, or a decimal or binary number after a certain number of digits. Also discussed under SERIES.

TS – *in underwater acoustics* abbreviation for TARGET STRENGTH. *Units* dB @ 1 m, or dB re 1 m^2.

tsunami – a destructive ocean wave generated by movement of the sea floor in mid-ocean (e.g. by underwater fault ruptures, subduction zone activity, or volcanic eruptions), or by underwater landslides. The period of the initiating disturbance can be as long as 1 hour, with associated wavelengths much greater than the ocean depth. Nonlinear steepening shortens and intensifies the wave as it approaches the shore, and heights of 20 m or more are typical at the shoreline.

TTS – abbreviation for TEMPORARY THRESHOLD SHIFT. *Units* dB.

tuned damper – a resonant VIBRATION ABSORBER that is lightly damped, and is therefore effective over a narrow frequency range around resonance.

turbulence – a chaotic type of eddying fluid flow that occurs at high REYNOLDS NUMBERS. It is characterized by rotational three-dimensional velocity fluctuations covering a wide frequency range, and a wide range of eddy sizes or length scales. Viscous dissipation of energy takes place at the smallest length scales, far removed from the larger scales that contain most of the kinetic energy.

turning point – *of a ray path in a stratified medium* an equivalent term for VERTEX.

two-dimensional spatial Fourier transform – an extension of the spatial Fourier transform to two dimensions, using cartesian coordinates (x, y). The Fourier transform of a field $u(x, y)$ is denoted by $U(k_x, k_y)$; it is given by

$$U(k_x, k_y) = \int_{-\infty}^{\infty} \int_{-\infty}^{\infty} u(x, y) \, e^{j(k_x x + k_y y)} \, dx \, dy,$$

with inverse

$$u(x, y) = \frac{1}{4\pi^2} \int_{-\infty}^{\infty} \int_{-\infty}^{\infty} U(k_x, k_y) \, e^{-j(k_x x + k_y y)} \, dk_x \, dk_y.$$

The wavenumbers k_x and k_y give the rate of change of phase with respect to distance in the x and y directions respectively.

Note (1): The two-dimensional Fourier transform is useful for propagating 3D single-frequency sound fields in the direction perpendicular to the x–y

plane. This involves transforming the field in the plane $z = z_1$ – denoted by $u(x, y, z_1)$ – with respect to x and y as above, and then propagating the wavenumber-domain field to the desired plane $z = z_2$. The inverse transform finally gives the required field $u(x, y, z_2)$.

Note (2): Two-dimensional spatial Fourier transforms are frequently used in acoustics, geophyics, and image processing. Practical problems can be caused by finite spatial windowing effects and by noisy data; noise is exponentially increased when the field is propagated back towards the source. See also HANKEL TRANSFORM.

two-point correlation – an abbreviation for the cross-correlation function between two fluctuating quantities measured at different positions. A typical example of a two-point correlation is the cross-correlation function between unsteady velocity components at two points, in a time-stationary turbulent flow.

two-port system – a system with one input and one output PORT (1). The linear response of the system is described by a TRANSFER MATRIX between the two ports.

tympanic membrane – the membrane that terminates the EXTERNAL EAR CANAL at its inner end. It vibrates in response to incident sound and transmits its vibration to the OSSICLES. Commonly referred to as the *eardrum*.

tympanometry – *in audiology* an OTOADMITTANCE TEST in which the acoustic admittance of the ear is measured over a range of values of applied mean pressure. Raising or lowering the mean pressure in the ear canal alters the stiffness of the eardrum.

U

ultraharmonic response – *of a nonlinear system* frequency components of the system response appearing at n/q times the driving frequency, where n and q are integers greater than 1 that do not have a common factor. Examples are $n/q = 3/2$ or $2/3$.

ultrasonic – refers to acoustic frequencies in the range (roughly) 20 kHz to 10^{13} Hz, and therefore normally inaudible. Hence ***ultrasonic frequency***, ***ultrasonic transducer***. Compare HYPERSONIC.

ultrasonic calorimeter – a device for measuring the total ultrasonic power radiated by a TRANSDUCER. It operates by absorbing the power in a thermally monitored container, and measuring the heating effect produced by dissipation.

ultrasonic interferometer – an instrument for determining ultrasonic attenuation and sound speed in fluids, as a function of frequency. It operates by measuring the standing waves set up between a TRANSDUCER and a movable reflector.

ultrasonics – the science and technology of ULTRASOUND.

ultrasound – sound at ULTRASONIC frequencies in any medium.

ultrasound Doppler flowmeter – a device that uses BACKSCATTERING of ultrasound from suspended particles in a flow to measure the flow velocity. The frequency of the backscattered signal from a moving particle is Doppler-shifted relative to the incident signal, and comparison of the two frequencies provides a measure of the particle's velocity component in the direction of the transducer.

ultrasound time-of-flight flowmeter – a device that uses time-of-flight measurements for ultrasound propagation between two transducers, in order to measure the average flow velocity along the insonated path. The acoustic travel time between the two transducers is longer for waves travelling against the flow than with the flow, and the time-of-flight difference provides a measure of the flow velocity component along the axis joining the transducers.

uncertainty principle – *for combined time and frequency localization of a transient signal* the principle that one cannot locate a signal with unlimited precision in both the time and frequency domains. Applied to spectral analysis, the uncertainty principle yields the well-known result that frequency resolution, Δf, and record duration or window length, T, are related by $\Delta f \approx 1/T$. Applied to time delay analysis, the corresponding result is $\Delta t \approx 1/B$; here Δt is the time resolution and B is the frequency bandwidth of the signal.

In mathematical form, the uncertainty principle is expressed by the inequality

$$\Delta\omega\,\Delta t \geq \frac{1}{2},$$

where $\Delta\omega$ and Δt respectively measure the spread of the signal in angular frequency and in time. The product $\Delta\omega\,\Delta t$ is sometimes called the **BT product** (bandwidth-time product) of the signal. For detailed definitions of $\Delta\omega$ and Δt, see the note below.

Note (1): Consider a real signal $x(t)$, with Fourier transform $X(f)$, whose total energy is finite:

$$\int_{-\infty}^{\infty} x^2(t)dt = \int_{-\infty}^{\infty} |X(f)|^2 df = E \qquad (E = \text{signal energy}).$$

Then the spread of the signal in time (Δt) and the spread in frequency (Δf) can be defined in terms of the second moments of $x^2(t)$ and $|X(f)|^2$ about their respective centroids:

$$(\Delta t)^2 = \frac{1}{E}\int_{-\infty}^{\infty}(t-\bar{t})^2 x^2(t)dt,$$

$$(\Delta f)^2 = \frac{1}{E}\int_{-\infty}^{\infty}(f-\bar{f})^2 |X(f)|^2 df.$$

The centroids $\bar{t}$ and $\bar{f}$, along the time and frequency axes, are given by

$$\bar{t} = \frac{1}{E}\int_{-\infty}^{\infty} t x^2(t)dt, \quad \bar{f} = \frac{1}{E}\int_{-\infty}^{\infty} f|X(f)|^2 df.$$

Note (2): As an example, the Gaussian pulse $x(t) = e^{-at^2}$ has a BT product of

$$\Delta\omega\,\Delta t = 2\pi\,\Delta f\,\Delta t = \frac{1}{2},$$

which is the minimum value possible.

uncorrelated – (1) *for signals* the opposite of correlated. See CORRELATED SIGNALS.

uncorrelated – (2) *for random variables* the same as orthogonal; see ORTHOGONAL RANDOM VARIABLES.

undamped natural frequency – *of a linear system* the natural frequency of the system with all damping removed. For example, a system whose displacement $x(t)$ is described by the equation

$$m\ddot{x} + R\dot{x} + kx = 0$$

has an undamped natural frequency f_0 given by

$$f_0 = \frac{1}{2\pi}\sqrt{\frac{k}{m}};$$

it is the frequency at which the system oscillates when the viscous damping coefficient R is set equal to zero. *Units* Hz.

underdamped – describes a system that repeatedly overshoots its equilibrium position after being displaced and released. See also DAMPING RATIO, CRITICAL DAMPING.

underexpanded jet – a supersonic gas jet that emerges from a nozzle at a higher pressure than that of the surrounding atmosphere, causing the formation of shock waves in the flow downstream of the nozzle. Compare IMPROPERLY EXPANDED JET.

undersampling – *in analogue-to-digital conversion* the use of too low a SAMPLING FREQUENCY for a particular ANALOGUE SIGNAL, so that the signal is ALIASED.

underwater acoustics – the science of sound propagation in the sea, and of sound radiation and scattering by underwater objects. Applications of underwater acoustics include ACOUSTICAL OCEANOGRAPHY and SONAR. From a fundamental point of view, the physics of sound in the ocean has areas in common with the physics of biomedical ultrasound (e.g. cavitation and scattering), and in both cases backscattered sound is used for imaging and detection of objects.

undisturbed-field equivalent sound level – ◆ *for an individual exposed to sound through headphones* the sound level of a diffuse sound field, incident on the head, that would produce the same level at the listener's eardrums as is produced by the headphones. *Units* dB re $(20\ \mu\text{Pa})^2$.

Note: The undisturbed-field equivalent sound level is calculated from the level at the eardrum by means of the diffuse-field HEAD-RELATED TRANSFER FUNCTION.

uniformity of sound – *in an auditorium* a good auditorium will have only small differences in acoustic quality between different audience seats. The term uniformity can refer to any attribute of sound in auditoria, but it is most commonly applied to sound level. Focusing by concave surfaces is one cause of non-uniformity.

unitary matrix – a square matrix **A** with the property that

$$\mathbf{A}^H \mathbf{A} = \mathbf{A}\mathbf{A}^H = \mathbf{I},$$

where $\mathbf{A}^H$ is the HERMITIAN TRANSPOSE of **A** and **I** is the IDENTITY MATRIX. Compare UNIT MATRIX.

unit circle – *in the complex plane* the circle $|z| = 1$; i.e. the set of complex numbers z with unit modulus.

unit impulse response function – *of a linear time-invariant system* an equivalent term for IMPULSE RESPONSE FUNCTION.

unit matrix – equivalent term for IDENTITY MATRIX. The unit matrix of size 2×2 has elements δ_{ij}, where δ_{ij} is the Kronecker delta.

unit normal – abbreviation for (*unit*) *normal vector*. See NORMAL VECTOR.

unit pulse function – see DIRAC DELTA FUNCTION, KRONECKER DELTA.

unit step function – equivalent term for HEAVISIDE UNIT STEP FUNCTION.

unit tensor – the unit tensor of order 2 has components δ_{ij}, where δ_{ij} is the Kronecker delta. See also TENSOR.

unit vector – a VECTOR with unit magnitude.

univariate – refers to STATISTICS that are formed from a single RANDOM VARIABLE. The term is used to distinguish them from JOINT STATISTICS, which are either BIVARIATE (two variables) or MULTIVARIATE (three or more variables).

universal gas constant – the GAS CONSTANT for one mole of gas; also known as the *molar gas constant*. Its value is 8.314 51 J mol^{-1} K^{-1}.

unwrapped phase – *of a system response function* a plot of the system PHASE (3) as a continuous function of frequency, obtained by keeping track of each 2π phase change. See also PHASE-SHIFT FUNCTION. *Units* rad.

V

van der Waals gas – a real-gas model, in the form of a P–ρ–T equation of state, that makes some allowance for departures from ideal-gas behaviour at finite densities. The van der Waals equation (**J D VAN DER WAALS 1873**) is

$$P = \frac{\rho RT}{1 - b\rho} - a\rho^2$$

where P is the gas pressure, ρ is the density, R is the specific gas constant, T is the absolute temperature and a, b are constants to be determined for each gas. By fitting this equation to the isotherm passing through the CRITICAL STATE, one obtains

$$a = \frac{27}{64}\frac{R^2 T_c^2}{P_c}, \quad b = \frac{RT_c}{8P_c} \quad \begin{cases} T_c = \text{critical temperature,} \\ P_c = \text{critical pressure.} \end{cases}$$

The speed of sound in a van der Waals gas is given by

$$\frac{c^2}{\tilde{\gamma} RT} = (1 - \tfrac{1}{8}\rho_r)^{-2} - \frac{27}{32}\frac{\rho_r}{\tilde{\gamma} T_r}$$

where $\tilde{\gamma}(T) = \gamma(T, P \rightarrow 0)$ is the specific-heat ratio in the low-pressure limit. The reduced temperature, T_r, and pseudo-reduced density, ρ_r, that appear in the sound speed equation are defined by

$$T_r = \frac{T}{T_c}, \quad \rho_r = \frac{\rho RT_c}{P_c}.$$

Note: The van der Waals parameter a actually varies significantly with ρ and T. An improved fit to real-gas data is obtained by allowing a to vary as $(1 + b\rho)\sqrt{T}$. However, the resulting sound speed equation is more complicated.

variance – *of a signal or random variable* the second moment, taken about the MEAN, of the numerical values taken by the signal or the variable. Thus a time-stationary signal $x(t)$ with mean $\bar{x}$ has variance

$$\text{var}\{x(t)\} = \langle [x(t) - \bar{x}]^2 \rangle = \langle x^2(t) \rangle - \bar{x}^2,$$

where the angle brackets denote a time average. See also COVARIANCE, SAMPLE VARIANCE.

variation – *of a functional* the derivative of the functional with respect to a parameter ε, evaluated at $\varepsilon = 0$; the parameter measures deviations of a function from a desired solution. For example, suppose the functional J is related to the function $y(x)$ by

$$J = \int_{x_1}^{x_2} F(x, y, y') \, dx.$$

The variation of J, relative to its value for a particular choice of $y(x)$, is defined in two stages. First a modified function $Y(x) = y(x) + \varepsilon\eta(x)$ is introduced, to give

$$J(\varepsilon) = \int_{x_1}^{x_2} F(x, Y, Y')\, dx.$$

Here $\eta(x)$ is an arbitrary function that vanishes at x_1 and x_2. Then the variation of J is defined by

$$\delta J = \left.\frac{dJ(\varepsilon)}{d\varepsilon}\right|_{\varepsilon=0} = 0.$$

See also CALCULUS OF VARIATIONS.

variational principle – *in mechanics* a statement that the behaviour of a system can be deduced either by requiring a specified FUNCTIONAL to be a maximum or minimum, or equivalently by requiring the VARIATION of the functional to be zero. A variational principle may be an independent postulate, such as HAMILTON'S PRINCIPLE, or may be derived from differential equations and boundary conditions governing the behaviour of a system. See also CALCULUS OF VARIATIONS.

VBS – *in underwater acoustics* abbreviation for the VOLUME BACKSCATTERING STRENGTH of a cloud of scatterers, expressed in decibels. *Units* dB re 1 m^{-1}.

vector – (1) a two- or three-dimensional quantity that has magnitude and direction, and whose component in the direction defined by a UNIT VECTOR **e** is the scalar product of the vector with **e**. Physical quantities such as force, velocity, DISPLACEMENT, ANGULAR VELOCITY and WAVE VECTOR are vectors in this sense.

vector – (2) an ordered set of N numbers, e.g. $(x_1, x_2, x_3, \ldots, x_N)$, is called a vector of length N. In MATRIX terminology, the vector $\mathbf{x} = \lfloor x_1, x_2, \ldots, x_N \rfloor$ is called a *row vector* and its TRANSPOSE $\mathbf{x}^T$ is a *column vector*.

vector potential – any SOLENOIDAL vector field **w** (representing displacement in a shear wave, for example) can be represented as the curl of another vector field **A**:

$$\mathbf{w} = \operatorname{curl} \mathbf{A} = \nabla \times \mathbf{A}.$$

The field **A** is then called the vector potential for **w**.

velocimeter – (1) an instrument for measuring the local speed of sound in a fluid. Typically, the travel time of a short acoustic pulse is measured between two transducers placed a known distance apart, one acting as a sound source and the other as a sound receiver.

velocimeter – (2) an instrument for measuring local velocities as a function of time. The quantity being measured may be the vibration velocity of a solid surface, or one or more components of the local flow-velocity vector at a point in a fluid. See also LASER VELOCIMETER.

velocity of sound – alternative term for SOUND SPEED. *Units* m s^{-1}.

velocity potential – if a fluid flow has a vector velocity field **u** that is irrotational and defined by

$$\mathbf{u} = \nabla \varphi, \quad \text{or alternatively} \quad \mathbf{u} = -\nabla \varphi,$$

then the scalar field φ is called the velocity potential of the flow. *Units* m^2 s^{-1}.

velocity profile – (1) *in fluid dynamics* a profile of fluid velocity as a function of a coordinate normal to the main flow direction. Examples are the ~ across a duct cross-section, or the ~ through a shear layer. The velocity is usually time-averaged.

Note: The velocity profile may be a function of either one or two coordinates transverse to the flow; it is called a *two-dimensional velocity profile* in the second case.

velocity profile – (2) *in underwater acoustics* the sound speed profile as a function of depth.

vertex – *of a ray path in a horizontally stratified medium* a point where the vertical coordinate of the ray trajectory is a maximum or minimum. An alternative name for a vertex is a *turning point*.

vestibular nerve – see AUDITORY NERVE.

vestibular system – the part of the human ear that senses translational and rotational acceleration of the head, and also its orientation with respect to gravity. Forming part of the labyrinth of the INNER EAR, it consists of the *semicircular canals* and the two vestibular sacs, all of which are filled with fluid.

Note: Angular or linear acceleration of the head induces fluid movement in the various cavities of the vestibular system. Sensors detect the movement and provide information about head motion to the central nervous system.

vibration – (1) any time-varying oscillation about a state of equilibrium. Vibration may be broadly classified as transient or steady-state, with a further subdivision into either deterministic or random vibration.

vibration – (2) *in mechanical engineering* the VIBRATION (1) of a structure or mechanical system.

vibration – (3) the science and technology of VIBRATION (2), particularly as applied to mechanical engineering. Also known as *vibrations*.

vibration absorber – a passive subsystem that is attached to a vibrating machine or structure in order to reduce its vibration amplitude over a specified frequency range. The absorber consists of a mass attached by a spring to the main system (for example a flexible cantilever beam, fixed to the structure at its root, with a mass fixed to its tip). An equivalent name is *vibration neutralizer*.

Note: At frequencies close to its own resonance, the vibration absorber works by applying a large local mechanical impedance to the main structure. A damped spring of stiffness K, carrying a lumped mass M at the free end, has a driving-point mechanical impedance at the opposite end whose value at the mass-spring resonance frequency is

$$R_{\text{res}} = \frac{1}{\eta}\sqrt{KM}, \qquad \text{(i.e. purely real)}.$$

Here η is the LOSS FACTOR of the mass–spring–damper system. The spring element sometimes has additional damping added; this increases the bandwidth over which the absorber operates, at the expense of a reduction in the absorber impedance at resonance.

vibrational energy – *of a structure* the energy associated with structural oscillations about a position of equilibrium. It consists of elastic and kinetic energy components. *Units* J.

vibration isolator – a resilient mounting system, or a resilient coupling between two pipes or structural elements, designed to reduce vibration transmission. The effectiveness of a vibration isolator is measured by its INSERTION LOSS, or alternatively by its ISOLATION EFFICIENCY.

vibration neutralizer – an equivalent term for VIBRATION ABSORBER.

vibroacoustic – an adjective used mainly by European acousticians; it refers to coupled fluid and structural waves in fluid-loaded structures. An equivalent term is *structural acoustic* or *structural–acoustical*.

vibroacoustics – the study of coupled fluid and structural waves in fluid-loaded structures.

vibroseis – a mechanical vibrator that excites SEISMIC WAVES by shaking the ground surface.

virtual acoustics – the science and technology of VIRTUAL ACOUSTIC SOURCE production. Signals are presented at the ears of a listener that produce the illusion of a source, whose position may be quite different from that of the transducers used to generate the signals.

virtual acoustic source – (1) *in psychoacoustics* an apparent source, perceived by a listener as being in a position where no actual source is present.

virtual acoustic source – (2) *in nonlinear acoustics* an equivalent source distribution that describes the nonlinear generation of sound at sum and difference frequencies from an interaction between two primary sound fields. Virtual sources of this type are used in PARAMETRIC ARRAYS.

virtual inertia – an equivalent term for ATTACHED MASS or VIRTUAL MASS. *Units* kg.

virtual mass – *of a rigid body in an incompressible fluid* the coefficient of proportionality between the force required to overcome fluid inertia, and the acceleration of the body. In general the force is in a different direction from the acceleration; this is accounted for by defining the virtual mass as a tensor with components m_{ik}, such that the force in the i direction is related to the acceleration in the k direction by $F_i = m_{ik} a_k$.

However, if the body has an axis of symmetry and accelerates along that axis, the virtual mass tensor is DIAGONAL, with $m_{ik} = m\delta_{ik}$. The force applied to the fluid, **F**, is then in the same direction as the acceleration, **a**, and the coefficient is a scalar, denoted by m:

$$\mathbf{F} = m\mathbf{a}.$$

An example is the fluid loading on a plane rigid piston vibrating in a baffle. *Units* kg.

Note (1): The incompressible virtual mass may be used to estimate inertial forces in a compressible fluid, provided the accelerating body is acoustically compact; the relative error is of order $(k_0 l)^2$, where l is a characteristic dimension of the body and k_0 is the acoustic wavenumber.

Note (2): Equivalent terms are **added mass**, **attached mass**, **entrained mass**, **induced mass**, and **virtual inertia**, but not APPARENT MASS.

virtual mode – *in underwater acoustics* a mode of propagation, in a stratified model of the ocean, whose radial eigenvalue or horizontal propagation wavenumber contains a significant imaginary component. Such modes correspond to steep-angle rays, propagating at grazing angles larger than the CRITICAL ANGLE FOR TRANSMISSION; they are therefore strongly attenuated in the radial direction. Compare EVANESCENT MODE, NORMAL-MODE FIELD.

Note: Because virtual modes do not show up as distinct peaks in a radial wavenumber decomposition of the field from a localized source (i.e. a HANKEL TRANSFORM of the source field), they are sometimes referred to as the *continuous spectrum* component of the field with respect to radial wavenumber. Another name for virtual modes is *leaky modes*.

viscosity – *of a fluid* in a wide range of fluids the viscous stress is linearly related to the rate of strain; such fluids are called *newtonian*. The constant of proportionality relating fluid stress and rate of strain is called the viscosity; it is the factor μ in the equation

$$\tau_{ij} = 2\mu(d_{ij} - \frac{1}{3}\Delta\delta_{ij}), \quad \text{where} \quad d_{ij} = \frac{1}{2}\left(\frac{\partial u_i}{\partial x_j} + \frac{\partial u_j}{\partial x_i}\right).$$

Here x_i are cartesian coordinates fixed in space ($i = 1, 2, 3$), and u_i are components of the vector $\mathbf{u}(\mathbf{x}, t)$ that represents the fluid velocity at position $\mathbf{x}$ and time t. Other symbols are: τ_{ij} for the components of the viscous stress; d_{ij} for the components of the local deformation rate; δ_{ij} for the Kronecker delta, and Δ for the DILATATION RATE d_{kk}. Also known as *dynamic viscosity*. *Units* Pa s.

viscothermal – *in a fluid* related to viscosity and heat conduction; for example, *viscothermal losses in a sound wave*.

viscothermal unsteady boundary layer – *produced by interaction of a sound field with a boundary* a combined term for the THERMAL UNSTEADY BOUNDARY LAYER and the VISCOUS UNSTEADY BOUNDARY LAYER; see BOUNDARY LAYER (2). The two boundary layers, characterized respectively by large gradients of unsteady fluid temperature and fluid velocity, are referred to jointly as the viscothermal unsteady boundary layer; in acoustical contexts, this is abbreviated to *viscothermal boundary layer*.

viscous damper rate – an alternative term for VISCOUS DAMPING COEFFICIENT. *Units* N s m^{-1}.

viscous damping – refers to any damping mechanism in which the amplitude of the damping force, over a range of frequencies, is modelled as being proportional to the velocity amplitude (rather than the displacement amplitude).

viscous damping coefficient – the coefficient of proportionality between the force output and relative-velocity input, for an ideal linear damping element or DASHPOT. The viscous damping coefficient is represented by the coefficient R in the equation

$$m\ddot{x} + R\dot{x} + kx = F\cos\omega t,$$

which describes the displacement response $x(t)$ of a linear single-degree-of-freedom system to a sinusoidal applied force. The coefficient R is sometimes referred to simply as the **damping coefficient** of the single-degree-of-freedom system. *Units* N s m^{-1}.

Note: The *critical damping coefficient*, defined by

$$R_{\text{crit}} = 2\sqrt{km},$$

is the value of R above which the system returns monotonically to equilibrium following a displacement.

viscous penetration depth, viscous skin depth – *for oscillatory viscous flow near a boundary* the quantity $\delta = \sqrt{2\nu/\omega}$, where ν is the KINEMATIC VISCOSITY of the fluid and ω is the angular frequency. See also BOUNDARY LAYER (2). Compare THERMAL PENETRATION DEPTH. *Units* m.

viscous unsteady boundary layer – *produced by interaction of a sound field with a boundary* a thin region of fluid adjacent to the boundary within which viscous diffusion of vorticity occurs. In this region, the oscillatory tangential velocity is the sum of an acoustic component (the value corresponding to an inviscid fluid), and a rotational component that cancels the acoustic component at the boundary; the relative velocity at the boundary is therefore zero.

Note: The boundary layer qualifies as thin provided $\delta \ll (R, \lambda)$, where R is the interface radius of curvature, λ is the acoustic wavelength and δ is the viscous boundary layer thickness. Since δ is of order $\sqrt{2\nu/\omega}$, where ν is the KINEMATIC VISCOSITY and ω is the angular frequency, this requires $\omega \nu/c^2$ to be much less than unity.

VLA – *in underwater acoustics* abbreviation for vertical line array.

vocal fold – a sharp-edged fold of mucous membrane overlying the vocal ligament, positioned at the top of the windpipe and enclosed by the larynx. The folds occur in two pairs one above the other; the folds in each pair project into the larynx from opposite sides, and produce sound when tensed and excited by airflow along the windpipe. The vocal folds are informally known as **vocal cords**.

Voigt model – *of a linear viscoelastic material with time-independent properties* a stress-strain relationship in which the deviatoric stress, σ, at any instant is a linear function of the deviatoric strain, ε, and its time derivative at that instant. In mathematical form,

$$\sigma = a\varepsilon + b\dot{\varepsilon} \qquad (a, b \text{ constants}).$$

Also known as a **Kelvin model** or **Kelvin–Voigt model** (**W VOIGT 1898**). Compare MAXWELL MODEL.

volume absorption – the part of the ROOM ABSORPTION contributed by acoustic energy dissipation in the main volume of the room (as opposed to that caused by absorption at the boundaries). In live rooms with many resonant modes excited, the field is approximately diffuse and $A_{\text{vol}} \approx 4mV$, where m is the ENERGY ATTENUATION COEFFICIENT for plane sound waves in the medium and V is the room volume. *Units* m².

Note: The Kelvin–Voigt model is a special case of the STANDARD LINEAR SOLID (or Maxwell–Voigt) model for viscoelastic solids. An even more general linear model is the RELAXATION MODEL.

volume absorption coefficient – *of an acoustic medium* an equivalent term for BULK ABSORPTION COEFFICIENT. *Units* m^{-1}.

volume acceleration – the time derivative of a VOLUME VELOCITY. An object immersed in an incompressible fluid, and displacing a time-varying volume X, generates a volume velocity $\dot{X}$ across a fixed surface that encloses the body, and a volume acceleration $\ddot{X}$. *Units* m³ s^{-2}.

Note: In a free field, a COMPACT object of time-varying volume generates an outgoing-wave pressure field given by

$$p(r, t) = \frac{\rho_0 \ddot{X}(t - r/c_0)}{4\pi r} \quad \text{(at distance } r \text{ and time } t\text{),}$$

provided r is large compared with the size of the object. Here ρ_0 and c_0 are the density and sound speed in the undisturbed fluid, assumed uniform. Compare FFOWCS WILLIAMS–HAWKINGS EQUATION (2).

volume backscattering strength – *of a volume distribution of scatterers in an acoustic medium* the SOURCE LEVEL of the backscattered field per unit scattering volume, minus the sound pressure level of the incident plane wave in dB re 1 μPa². In equation form, the volume backscattering strength VBS is given by

$$\text{VBS} = 10 \log_{10} \left[\frac{r_{\text{ref}} \bar{r}^2 p_{\text{back}}^2(\bar{r})}{V p_{\text{inc}}^2} \right],$$

where $p_{\text{back}}(\bar{r})$ is the rms backscattered pressure measured at distance $\bar{r}$ from the scattering region and corrected for absorption, r_{ref} is the reference distance of 1 m, V is the scattering volume, and p_{inc} is the rms pressure of the incident wave. Compare BISTATIC VOLUME SCATTERING STRENGTH. *Units* dB re 1 m^{-1}.

Note: Provided consistent units are used, the volume backscattering strength is numerically related to the DIFFERENTIAL SCATTERING CROSS-

SECTION s_d of the scattering region (with dimensions of area per unit solid angle) by the equation

$$\text{VBS} = 10\log_{10}(r_{\text{ref}}s_d/V).$$

volume compliance – *of a compliant acoustic lumped element* the quantity

$$C = -d\mathcal{V}/dP,$$

assumed to be independent of frequency at low frequencies. Here $\mathcal{V}$ is the volume of the compliant acoustic element and P is the pressure applied to it. In the case of a Helmholtz resonator or similar rigid cavity with an opening, the definition is

$$C = X_{\text{in}}/p,$$

where X_{in} is the volume displacement of fluid into the cavity through the opening, and p is the incremental pressure. *Units* $\text{m}^3\,\text{Pa}^{-1}$.

Note: The associated ACOUSTIC ADMITTANCE of the lumped element is $j\omega C$.

volume diffusers – *in room acoustics* scattering objects that are suspended in a room and redistribute the energy of incident sound waves in all directions; the aim is to produce a more diffuse sound field. Such diffusers are often used in reverberation chambers. Compare ROTATING DIFFUSER.

volume flowrate – *across a given surface* the quantity

$$q = \int_S \mathbf{u}\cdot\mathbf{n}\,dS,$$

evaluated over the given surface S. Here $\mathbf{n}$ is the unit normal to S, and $\mathbf{u}$ is the fluid velocity vector. Compare VOLUME VELOCITY, which is the term used in acoustics for the same quantity. *Units* $\text{m}^3\,\text{s}^{-1}$.

volume scattering coefficient – *of a volume distribution of scatterers in an acoustic medium* an equivalent term for the BULK SCATTERING COEFFICIENT of the medium, i.e. the total scattering cross-section per unit volume. *Units* m^{-1}.

volume scattering strength (VSS) – *for a continuous volume distribution of scatterers in an acoustic medium* a logarithmic measure of the scattering strength in a specified direction, normalized to unit scattering volume. For underwater acoustics, see VOLUME BACKSCATTERING STRENGTH, BISTATIC VOLUME SCATTERING STRENGTH. If the scattering is isotropic, the volume scattering strength in any direction is related to the VOLUME SCATTERING COEFFICIENT, m_v, by

$$\text{VSS} = 10\log_{10}\frac{m_v}{4\pi}.$$

Units dB re 1 m^{-1}.

volume source – a MONOPOLE SOURCE, i.e. a multipole source distribution of order one.

volume stiffness – *of a compliant acoustic lumped element* the reciprocal of the VOLUME COMPLIANCE. *Units* Pa m^{-3}.

volume strain – *in a fluid or solid that is deformed from its initial state* the increase in volume of an element of material, per unit initial volume. Also known as *volumetric strain*. *Units* none.

volume thermal expansivity – the volumetric coefficient of thermal expansion of a fluid or solid, defined by

$$\alpha = \frac{1}{V}\left(\frac{\partial V}{\partial T}\right)_P.$$

It is the proportional increase in volume caused by unit rise in temperature at constant pressure; an equivalent term is *volume thermal expansion coefficient*. In the equation above, α is the thermal expansivity; other symbols used are V for specific volume, T for absolute temperature, and P for pressure. *Units* K^{-1}.

Note: The volume thermal expansivity of a fluid is related to the SPECIFIC-HEAT RATIO, γ, by

$$\alpha^2 = \frac{(\gamma-1)C_p}{c^2 T},$$

where C_p is the specific heat at constant pressure and c is the speed of sound. It also appears in the differential relation for the density, ρ, as a function of P and the specific entropy s:

$$d\rho = \frac{1}{c^2}dP + \left(\frac{\partial\rho}{\partial s}\right)_P ds, \quad \text{where} \quad \left(\frac{\partial\rho}{\partial s}\right)_P = -\frac{T\alpha\rho}{C_p}.$$

volume velocity – *across a surface in a sound field* the integral $\int u_n\, dS$ of the PARTICLE VELOCITY normal to a specified surface S, evaluated over the surface (usually fixed in space). *Units* m^3 s^{-1}.

Note: The term *volume velocity* usually denotes a fluctuating acoustic quantity, whereas *volume flowrate* (defined in exactly the same way) is used in mechanical engineering to denote a non-acoustic quantity.

vortex sheet – *in inviscid fluid dynamics* a tangential surface of discontinuity, with the fluid velocity parallel to the surface having different values on the two sides. Compare SHOCK WAVE, which is a surface of discontinuity in the fluid velocity normal to the surface.

Note: Vortex sheets are unstable; see KELVIN–HELMHOLTZ INSTABILITY. Furthermore in a viscous fluid, a true vortex sheet can exist only instantaneously since the vorticity diffuses into the fluid on either side of the sheet.

vorticity – a measure of local fluid rotation, defined by

$$\omega = \nabla \times \mathbf{u} \quad (\mathbf{u} = \text{fluid velocity vector}).$$

Units s^{-1}.

W

warble tone – a signal that varies as $\cos\{[\omega_0 + f(t)]t + \alpha\}$, where t is time, $f(t)$ is a periodic function, and (α, ω_0) are real constants.

warmth – the sensation for the listener created by a rich bass sound in concert halls. In a concert hall with warmth, bass instruments – especially cellos and double basses – are easily heard. See also BASS RATIO.

water hammer – a rapid transient pressure pulse that propagates in liquid-filled pipes and is caused by a sudden change in flowrate at some point along the pipe.

wave – a travelling disturbance. See MECHANICAL WAVE PROPAGATION, LONGITUDINAL WAVES, TRANSVERSE WAVES.

wave equation – an equation of the form

$$\nabla^2 u - \frac{1}{c^2}\frac{\partial^2 u}{\partial t^2} = 0,$$

where c is the *wave speed* and t denotes time. The *wave variable* u is either a scalar (e.g. acoustic pressure) or a vector (e.g. particle displacement in a solid). In acoustics, the number of spatial dimensions is typically 3, but may be reduced to 2 or 1 in idealized problems. See also HELMHOLTZ EQUATION.

waveform – *of a signal* a continuous record of the signal as a function of time. If the signal is periodic, the waveform refers to one period.

wavefront – (1) a propagating surface of discontinuity across which an acoustic variable (e.g. pressure) changes by a vanishingly small amount. See HUYGENS' PRINCIPLE.

wavefront – (2) a surface of constant phase, in a single-frequency progressive wave field. The speed of advance of a plane wavefront defines the local PHASE SPEED (1) of free waves at that frequency.

wavefront – (3) *in geometrical acoustics* any surface on which the EIKONAL $\tau(\mathbf{x})$ is constant. The wavefronts advance in the direction of τ increasing, at a speed $|\nabla\tau|^{-1}$.

wavefront reversal – *of an acoustic field* an equivalent term for PHASE CONJUGATION.

wavefront speed – *for linear progressive waves of any type* the limiting high-frequency value of the PHASE SPEED (1). If no finite limit exists, the wavefront speed is not defined. *Units* m s^{-1}.

Note (1): The direction of propagation must be specified, if the medium is anisotropic.

Note (2): If a finite wavefront speed exists, it also coincides with the high-frequency limit of the group velocity in the propagation direction.

waveguide mode – *in a uniform duct or channel containing an acoustic medium* a transverse pattern of spatial pressure dependence that corresponds to one of the EIGENFUNCTIONS of the duct cross-section. Such a pattern, when excited at a single frequency at one axial location, propagates along the waveguide with a characteristic wavenumber $k(\omega)$ that is determined by

$$k(\omega) = \pm\sqrt{k_0^2 - \kappa^2}.$$

Here k_0 is the ACOUSTIC WAVENUMBER at angular frequency ω, and κ is the mode CUTOFF WAVENUMBER. See also CHARACTERISTIC WAVENUMBER.

wave impedance – the SPECIFIC RADIATION IMPEDANCE that a homogeneous fluid layer presents to an infinite flat surface, when the surface normal velocity takes the form of a sinusoidal travelling wave. The wave impedance as defined here is a complex quantity that is the same at all points on the surface. *Units* Pa s m^{-1}.

Note (1): The term wave impedance is commonly used for the characteristic impedance of plane waves; i.e. (in a fluid) the product ρc.

Note (2): If the surface lies in the plane $y = 0$, and a uniform acoustic medium extends from $y = 0$ to $y = \infty$, the wave impedance equals the CHARACTERISTIC IMPEDANCE

$$z_{y,\text{char}} = \frac{\omega \rho}{k_y} = \rho c \frac{k_0}{k_y}, \quad (k_0 = \omega/c).$$

Here ρ and c are the density and sound speed of the medium, ω is the angular frequency, and k_y is the propagation wavenumber (defined such that the radiated field varies with y as $e^{-jk_y y}$). The propagation wavenumber is related to the surface wavenumber, k_t, by

$$k_y = \begin{cases} \sqrt{k_0^2 - k_t^2} & (\text{real } k_y), \\ -j\sqrt{k_t^2 - k_0^2} & (\text{imaginary } k_y). \end{cases}$$

Note (3): The requirement of homogeneity applies only in directions parallel to the surface. The acoustic properties of the layer can vary in the normal direction while still leaving the wave impedance independent of position.

wavelength – (1) *of a one-dimensional periodic spatial pattern* the spatial period, i.e. the distance over which a complete cycle occurs. The spatial analogue of the PERIOD of a time-harmonic signal. *Units* m.

wavelength – (2) *of a 2-dimensional or 3-dimensional field that is spatially periodic in all directions* the spatial period in a specified direction; or if no direction is specified, the spatial period in the direction for which it is shortest. *Units* m.

wavelength – (3) *for sound at a specified frequency in a given medium* the quantity c/f, where f is the frequency and c is the sound speed in the medium. Equivalent terms are **wavelength of sound** and ACOUSTIC WAVELENGTH. *Units* m.

wavelet – ◆ an infinitesimal propagating pressure disturbance; a wavefront is composed of wavelets, which may interact (as in nonlinear acoustics) or not (as in linear acoustics).

wavenormal – *at any point on a wavefront* the unit vector pointing normal to the wavefront, in the direction of advance.

wavenumber, wave number – (1) *for a quantity in acoustics that varies sinusoidally along a specified axis* 2π times the number of cycles per unit distance; also known as ANGULAR WAVENUMBER. The wavenumber in this sense may be considered as the spatial analogue of ANGULAR FREQUENCY. Compare WAVENUMBER VECTOR (1), MODAL WAVENUMBER, COMPLEX WAVENUMBER. *Units* rad m^{-1}.

wavenumber, wave number – (2) *at a specified frequency in a lossless acoustic medium* the quantity $2\pi f/c$, where f is the frequency and c is the sound speed in the medium. A preferred term is ACOUSTIC WAVENUMBER. *Units* rad m^{-1}.

Note: In optics, the term *wavenumber* (also called *repetency*) means $1/\lambda$ where λ is the wavelength. The terms *angular wavenumber* or *circular wavenumber* (corresponding to angular or circular frequency), or else *propagation number*, are used for wavenumber in the senses defined above (c being interpreted as the speed of light). As yet, this terminology is not widely adopted in acoustics.

wavenumber, wave number – (3) *of single-frequency progressive waves propagating in a specified direction* the propagation wavenumber of the wave, i.e. its CHARACTERISTIC WAVENUMBER in the direction of propagation. The wavenumber in this sense is generally a complex function of frequency, $k(\omega)$, that depends on the properties and geometry of the medium; compare

DISPERSION RELATION. *Units* (real part) rad m^{-1}; (imaginary part) Np m^{-1}.

wavenumber kernel – *in underwater acoustics* an alternative term for the DEPTH-DEPENDENT GREEN'S FUNCTION. *Units* m.

wavenumber vector – (1) *of an acoustic field variable that varies sinusoidally with position* a vector whose component in any direction equals the WAVENUMBER in that direction. For example, if a two-dimensional field has wavenumbers k_x, k_y in the x, y coordinate directions, the 2D wavenumber vector $\mathbf{k} = (k_x, k_y)$. *Units* rad m^{-1}.

wavenumber vector – (2) *of single-frequency plane progressive waves in a lossless medium* the vector $\mathbf{k} = k\mathbf{n}$, where $\mathbf{n}$ is the unit vector in the WAVENORMAL direction, and k is the propagation wavenumber or CHARACTERISTIC WAVENUMBER. *Units* rad m^{-1}.

Note: In optics, the term *propagation vector* is used for wavenumber vector as defined above, and *wave vector* is used for $(1/2\pi)$ times this quantity.

wave-slowness vector – an equivalent term for SLOWNESS VECTOR. *Units* s m^{-1}.

wave vector – *in acoustics* alternative term (not recommended) for WAVENUMBER VECTOR. *Units* rad m^{-1}.

Note: In optics, the term *wave vector* has a different meaning; see the note under WAVENUMBER VECTOR.

weak shock – a SHOCK WAVE across which the relative change in density is small. In symbols, with subscripts 1, 2 denoting conditions ahead of and behind the shock, the **weak-shock criterion** is

$$|\Delta\rho| = |\rho_2 - \rho_1| \ll \rho_1, \quad (\rho = \text{fluid density}).$$

An important feature of weak shocks is the rapidity with which Δs, the increase in specific entropy through the shock, tends to zero as the shock strength $\Delta\rho/\rho$ is reduced: $\Delta s \propto (\Delta\rho)^3$. See also WEAK-SHOCK RELATIONS, WEAK-SHOCK THEORY.

weak-shock relations – approximate relations, asymptotically valid in the limiting case of weak shocks, that connect quantities on the two sides of a shock wave. An example is the equation that relates the increase in specific entropy, s, through a shock to the change in pressure:

$$\Delta s = s_2 - s_1 = \frac{h_{PPP}}{12 h_s}(P_2 - P_1)^3 + O(P_2 - P_1)^4,$$

where h is the specific enthalpy, P is the fluid pressure, and subscripts 1, 2 label conditions ahead of and behind the shock wave. Partial derivatives of the function $h(P, s)$ are denoted by subscripts. An equivalent nondimensional version of this equation is

$$c_1^{-2} T_1 \Delta s = \frac{1}{6}\beta \Pi^3, \qquad \Pi = \frac{P_2 - P_1}{\rho_1 c_1^2},$$

where T is the absolute temperature, β is the COEFFICIENT OF NONLINEARITY, and ρ, c are the fluid density and sound speed.

Note: Two other useful weak-shock relations, involving the fluid velocity, w, relative to the shock, are given below. They relate the decrease in w through the shock to the pressure-rise parameter Π and to the shock Mach number, $w_1/c_1 = M_1$:

$$\frac{1}{c_1}(w_1 - w_2) = \Pi - \frac{1}{2}\beta\Pi^2 + O(\Pi^3)$$
$$= \frac{1}{\beta}(M_1 - 1) + O(M_1 - 1)^2.$$

weak-shock theory – *in nonlinear acoustics* an approximate description of progressive-wave propagation at finite amplitudes, based on a lossless-medium model in which dissipation is confined to shocks. Weak-shock theory discards terms beyond second order in the acoustic variables, with the result that backward-travelling waves do not appear when shocks form and decay.

Webster's horn equation – a one-dimensional equation that provides an approximate description of sound transmission along a slowly-tapering horn, for wavelengths that are long compared with the transverse dimensions of the horn. For the specific case of an ACOUSTIC HORN of cross-sectional area $A(x)$, filled with uniform fluid of sound speed c and with rigid walls, the Webster equation is

$$\frac{1}{A}\frac{\partial}{\partial x}\left(A\frac{\partial p}{\partial x}\right) - \frac{1}{c^2}\frac{\partial^2 p}{\partial t^2} = 0$$

(A G WEBSTER 1919). Here $p(x, t)$ is the acoustic pressure averaged over the horn cross-section, t is time, and x is distance measured along the horn axis.

weighted mean – if a set of numbers x_i ($i = 1$ to N) has an associated set of WEIGHTING FACTORS w_i, the weighted mean is defined as

$$\bar{x} = \frac{\sum w_i x_i}{\sum w_i}.$$

weighted sound pressure level – *for airborne sounds* a sound pressure level based on applying a standard frequency weighting to an acoustic pressure signal, followed by exponentially-weighted time averaging of the squared signal. The type of exponential weighting needs to be specified. *Units* dB re $(20\ \mu Pa)^2$.

Note: The abbreviation **sound level** is commonly used, particularly when the weighted sound pressure level is based on fast (F) time weighting and A frequency weighting. See also the entry for SOUND LEVEL.

weighted sound reduction index – *of a partition* a method of rating the airborne sound insulation of a partition, based on SOUND INSULATION measurements in one-third octave bands from 50 Hz to 3.15 kHz. A standard-shape curve is fitted to the data, and the nominal sound insulation at 500 Hz defines the weighted index. *Units* dB.

weighting – *applied to a data sample* premultiplication of the data by a set of WEIGHTING FACTORS.

weighting factors – multiplying factors that are applied to numerical data values, or to the FREQUENCIES associated with data BINS, prior to calculating aggregate properties or STATISTICS such as a WEIGHTED MEAN. Also known as *statistical weights*.

weighting function – a set of WEIGHTING FACTORS in the form of a continuous function, e.g. of time or frequency. The weighting function of an ARRAY is the vector of weighting factors applied to the outputs from each array element.

Weyl integral – a generic term for an integral representation of spherical outgoing waves in terms of plane waves (**H WEYL 1919**). The Weyl integral for outgoing waves with spherical symmetry is

$$\frac{e^{-jkR}}{R} = \frac{1}{2\pi j} \int_{-\infty}^{\infty} dk_y \int_{-\infty}^{\infty} \frac{1}{\gamma} e^{-j\gamma|z|} \exp[-j(k_x x + k_y y)] dk_x$$

where (x, y, z) are cartesian coordinates; here the monopole is located at the origin, and $R = \sqrt{x^2 + y^2 + z^2}$ is the spherical radius. The complex numbers k (acoustic wavenumber) and γ (vertical wavenumber) are related by $j\gamma = \sqrt{k_x^2 + k_y^2 - k^2}$, and their phases are subject to the constraints

$\text{Re}(jk) \geq 0$ (k corresponds to non-amplifying waves),

$$-\frac{\pi}{2} < \arg \sqrt{k_x^2 + k_y^2 - k^2} \leq \frac{\pi}{2}$$

(γ corresponds to waves that decay in the $\pm z$ directions).

Compare SOMMERFELD INTEGRAL.

whispering gallery modes – *in a cylindrical room* transverse cylindrical waves of high azimuthal order, equivalent to rays that strike the curved interior surface almost tangentially in planes normal to the axis of the cylinder. Such rays remain close to the surface at all times as they travel round the room, with the result that the energy of the sound field is confined to a thin layer near the cylindrical boundary.

white noise – a STATIONARY, RANDOM, CONTINUOUS SIGNAL with a flat power spectrum. If the single-sided autospectral density of such a signal is N_0, the AUTOCORRELATION FUNCTION is $\frac{1}{2}N_0$ multiplied by a delta function $\delta(\tau)$, where τ is the time shift. Signal values separated by any finite delay τ are therefore uncorrelated.

Note (1): The statistical properties of white noise do not necessarily follow a GAUSSIAN DISTRIBUTION. GAUSSIAN NOISE, likewise, does not have to be white.

Note (2): White noise in the strict sense defined above is a useful theoretical concept, but cannot be realized physically. BAND-LIMITED WHITE NOISE is a practical approximation; for example, if the autospectral density of the noise signal drops to zero beyond $|f| = B$ its autocorrelation function is $BN_0 \operatorname{sinc} 2B\tau$.

wideband – covering all frequencies, or (in practice) all frequencies within the bandwidth of a measurement system. Equivalent to OVERALL.

Wiener–Hopf equation – an integral equation in one dimension (specifically, on a half-line) that involves a convolution and can be solved by Fourier transform methods (**N WIENER, E HOPF 1931**). The general form of the equation is

$$u(x) - \int_0^\infty K(x-s)\,u(s)\,ds = f(x), \qquad 0 \leq x < \infty;$$

here $u(x)$ is the unknown function, and the kernel K is specified along with the input function $f(x)$.

Wiener–Hopf technique – a mathematical method for solving boundary-value problems and integral equations of a specific type particularly associated with diffraction phenomena; see WIENER–HOPF EQUATION. The technique, originally devised in connection with neutron transport theory, was later applied to diffraction problems involving electromagnetic and acoustic waves. It depends on a special factorization of a function of a complex variable arising in an integral transform. For solving the harder boundary-value problems in acoustics and aeroacoustics, the technique has no equal (**N WIENER, E HOPF 1931**).

Wiener–Khinchin theorem – states that the AUTOCORRELATION FUNCTION and the AUTOSPECTRAL DENSITY of a stationary random real signal form a Fourier transform pair (**N WIENER 1930, A YA KHINCHIN 1934**):

$$R_{xx}(\tau) \leftrightarrow S_{xx}(f).$$

Here $x(t)$ is the signal as a function of time t, $R_{xx}(\tau)$ is its autocorrelation function, and τ is a time shift or delay. Likewise $S_{xx}(f)$ denotes the signal autospectral density at frequency f. Compare CONVOLUTION THEOREM.

wind noise – *in underwater acoustics* underwater ambient noise with a broadband downward-sloping spectrum, typically observed in the frequency range 500 Hz to 50 kHz. The spectrum level of wind noise increases with wind speed; for each doubling of wind speed, the level increases by 6 to 7 dB.

Note: The typical spectrum slope of wind noise in deep oceans is between -5 and -6 dB per octave. The likely mechanism of noise generation is wave breaking, with associated formation and collapse of bubbles.

windowing – *in signal processing* application of a WEIGHTING FUNCTION, called a *window*, to a signal. A key property of a window function is that it is zero outside a finite interval; the windowed signal is therefore of finite length.

WKB method, WKBJ method – a procedure for finding approximate analytical solutions to differential equations of the form

$$\frac{d^2 u}{dx^2} + q^2(x)\, u = 0.$$

Typically, x is a spatial coordinate, so $q(x)$ may be interpreted as a local wavenumber. The WKBJ method leads to the result

$$u(x) \approx q^{-1/2} \left(A \exp[-j \int_0^x q(x')\, dx'] + B \exp[+j \int_0^x q(x')\, dx']\right),$$

as a first approximation (**G WENTZEL, H A KRAMERS, L N BRILLOUIN 1926; H JEFFREYS 1923**).

Note: The WKBJ method becomes more accurate, the more slowly $q(x)$ varies with x. The fundamental requirement is $|q^2| \gg |q'(x)|$, which means that the relative change in q over one wavelength is small.

wolf note, wolf tone – *in bowed-string musical instruments* an audible beating phenomenon that is associated with nonlinear coupling between vibration of the instrument body, and wave motion in the bowed string. It occurs when the driving-point admittance of one of the string terminations has a resonance peak whose period almost coincides with the round-trip time for a wave travelling twice the full length of the string, with one reflection at each end. The nonlinear coupling probably also involves the nonlinear stick-slip dynamics of the bowing process.

Womersley parameter – *for periodic pulsatile flow in a tube* a dimensionless frequency parameter, related to the period T of the oscillatory flow by

$$W = a\left(\frac{2\pi}{\nu T}\right)^{1/2},$$

where a is the tube internal radius and ν is the kinematic viscosity of the fluid. The term is used in biological fluid dynamics. *Units* none.

Note: The Womersley parameter is the square root of the STOKES NUMBER based on a; it is equivalent to the SHEAR NUMBER.

work – a transfer of ENERGY between a system and its surroundings, by means other than HEAT. The ***work done on*** a system is the amount of energy transferred to the system. For example, the work done by a force $\mathbf{F}(t)$, applied to a point on the system that moves with velocity $\mathbf{v}(t)$, is

$$W_{12} = \int_{t_1}^{t_2} \mathbf{F}\cdot\mathbf{v}\,dt \qquad \text{(between times } t_1 \text{ and } t_2\text{)}.$$

Units J.

WOTAN – *in underwater acoustics* abbreviation for weather observation through ambient noise.

Wronskian – *of a set of functions of a given independent variable* a determinant, formed from the n functions $h_1(x)$, $h_2(x)$, ..., $h_n(x)$ and their derivatives up to order $(n-1)$, that is used to test whether the functions form a LINEARLY INDEPENDENT SET over a specified range of the variable x. The Wronskian W is defined by the determinant below, in which the first row consists of the n functions, the second row consists of their first derivatives, and so on up to the $(n-1)$-order derivatives.

$$W = \begin{vmatrix} h_1 & h_2 & \cdots & h_n \\ h'_1 & h'_2 & \cdots & h'_n \\ \cdots & \cdots & \cdots & \cdots \\ h_1^{(n-1)} & h_2^{(n-1)} & \cdots & h_n^{(n-1)} \end{vmatrix}.$$

Also known as the ***Wronskian determinant*** (J HOENE-WRONSKI 1812).

Note: If within the specified range of x there is at most a finite number of points where W vanishes, then the set of functions $h_1(x)$, $h_2(x)$, ..., $h_n(x)$ is linearly independent.

X

XBT – *in underwater acoustics* abbreviation for expendable BATHYTHERMOGRAPH.

XSV – *in underwater acoustics* abbreviation for expendable sound VELOCIMETER.

X wave – a type of LOCALIZED WAVE that propagates as an axisymmetric pulse, with a wavefront cross-section (in a plane through the beam axis) that resembles the letter X. Because of their ability to remain focused while propagating over distances comparable with the Rayleigh distance, such waves have acoustical applications in nondestructive testing and medical imaging.

Note: An example of an acoustic X wave is the complex pressure field

$$p(r,z,t) \propto \frac{1}{\sqrt{r^2 + [a_1 + j\gamma(z - v_1 t)]^2}}.$$

Here r, z are radial and axial CYLINDRICAL COORDINATES; t is time; a_1 and v_1 are parameters, and $\gamma = [(v_1/c)^2 - 1]^{1/2}$, where c is the speed of sound. The actual transient pressure field is taken as the real part of this expression.

Y

Young's modulus – *of a solid* the ratio of the longitudinal stress to the longitudinal strain in the same direction, when a rod of the material is stretched by a small amount along its axis and allowed to contract freely transverse to the axis. Also known as *elastic modulus*. Compare COMPLEX MODULUS. Symbol E. *Units* Pa.

Young waves – LONGITUDINAL WAVES that propagate in a uniform bar of solid material with stress-free sides. The speed of propagation is a property of the rod material, given by $c_L = \sqrt{E/\rho}$ where E is the YOUNG'S MODULUS and ρ is the density.

Z

zero matrix – an alternative term for NULL MATRIX.

zero-padding – *applied to a finite-length discrete sequence* the addition of dummy samples (of value 0) at the beginning or end of the sequence, in order to increase the length of the sequence up to some desired value. For typical uses of zero-padding, see DISCRETE FOURIER TRANSFORM, CIRCULAR CONVOLUTION (2).

zeros – (1) *of a given function* the zeros of the function $f(x)$ are the values of x for which the function is zero, i.e. the ROOTS of $f(x) = 0$.

zeros – (2) *of a linear time-invariant system* the points in the S-PLANE where the system transfer function has zero modulus.

zone of silence – *for a given source* a region into which acoustic RAYS cannot penetrate; also known as a *sound shadow*. In outdoor sound propagation, upward refraction due to a decrease in temperature with height can combine with the ground to produce a zone of silence near the ground, beyond a certain radius from the source. Compare ABNORMAL AUDIBILITY ZONE.

z-plane – the complex plane in which the Z-TRANSFORM is represented.

z-transform – *of a digital signal* a transform that converts a digital signal or sequence into a continuous function of the complex variable z. The direct z-transform of sequence $x[n]$ is

$$X(z) = \sum_{n=-\infty}^{\infty} x[n]\, z^{-n}.$$

The inverse transform is

$$x[n] = \frac{1}{2\pi j} \oint_C X(z)\, z^{n-1}\, dz,$$

where the integration contour C in the complex z-plane must encircle the origin and be in the convergence region of $X(z)$.

Note: The z-transform is the discrete-time equivalent of the LAPLACE TRANSFORM, with z representing the complex phase factor for one time step. The complex variable z is the counterpart of the factor e^{-st} in the Laplace transform. There is also a close parallel with the DISCRETE-TIME FOURIER TRANSFORM.

Zwicker phons – a single-number scale for rating the loudness of complex sounds. It is based on a model that simulates the nonlinear operation of the human ear, in contrast to simple frequency-weighting procedures such as A-weighting (**E ZWICKER 1958**).

Select Bibliography

Sources that have been particularly useful in compiling this dictionary are listed below, along with references for further reading on the topics covered. The latter represent a personal choice, and what follows is by no means a complete listing of the many excellent books available.

Dictionaries

———————————————————— (1995) *Longman Dictionary of the English Language*. London, UK, Viking.
Clapham, C. (1996) *Concise Oxford Dictionary of Mathematics* (2nd edn). Oxford, UK, Oxford University Press.
Hazewinkel, M. (ed.) (1988) *Encyclopaedia of Mathematics*: Annotated translation of Soviet Mathematical Encyclopedia (I M Vinogradov, editor-in-chief). Dordrecht, NL, Reidel (Kluwer Academic).
Isaacs, A., Daintith, J., Martin, E. (eds) (1991) *The Oxford Dictionary for Scientific Writers and Editors*. Oxford, UK, Clarendon Press.
Lafferty, P., Rowe, J. (eds) (1998) *The Hutchinson Dictionary of Science* (2nd edn). Oxford, UK, Helicon.
Martin, E. (1998) *Concise Oxford Medical Dictionary* (5th edn). Oxford, UK, Oxford University Press.
Mendel, L.L., Danhauer, J.L., Singh, S. (1999) *Singular's Illustrated Dictionary of Audiology*. San Diego, CA; London, UK, Singular Publishing.
Parker, S.F. (ed.) (1997) *McGraw-Hill Dictionary of Mathematics*. New York, NY, McGraw-Hill.
Parker, S.F. (ed.) (1994) *McGraw-Hill Dictionary of Scientific and Technical Terms* (5th edn). New York, NY, McGraw-Hill.
Schwartzman, S. (1994) *The Words of Mathematics*. Washington, DC, Mathematical Association of America.
Weissing, H. (1992) *Fachwörterbuch Akustik* (Specialist Dictionary of Acoustics): English–German, German–English. Berlin, Germany; Paris, France, Verlag Alexandre Hatier.

Subject Classification and Standards

Acoustical Society of America (1994) American National Standard: Acoustical Terminology. ANSI S1.1-1994 (ASA 111-1994). New York, NY, Standards Secretariat, ASA.
Editor-in-Chief, *Journal of the Acoustical Society of America* (updated) Physics and Astronomy Classification Scheme, Section 43: Acoustics. Published in

June and December issues of the Journal. Melville, NY, Acoustical Society of America.
International Electrotechnical Commission (1994) *International Electrotechnical Vocabulary*, Chapter 801: Acoustics and electroacoustics (2nd edn). IEC 50(801): 1994. Geneva, CH, Bureau Central, CEI/IEC.

Books

Allard, J.-F. (1993) *Propagation of Sound in Porous Media: Modelling Sound Absorbing Materials*. London, UK, Elsevier Applied Science.
Barron, M. (1993) *Auditorium Acoustics and Architectural Design*. London, UK, E & F N Spon.
Bendat, J.S., Piersol, A.G. (1986) *Random Data: Analysis and Measurement Procedures* (2nd edn). New York, NY, Wiley-Interscience.
Beranek, L.L., Vér, I.L. (eds) (1992) *Noise and Vibration Control Engineering*. New York, NY, J Wiley.
Beranek, L. (1996) *Concert and Opera Halls: How they sound*. Woodbury, NY, Acoustical Society of America.
Biering, H., Pedersen, O.Z. (1983) *System Analysis and Time Delay Spectrometry*, Parts I and II (Technical Review, 1 – 1983 and 2 – 1983). Nærum, DK, Brüel & Kjær.
Boas, M.L. (1983) *Mathematical Methods in the Physical Sciences* (2nd edn). New York, NY, J Wiley.
Brekhovskikh, L.M. (1980) *Waves in Layered Media* (2nd edn) (transl. Robert T. Beyer). New York, NY; London, UK, Academic Press.
Brekhovskikh, L.M., Godin, O.A. (1999) *Acoustics of Layered Media II: Point Sources and Bounded Beams* (2nd edn). Berlin, Germany; New York, NY, Springer-Verlag.
Brekhovskikh, L.M., Lysanov, Y.P. (1991) *Fundamentals of Ocean Acoustics* (2nd edn). Berlin, Germany; New York, NY, Springer-Verlag.
Brook, D., Wynne, R.J. (1988) *Signal Processing: Principles and Applications*. London, UK, Edward Arnold.
Buckingham, M.J. (1983) *Noise in Electronic Devices and Systems*. Chichester, UK; New York, NY, Ellis Horwood/J Wiley.
Crighton, D.G. et al. (1990) *Modern Methods in Analytical Acoustics*. Berlin, Germany, Springer.
Crocker, M.J. (1997) *Encyclopedia of Acoustics* (4 vols.). New York, NY, J Wiley.
Davis, D., Davis, C. (1987) *Sound System Engineering* (2nd edn). Indianapolis, IN, Macmillan/Howard W Sams.
Drumheller, D.S. (1998) *Introduction to Wave Propagation in Nonlinear Fluids and Solids*. New York, NY; Cambridge, UK, Cambridge University Press.
Duck F.A., Baker, A.C., Starritt, H.C. (eds) (1998) *Ultrasound in Medicine*. Bristol, UK; Philadelphia, PA, Institute of Physics Publishing.

Fahy, F.J. (1985) *Sound and Structural Vibration: Radiation, Transmission and Response*. London, UK, Academic Press.
Fahy, F.J. (1995) *Sound Intensity* (2nd edn). London, UK; New York, NY, E & F N Spon/Routledge.
Fahy, F.J., Walker, J. (eds) (1998) *Fundamentals of Noise and Vibration*. London, UK; New York, NY, E & F N Spon/Routledge.
Frisk, G.V. (1994) *Ocean and Seabed Acoustics: A Theory of Wave Propagation*. Englewood Cliffs, NJ, Prentice-Hall.
Fuller, C.R., Elliott, S.J., Nelson, P.A. (1996) *Active Control of Vibration*. San Diego, CA; London, UK, Academic Press.
Gill, A.E. (1982) *Atmosphere–Ocean Dynamics*. San Diego, CA; London, UK, Academic Press.
Goldstein, M.E. (1976) *Aeroacoustics*. New York, NY, McGraw-Hill.
Griffin, M.J. (1990) *Handbook of Human Vibration*. London, UK; San Diego, CA, Academic Press.
Hamilton, M.F., Blackstock, D.T. (eds) (1998) *Nonlinear Acoustics*. San Diego, CA, Academic Press.
Harris, C.M. (ed.) (1996) *Shock and Vibration Handbook* (4th edn). New York, NY, McGraw-Hill.
Hsu, H.P. (1995) *Theory and Problems of Signals and Systems* (Schaum's Outline Series). New York, NY, McGraw-Hill.
Howe, M.S. (1998) *Acoustics of Fluid–Structure Interactions*. New York, NY; Cambridge, UK, Cambridge University Press.
Ingard, U. (1994) *Notes on Sound Absorption Technology*. Poughkeepsie, NY, Noise Control Foundation.
Jensen, F.B., Kuperman, W.A., Porter, M.B., Schmidt, H. (1994) *Computational Ocean Acoustics*. Woodbury, NY, AIP Press.
Jones, D.S. (1986) *Acoustic and Electromagnetic Waves*. Oxford, UK, Clarendon Press.
Junger, M.C., Feit, D. (1986) *Sound, Structures and their Interaction* (2nd edn). Cambridge, MA; London, UK, MIT Press.
Kuttruff, H. (2000) *Room Acoustics* (4th edn). London, UK; New York, NY, E & F N Spon/Routledge.
Leighton, T.G. (1994) *The Acoustic Bubble*. London, UK; San Diego, CA, Academic Press.
Lockett, F.J. (1972) *Nonlinear Viscoelastic Solids*. London, UK; New York, NY, Academic Press.
Mead, D.J. (1998) *Passive Vibration Control*. Chichester, UK; New York, NY, J Wiley.
Meirovitch, L. (1986) *Elements of Vibration Analysis* (2nd edn). New York, NY, McGraw-Hill.
Moore, B.C.J. (1997) *An Introduction to the Psychology of Hearing* (4th edn). San Diego, CA; London, UK, Academic Press.
Nayfeh, A.H., Mook, D.T. (1979) *Nonlinear Oscillations*. New York, NY, Wiley-Interscience.

Nelson, P.A., Elliott, S.J. (1992) *Active Control of Sound*. San Diego, CA; London, UK, Academic Press.
Oppenheim, A.V., Willsky, A.S., Nawab, S.H. (1997) *Signals and Systems* (2nd edn). London, UK, Prentice-Hall International.
Papoulis, A. (1984) *Signal Analysis*. New York, NY, McGraw-Hill.
Pierce, A.D. (1989) *Acoustics: An Introduction to its Physical Principles and Applications* (1st edn of 1981, reprinted with corrections). Woodbury, NY, Acoustical Society of America.
Rayleigh, Lord (1945) *The Theory of Sound* (2 vols.) (2nd edn of 1896, reprinted with corrections). New York, NY, Dover.
Thompson, P.A. (1984) *Compressible-Fluid Dynamics* (1st edn of 1979, reprinted by the author). New York, NY, McGraw-Hill (1st edn).
Tohyama, M., Suzuki, H., Ando, Y. (1995) *The Nature and Technology of Acoustic Space*. San Diego, CA; London, UK, Academic Press.
Trusler, J.P.M. (1991) *Physical Acoustics and Metrology of Fluids*. Bristol, UK; Philadelphia, PA, Institute of Physics Publishing/Adam Hilger.
Urick, R.J. (1983) *Principles of Underwater Sound* (3rd edn). New York, NY, McGraw-Hill.
Wark, K. Jr. (1988) *Thermodynamics* (5th edn). New York, NY, McGraw-Hill.
Ziemer, R.E., Tranter, W.H. (1990) *Principles of Communication: Systems, Modulation and Noise* (3rd edn). Boston, MA, Houghton Mifflin.

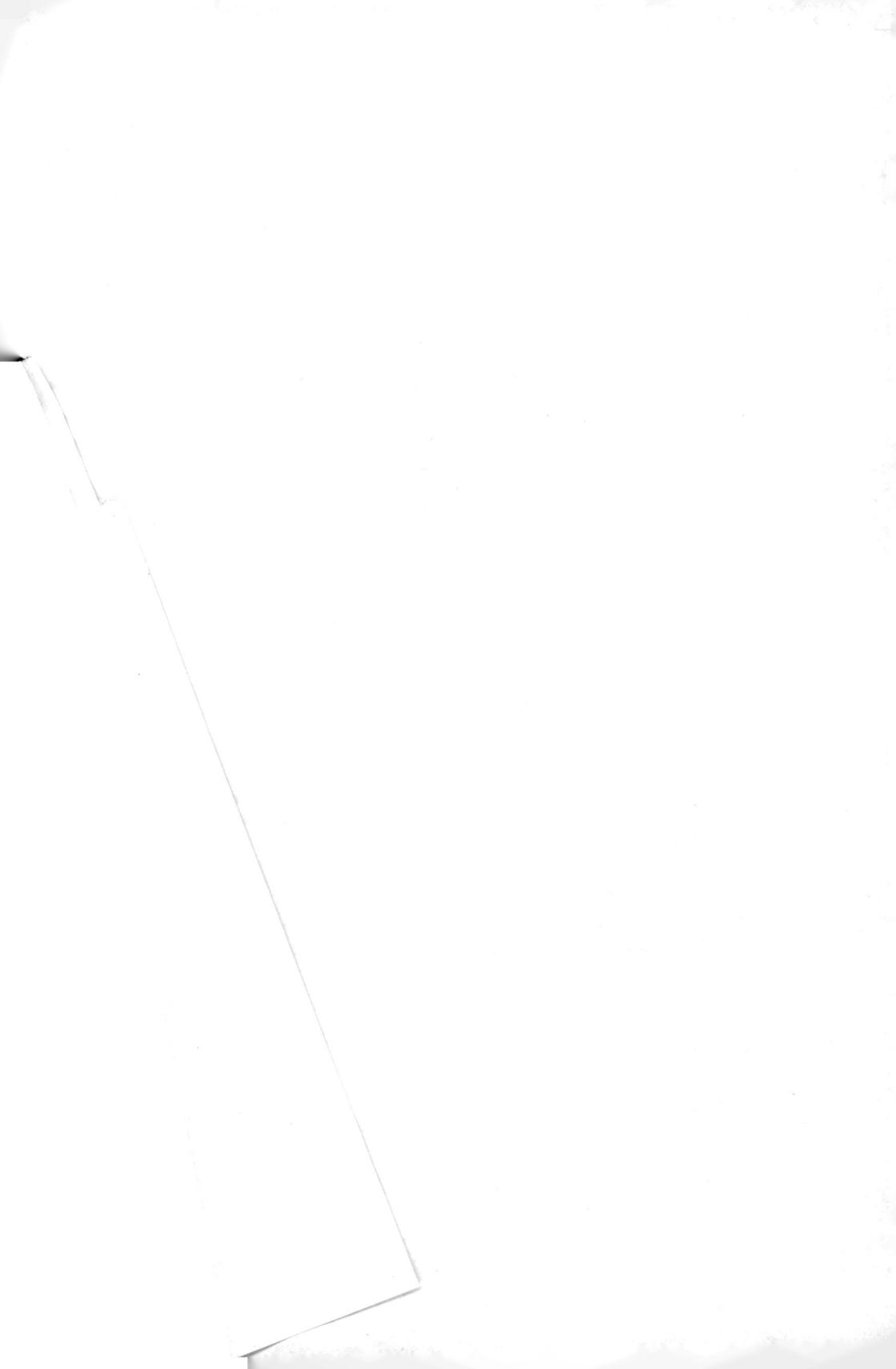